proximated using computers. The theory of dynamical systems puts emphasis on qualitative anal‐
ysis of systems described by differential equations, while many numerical methods have been de‐
veloped to determine solutions with a given degree of accuracy.

History

Differential equations first came into existence with the invention of calculus by Newton and Leib‐
niz. In Chapter 2 of his 1671 work "Methodus fluxionum et Serierum Infinitarum", Isaac Newton
listed three kinds of differential equations:

$$\frac{dy}{dx} = f(x)$$

$$\frac{dy}{dx} = f(x, y)$$

$$x_1 \frac{\partial y}{\partial x_1} + x_2 \frac{\partial y}{\partial x_2} = y$$

He solves these examples and others using infinite series and discusses the non-uniqueness of
solutions.

Jacob Bernoulli proposed the Bernoulli differential equation in 1695. This is an ordinary differen‐
tial equation of the form

$$y' + P(x)y = Q(x)y^n$$

for which the following year Leibniz obtained solutions by simplifying it.

Historically, the problem of a vibrating string such as that of a musical instrument was studied by
Jean le Rond d'Alembert, Leonhard Euler, Daniel Bernoulli, and Joseph-Louis Lagrange. In 1746,
d'Alembert discovered the one-dimensional wave equation, and within ten years Euler discovered
the three-dimensional wave equation.

The Euler–Lagrange equation was developed in the 1750s by Euler and Lagrange in connection with
their studies of the tautochrone problem. This is the problem of determining a curve on which a
weighted particle will fall to a fixed point in a fixed amount of time, independent of the starting point.

Lagrange solved this problem in 1755 and sent the solution to Euler. Both further developed
Lagrange's method and applied it to mechanics, which led to the formulation of Lagrangian
mechanics.

Fourier published his work on heat flow in *Théorie analytique de la chaleur* (The Analytic
Theory of Heat), in which he based his reasoning on Newton's law of cooling, namely, that the
flow of heat between two adjacent molecules is proportional to the extremely small difference
of their temperatures. Contained in this book was Fourier's proposal of his heat equation for
conductive diffusion of heat. This partial differential equation is now taught to every student
of mathematical physics.

Example

For example, in classical mechanics, the motion of a body is described by its position and velocity as the time value varies. Newton's laws allow (given the position, velocity, acceleration and various forces acting on the body) one to express these variables dynamically as a differential equation for the unknown position of the body as a function of time.

In some cases, this differential equation (called an equation of motion) may be solved explicitly.

An example of modelling a real world problem using differential equations is the determination of the velocity of a ball falling through the air, considering only gravity and air resistance. The ball's acceleration towards the ground is the acceleration due to gravity minus the acceleration due to air resistance.

Gravity is considered constant, and air resistance may be modeled as proportional to the ball's velocity. This means that the ball's acceleration, which is a derivative of its velocity, depends on the velocity (and the velocity depends on time). Finding the velocity as a function of time involves solving a differential equation and verifying its validity.

Types

Differential equations can be divided into several types. Apart from describing the properties of the equation itself, these classes of differential equations can help inform the choice of approach to a solution. Commonly used distinctions include whether the equation is: Ordinary/Partial, Linear/Non-linear, and Homogeneous/Inhomogeneous. This list is far from exhaustive; there are many other properties and subclasses of differential equations which can be very useful in specific contexts.

Ordinary Differential Equations

An ordinary differential equation (*ODE*) is an equation containing a function of one independent variable and its derivatives. The term "*ordinary*" is used in contrast with the term partial differential equation which may be with respect to *more than* one independent variable.

Linear differential equations, which have solutions that can be added and multiplied by coefficients, are well-defined and understood, and exact closed-form solutions are obtained. By contrast, ODEs that lack additive solutions are nonlinear, and solving them is far more intricate, as one can rarely represent them by elementary functions in closed form: Instead, exact and analytic solutions of ODEs are in series or integral form. Graphical and numerical methods, applied by hand or by computer, may approximate solutions of ODEs and perhaps yield useful information, often sufficing in the absence of exact, analytic solutions.

Partial Differential Equations

A partial differential equation (*PDE*) is a differential equation that contains unknown multivariable functions and their partial derivatives. (This is in contrast to ordinary differential equations, which deal with functions of a single variable and their derivatives.) PDEs are used to formulate problems involving functions of several variables, and are either solved in closed form, or used to create a relevant computer model.

Ordinary differential equations (ODEs) arise in many contexts of mathematics and science (social as well as natural). Mathematical descriptions of change use differentials and derivatives. Various differentials, derivatives, and functions become related to each other via equations, and thus a differential equation is a result that describes dynamically changing phenomena, evolution, and variation. Often, quantities are defined as the rate of change of other quantities (for example, derivatives of displacement with respect to time), or gradients of quantities, which is how they enter differential equations.

Specific mathematical fields include geometry and analytical mechanics. Scientific fields include much of physics and astronomy (celestial mechanics), meteorology (weather modelling), chemistry (reaction rates), biology (infectious diseases, genetic variation), ecology and population modelling (population competition), economics (stock trends, interest rates and the market equilibrium price changes).

Many mathematicians have studied differential equations and contributed to the field, including Newton, Leibniz, the Bernoulli family, Riccati, Clairaut, d'Alembert, and Euler.

A simple example is Newton's second law of motion — the relationship between the displacement x and the time t of an object under the force F, is given by the differential equation

$$m\frac{d^2 x(t)}{dt^2} = F(x(t))$$

which constrains the motion of a particle of constant mass m. In general, F is a function of the position $x(t)$ of the particle at time t. The unknown function $x(t)$ appears on both sides of the differential equation, and is indicated in the notation $F(x(t))$.

Definitions

In what follows, let y be a dependent variable and x an independent variable, and $y = f(x)$ is an unknown function of x. The notation for differentiation varies depending upon the author and upon which notation is most useful for the task at hand. In this context, the Leibniz's notation $(dy/dx, d^2y/dx^2, \ldots d^ny/dx^n)$ is more useful for differentiation and integration, whereas Newton's and Lagrange's notation $(y', y'', \ldots y^{(n)})$ is more useful for representing *derivatives* of any order compactly.

General Definition of an ODE

Given F, a function of x, y, and derivatives of y. Then an equation of the form

$$F(x, y, y', \cdots y^{(n-1)}) = y^{(n)}$$

is called an explicit *ordinary differential equation* of *order n*.

More generally, an *implicit* ordinary differential equation of order n takes the form:

$$F\left(x, y, y', y'', \cdots, y^{(n)}\right) = 0$$

There are further classifications:

Autonomous

A differential equation not depending on x is called *autonomous*.

Linear

A differential equation is said to be *linear* if F can be written as a linear combination of the derivatives of y:

$$y^{(n)} = \sum_{i=0}^{n-1} a_i(x)y^{(i)} + r(x)$$

where $a_i(x)$ and $r(x)$ are continuous functions in x. The function $r(x)$ is called the *source term*, leading to two further important classifications:

Homogeneous

If $r(x) = 0$, and consequently one "automatic" solution is the trivial solution, $y = 0$. The solution of a linear homogeneous equation is a complementary function, denoted here by y_c.

Nonhomogeneous (or inhomogeneous)

If $r(x) \neq 0$. The additional solution to the complementary function is the particular integral, denoted here by y_p.

The general solution to a linear equation can be written as $y = y_c + y_p$.

Non-linear

A differential equation that cannot be written in the form of a linear combination.

System of ODEs

A number of coupled differential equations form a system of equations. If y is a vector whose elements are functions; $y(x) = [y_1(x), y_2(x),..., y_m(x)]$, and F is a vector-valued function of y and its derivatives, then

$$\mathbf{y}^{(n)} = \mathbf{F}\left(x, \mathbf{y}, \mathbf{y}', \mathbf{y}'', \cdots \mathbf{y}^{(n-1)}\right)$$

is an *explicit system of ordinary differential equations* of *order or dimension m*. In column vector form:

$$\begin{pmatrix} y_1^{(n)} \\ y_2^{(n)} \\ \vdots \\ y_m^{(n)} \end{pmatrix} = \begin{pmatrix} f_1\left(x, \mathbf{y}, \mathbf{y}', \mathbf{y}'', \cdots \mathbf{y}^{(n-1)}\right) \\ f_2\left(x, \mathbf{y}, \mathbf{y}', \mathbf{y}'', \cdots \mathbf{y}^{(n-1)}\right) \\ \vdots \\ f_m\left(x, \mathbf{y}, \mathbf{y}', \mathbf{y}'', \cdots \mathbf{y}^{(n-1)}\right) \end{pmatrix}$$

These are not necessarily linear. The *implicit* analogue is:

$$\mathbf{F}\left(x,\mathbf{y},\mathbf{y}',\mathbf{y}'',\cdots\mathbf{y}^{(n)}\right)=\mathbf{0}$$

where $\mathbf{0} = (0, 0,\ldots 0)$ is the zero vector. In matrix form

$$\begin{pmatrix} f_1(x,\mathbf{y},\mathbf{y}',\mathbf{y}'',\cdots\mathbf{y}^{(n)}) \\ f_2(x,\mathbf{y},\mathbf{y}',\mathbf{y}'',\cdots\mathbf{y}^{(n)}) \\ \vdots \\ f_m(x,\mathbf{y},\mathbf{y}',\mathbf{y}'',\cdots\mathbf{y}^{(n)}) \end{pmatrix} = \begin{pmatrix} 0 \\ 0 \\ \vdots \\ 0 \end{pmatrix}$$

For a system of the form $\mathbf{F}\left(x,\mathbf{y},\mathbf{y}'\right)=\mathbf{0}$, some sources also require that the Jacobian matrix $\dfrac{\partial\mathbf{F}(x,\mathbf{u},\mathbf{v})}{\partial\mathbf{v}}$ be non-singular in order to call this an implicit ODE [system]; an implicit ODE system satisfying this Jacobian non-singularity condition can be transformed into an explicit ODE system. In the same sources, implicit ODE systems with a singular Jacobian are termed differential algebraic equations (DAEs). This distinction is not merely one of terminology; DAEs have fundamentally different characteristics and are generally more involved to solve than (nonsigular) ODE systems. Presumably for additional derivatives, the Hessian matrix and so forth are also assumed non-singular according to this scheme, although note that any ODE of order greater than one can be [and usually is] rewritten as system of ODEs of first order, which makes the Jacobian singularity criterion sufficient for this taxonomy to be comprehensive at all orders.

Solutions

Given a differential equation

$$F\left(x,y,y',\cdots,y^{(n)}\right)=0$$

a function $u: I \subset \mathrm{R} \to \mathrm{R}$ is called the *solution* or integral curve for F, if u is n-times differentiable on I, and

$$F(x,u,u',\cdots,u^{(n)})=0 \quad x \in I.$$

Given two solutions $u: J \subset \mathrm{R} \to \mathrm{R}$ and $v: I \subset \mathrm{R} \to \mathrm{R}$, u is called an *extension* of v if $I \subset J$ and

$$u(x)=v(x) \quad x \in I.$$

A solution that has no extension is called a *maximal solution*. A solution defined on all of R is called a *global solution*.

A *general solution* of an nth-order equation is a solution containing n arbitrary independent constants of integration. A *particular solution* is derived from the general solution by setting the constants to particular values, often chosen to fulfill set 'initial conditions or boundary conditions'. A singular solution is a solution that cannot be obtained by assigning definite values to the arbitrary constants in the general solution.

Theories of ODEs

Singular Solutions

The theory of singular solutions of ordinary and partial differential equations was a subject of research from the time of Leibniz, but only since the middle of the nineteenth century did it receive special attention. A valuable but little-known work on the subject is that of Houtain (1854). Darboux (starting in 1873) was a leader in the theory, and in the geometric interpretation of these solutions he opened a field worked by various writers, notable ones being Casorati and Cayley. To the latter is due (1872) the theory of singular solutions of differential equations of the first order as accepted circa 1900.

Reduction to Quadratures

The primitive attempt in dealing with differential equations had in view a reduction to quadratures. As it had been the hope of eighteenth-century algebraists to find a method for solving the general equation of the nth degree, so it was the hope of analysts to find a general method for integrating any differential equation. Gauss (1799) showed, however, that the differential equation meets its limitations very soon unless complex numbers are introduced. Hence, analysts began to substitute the study of functions, thus opening a new and fertile field. Cauchy was the first to appreciate the importance of this view. Thereafter, the real question was to be not whether a solution is possible by means of known functions or their integrals but whether a given differential equation suffices for the definition of a function of the independent variable or variables, and, if so, what are the characteristic properties of this function.

Fuchsian Theory

Two memoirs by Fuchs (*Crelle*, 1866, 1868), inspired a novel approach, subsequently elaborated by Thomé and Frobenius. Collet was a prominent contributor beginning in 1869, although his method for integrating a non-linear system was communicated to Bertrand in 1868. Clebsch (1873) attacked the theory along lines parallel to those followed in his theory of Abelian integrals. As the latter can be classified according to the properties of the fundamental curve that remains unchanged under a rational transformation, so Clebsch proposed to classify the transcendent functions defined by the differential equations according to the invariant properties of the corresponding surfaces $f = 0$ under rational one-to-one transformations.

Lie's Theory

From 1870, Sophus Lie's work put the theory of differential equations on a more satisfactory foundation. He showed that the integration theories of the older mathematicians can, by the introduction of what are now called Lie groups, be referred to a common source, and that ordinary differential equations that admit the same infinitesimal transformations present comparable difficulties of integration. He also emphasized the subject of transformations of contact.

Lie's group theory of differential equations has been certified, namely: (1) that it unifies the many ad hoc methods known for solving differential equations, and (2) that it provides powerful new ways to find solutions. The theory has applications to both ordinary and partial differential equations.

A general approach to solve DEs uses the symmetry property of differential equations, the continuous infinitesimal transformations of solutions to solutions (Lie theory). Continuous group theory, Lie algebras, and differential geometry are used to understand the structure of linear and nonlinear (partial) differential equations for generating integrable equations, to find its Lax pairs, recursion operators, Bäcklund transform, and finally finding exact analytic solutions to the DE.

Symmetry methods have been recognized to study differential equations, arising in mathematics, physics, engineering, and many other disciplines.

Sturm–Liouville Theory

Sturm–Liouville theory is a theory of a special type of second order ordinary differential equations. Their solutions are based on eigenvalues and corresponding eigenfunctions of linear operators defined in terms of second-order homogeneous linear equations. The problems are identified as Sturm-Liouville Problems (SLP) and are named after J.C.F. Sturm and J. Liouville, who studied such problems in the mid-1800s. The interesting fact about regular SLPs is that they have an infinite number of eigenvalues, and the corresponding eigenfunctions form a complete, orthogonal set, which makes orthogonal expansions possible. This is a key idea in applied mathematics, physics, and engineering. SLPs are also useful in the analysis of certain partial differential equations.

Existence and Uniqueness of Solutions

There are several theorems that establish existence and uniqueness of solutions to initial value problems involving ODEs both locally and globally. The two main theorems are

Theorem	Assumption	Conclusion
Peano existence theorem	F continuous	local existence only
Picard–Lindelöf theorem	F Lipschitz continuous	local existence and uniqueness

which are both local results.

Note that uniqueness theorems like the Lipschitz one above do not apply to DAE systems, which may have multiple solutions stemming from their (non-linear) algebraic part alone.

Local Existence and Uniqueness Theorem Simplified

The theorem can be stated simply as follows. For the equation and initial value problem:

$$y' = F(x, y), \quad y_0 = y(x_0)$$

if F and $\partial F/\partial y$ are continuous in a closed rectangle

$$R = [x_0 - a, x_0 + a] \times [y_0 - b, y_0 + b]$$

in the x-y plane, where a and b are real (symbolically: $a, b \in \mathbb{R}$) and × denotes the cartesian product, square brackets denote closed intervals, then there is an interval

$$I = [x_0 - h, x_0 + h] \subset [x_0 - a, x_0 + a]$$

for some $h \in \mathbb{R}$ where *the* solution to the above equation and initial value problem can be found. That is, there is a solution and it is unique. Since there is no restriction on F to be linear, this applies to non-linear equations that take the form F(x, y), and it can also be applied to systems of equations.

Global Uniqueness and Maximum Domain of Solution

When the hypotheses of the Picard–Lindelöf theorem are satisfied, then local existence and uniqueness can be extended to a global result. More precisely:

For each initial condition (x_o, y_o) there exists a unique maximum (possibly infinite) open interval

$$I_{max} = (x_-, x_+), x_\pm \in \mathbb{R}, x_0 \in I_{max}$$

such that any solution that satisfies this initial condition is a restriction of the solution that satisfies this initial condition with domain I_{max}.

In the case that $x_\pm \nrightarrow \pm\infty$, there are exactly two possibilities

- explosion in finite time: $\limsup\limits_{x \to x_\pm} \| y(x) \| \to \infty$

- leaves domain of definition: $\lim\limits_{x \to x_\pm} y(x) \in \partial\overline{\Omega}$

where Ω is the open set in which F is defined, and $\partial\overline{\Omega}$ is its boundary.

Note that the maximum domain of the solution

- is always an interval (to have uniqueness)

- may be smaller than $\mathbb{R}$

- may depend on the specific choice of (x_o, y_o).

Example

$$y' = y^2$$

This means that F(x, y) = y², which is C^1 and therefore locally Lipschitz continuous, satisfying the Picard–Lindelöf theorem.

Even in such a simple setting, the maximum domain of solution cannot be all R, since the solution is

$$y(x) = \frac{y_0}{(x_0 - x)y_0 + 1}$$

which has maximum domain:

$$
\begin{cases}
\mathbb{R} & y_0 = 0 \\[2mm]
(-\infty, x_0 + \dfrac{1}{y_0}) & y_0 > 0 \\[3mm]
(x_0 + \dfrac{1}{y_0}, +\infty) & y_0 < 0
\end{cases}
$$

This shows clearly that the maximum interval may depend on the initial conditions. The domain of y could be taken as being $\mathbf{R} \setminus (x_0 + 1/y_0)$, but this would lead to a domain that is not an interval, so that the side opposite to the initial condition would be disconnected from the initial condition, and therefore not uniquely determined by it.

The maximum domain is not $\mathbb{R}$ because

$$\lim_{x \to x_\pm} \| y(x) \| \to \infty,$$

which is one of the two possible cases according to the above theorem.

Reduction of Order

Differential equations can usually be solved more easily if the order of the equation can be reduced.

Reduction to a First-order System

Any explicit differential equation of order n,

$$F\left(x, y, y', y'', \cdots, y^{(n-1)}\right) = y^{(n)}$$

can be written as a system of n first-order differential equations by defining a new family of unknown functions

$$y_i = y^{(i-1)}.$$

for $i = 1, 2, \dots n$. The n-dimensional system of first-order coupled differential equations is then

$$
\begin{aligned}
y_1' &= y_2 \\
y_2' &= y_3 \\
&\;\;\vdots \\
y_{n-1}' &= y_n \\
y_n' &= F(x, y_1, \cdots, y_n).
\end{aligned}
$$

more compactly in vector notation:

$$\mathbf{y}' = \mathbf{F}(x, \mathbf{y})$$

where

$$= (y_1, \cdots, y_n), \quad \mathbf{F}(x, y_1, \cdots, y_n) = (y_2, \cdots, y_n, F(x, y_1, \cdots, y_n)).$$

Summary of Exact Solutions

Some differential equations have solutions that can be written in an exact and closed form. Several important classes are given here.

In the table below, $P(x)$, $Q(x)$, $P(y)$, $Q(y)$, and $M(x,y)$, $N(x,y)$ are any integrable functions of x, y, and b and c are real given constants, and C_1, C_2,... are arbitrary constants (complex in general). The differential equations are in their equivalent and alternative forms that lead to the solution through integration.

In the integral solutions, λ and ε are dummy variables of integration (the continuum analogues of indices in summation), and the notation $\int^x F(\lambda)d\lambda$ just means to integrate $F(\lambda)$ with respect to λ, then *after* the integration substitute $\lambda = x$, without adding constants (explicitly stated).

Differential equation	Solution method	General solution
Separable equations		
First-order, separable in x and y (general case, see below for special cases) $$P_1(x)Q_1(y) + P_2(x)Q_2(y)\frac{dy}{dx} = 0$$ $$P_1(x)Q_1(y)dx + P_2(x)Q_2(y)dy = 0$$	Separation of variables (divide by P_2Q_1).	$$\int^x \frac{P_1(\lambda)}{P_2(\lambda)}d\lambda + \int^y \frac{Q_2(\lambda)}{Q_1(\lambda)}d\lambda = C$$
First-order, separable in x $$\frac{dy}{dx} = F(x)$$ $$dy = F(x)dx$$	Direct integration.	$$y = \int^x F(\lambda)d\lambda + C$$
First-order, autonomous, separable in y $$\frac{dy}{dx} = F(y)$$ $$dy = F(y)dx$$	Separation of variables (divide by F).	$$x = \int^y \frac{d\lambda}{F(\lambda)} + C$$
First-order, separable in x and y $$P(y)\frac{dy}{dx} + Q(x) = 0$$ $$P(y)dy + Q(x)dx = 0$$	Integrate throughout.	$$\int^y P(\lambda)d\lambda + \int^x Q(\lambda)d\lambda = C$$

General first-order equations		
First-order, homogeneous $$\frac{dy}{dx} = F\left(\frac{y}{x}\right)$$	Set y = ux, then solve by separation of variables in u and x.	$$\ln(Cx) = \int^{y/x} \frac{d\lambda}{F(\lambda) - \lambda}$$
First-order, separable $$yM(xy) + xN(xy)\frac{dy}{dx} = 0$$ $$yM(xy)dx + xN(xy)dy = 0$$	Separation of variables (divide by xy).	$$\ln(Cx) = \int^{xy} \frac{N(\lambda)d\lambda}{\lambda[N(\lambda) - M(\lambda)]}$$ If N = M, the solution is xy = C.
Exact differential, first-order $$M(x,y)\frac{dy}{dx} + N(x,y) = 0$$ $$M(x,y)dy + N(x,y)dx = 0$$ where $\dfrac{\partial M}{\partial x} = \dfrac{\partial N}{\partial y}$	Integrate throughout.	$$F(x,y) = \int^{y} M(x,\lambda)d\lambda + \int^{x} N(\lambda,y)d\lambda$$ $$+ Y(y) + X(x) = C$$ where Y(y) and X(x) are functions from the integrals rather than constant values, which are set to make the final function F(x, y) satisfy the initial equation.
Inexact differential, first-order $$M(x,y)\frac{dy}{dx} + N(x,y) = 0$$ $$M(x,y)dy + N(x,y)dx = 0$$ where $\dfrac{\partial M}{\partial x} \neq \dfrac{\partial N}{\partial y}$	Integration factor $\mu(x, y)$ satisfying $$\frac{\partial(\mu M)}{\partial x} = \frac{\partial(\mu N)}{\partial y}$$	If μ(x, y) can be found: $$F(x,y) = \int^{y}\mu(x,\lambda)M(x,\lambda)d\lambda + \int^{x}\mu(\lambda,y)N(\lambda,y)d\lambda$$ $$+ Y(y) + X(x) = C$$
General second-order equations		
Second-order, autonomous $$\frac{d^2 y}{dx^2} = F(y)$$	Multiply equation by $2dy/dx$, substitute $2\dfrac{dy}{dx}\dfrac{d^2y}{dx^2} = \dfrac{d}{dx}\left(\dfrac{dy}{dx}\right)^2$ then integrate twice.	$$x = \pm\int^{y} \frac{d\lambda}{\sqrt{2\int^{\lambda}F(\epsilon)d\epsilon + C_1}} + C_2$$
Linear equations (up to nth order)		
First-order, linear, inhomogeneous, function coefficients $$\frac{dy}{dx} + P(x)y = Q(x)$$	Integrating factor: $e^{\int^{x}P(\lambda)d\lambda}$	$$y = e^{-\int^{x}P(\lambda)d\lambda}\left[\int^{x}e^{\int^{\lambda}P(\epsilon)d\epsilon}Q(\lambda)d\lambda + C\right]$$
Second-order, linear, inhomogeneous, constant coefficients $$\frac{d^2 y}{dx^2} + b\frac{dy}{dx} + cy = r(x)$$	Complementary function y_c: assume $y_c = e^{\alpha x}$, substitute and solve polynomial in α, to find the linearly independent functions $e^{\alpha_j x}$ Particular integral yp: in general the method of variation of parameters, though for very simple $r(x)$ inspection may work.	$$y = y_c + y_p$$ If b2 > 4c, then: $$y_c = C_1 e^{\left(-b + \sqrt{b^2 - 4c}\right)\frac{x}{2}} + C_2 e^{-\left(b + \sqrt{b^2 - 4c}\right)\frac{x}{2}}$$ If b2 = 4c, then: $$y_c = (C_1 x + C_2)e^{-\frac{1}{2}bx}$$ If b2 < 4c, then: $$y_c = e^{-b\frac{x}{2}}\left[C_1\sin\left(\sqrt{b^2-4c}\,\frac{x}{2}\right) + C_2\cos\left(\sqrt{b^2-4c}\,\frac{x}{2}\right)\right]$$

| nth-order, linear, inhomogeneous, constant coefficients $$\sum_{j=0}^{n} b_j \frac{d^j y}{dx^j} = r(x)$$ | Complementary function yc: assume yc = eαx, substitute and solve polynomial in α, to find the linearly independent functions $e^{\alpha_j x}$

 Particular integral yp: in general the method of variation of parameters, though for very simple r(x) inspection may work. | $y = y_c + y_p$

 Since α_j are the solutions of the polynomial of degree n: $\prod_{j=1}^{n}(\alpha - \alpha_j) = 0$ for αj all different,

 $$y_c = \sum_{j=1}^{n} C_j e^{\alpha_j x}$$

 for each root αj repeated kj times,

 $$y_c = \sum_{j=1}^{n} \left(\sum_{\ell=1}^{k_j} C_\ell x^{\ell-1} \right) e^{\alpha_j x}$$

 for some αj complex, then setting α = χj + iγj, and using Euler's formula, allows some terms in the previous results to be written in the form

 $C_j e^{\alpha_j x} = C_j e^{\chi_j x} \cos(\gamma_j x + \phi_j)$

 where ϕ_j is an arbitrary constant (phase shift). |

Software for ODE Solving

- Maxima computer algebra system (GPL)

- COPASI a free (Artistic License 2.0) software package for the integration and analysis of ODEs.

- MATLAB a Technical Computing Software (MATrix LABoratory)

- GNU Octave a high-level language, primarily intended for numerical computations.

- Scilab open source software for numerical computation.

- Maple

- Mathematica

- Julia (programming language)

- SciPy a Python package that includes an ODE integration module.

- Chebfun an open-source package, written in MATLAB, for computing with functions to 15-digit accuracy.

- GNU R an open source computational environment primarily intended for statistics, which includes package for ODE solving.

- EROS.NET a free ODE solver for .NET.

Method of Undetermined Coefficients

In mathematics, the method of undetermined coefficients is an approach to finding a particular solution to certain inhomogeneous ordinary differential equations and recurrence relations. It is closely related to the annihilator method, but instead of using a particular kind of differential operator (the annihilator) in order to find the best possible form of the particular solution, a "guess" is made as to the appropriate form, which is then tested by differentiating the resulting equation. For complex equations, the annihilator method or variation of parameters is less time consuming to perform.

Undetermined coefficients is not as general a method as variation of parameters, since it only works for differential equations that follow certain forms.

Description of the Method

Consider a linear non-homogeneous ordinary differential equation of the form

$$\sum_{i=0}^{n} c_i y^{(i)} + y^{(n+1)} = g(x)$$

... where $y^{(i)}$ denotes the i-th derivate of y, and c_i denotes a function of x

The method consists of finding the general homogeneous solution y_c for the complementary linear homogeneous differential equation

$$\sum_{i=0}^{n} c_i y^{(i)} + y^{(n+1)} = 0,$$

and a particular integral y_p of the linear non-homogeneous ordinary differential equation based on $g(x)$. Then the general solution y to the linear non-homogeneous ordinary differential equation would be

$$y = y_c + y_p.$$

If $g(x)$ consists of the sum of two functions $h(x) + w(x)$ and we say that y_{p_1} is the solution based on $h(x)$ and y_{p_2} the solution based on $w(x)$. Then, using a superposition principle, we can say that the particular integral y_p is

$$y_p = y_{p_1} + y_{p_2}.$$

Typical Forms of the Particular Integral

In order to find the particular integral, we need to 'guess' its form, with some coefficients left as variables to be solved for. This takes the form of the first derivative of complementary function. Below is a table of some typical functions and the solution to guess for them.

Function of x	Form for y
ke^{ax}	Ce^{ax}

$kx^n, n = 0, 1, 2, \ldots$

$$\sum_{i=0}^{n} K_i x^i$$

$k \cos(ax)$ or $k \sin(ax)$

$K \cos(ax) + M \sin(ax)$

$ke^{ax} \cos(bx)$ or $ke^{ax} \sin(bx)$

$e^{ax}(K \cos(bx) + M \sin(bx))$

$\left(\sum_{i=0}^{n} k_i x^i \right) \cos(bx)$ or $\left(\sum_{i=0}^{n} k_i x^i \right) \sin(bx)$

$\left(\sum_{i=0}^{n} Q_i x^i \right) \cos(bx) + \left(\sum_{i=0}^{n} R_i x^i \right) \sin(bx)$

$\left(\sum_{i=0}^{n} k_i x^i \right) e^{ax} \cos(bx)$ or $\left(\sum_{i=0}^{n} k_i x^i \right) e^{ax} \sin(bx)$

$e^{ax} \left(\left(\sum_{i=0}^{n} Q_i x^i \right) \cos(bx) + \left(\sum_{i=0}^{n} R_i x^i \right) \sin(bx) \right)$

If a term in the above particular integral for y appears in the homogeneous solution, it is necessary to multiply by a sufficiently large power of x in order to make the solution independent. If the function of x is a sum of terms in the above table, the particular integral can be guessed using a sum of the corresponding terms for y.

Examples

Example 1

Find a particular integral of the equation

$$y'' + y = t \cos t.$$

The right side $t \cos t$ has the form

$$P_n e^{\alpha t} \cos \beta t$$

with $n = 1$, $\alpha = 0$, and $\beta = 1$.

Since $\alpha + i\beta = i$ is a *simple root* of the characteristic equation

$$\lambda^2 + 1 = 0$$

we should try a particular integral of the form

$$\begin{aligned}
y_p &= t[F_1(t)e^{\alpha t} \cos \beta t + G_1(t)e^{\alpha t} \sin \beta t] \\
&= t[F_1(t) \cos t + G_1(t) \sin t] \\
&= t[(A_0 t + A_1) \cos t + (B_0 t + B_1) \sin t] \\
&= (A_0 t^2 + A_1 t) \cos t + (B_0 t^2 + B_1 t) \sin t.
\end{aligned}$$

Substituting y_p into the differential equation, we have the identity

$$t\cos t = y_p'' + y_p$$
$$= [(A_0 t^2 + A_1 t)\cos t + (B_0 t^2 + B_1 t)\sin t]''$$
$$+ [(A_0 t^2 + A_1 t)\cos t + (B_0 t^2 + B_1 t)\sin t]$$
$$= [2A_0 \cos t + 2(2A_0 t + A_1)(-\sin t) + (A_0 t^2 + A_1 t)(-\cos t)]$$
$$+ [2B_0 \sin t + 2(2B_0 t + B_1)\cos t + (B_0 t^2 + B_1 t)(-\sin t)]$$
$$+ [(A_0 t^2 + A_1 t)\cos t + (B_0 t^2 + B_1 t)\sin t]$$
$$= [4B_0 t + (2A_0 + 2B_1)]\cos t + [-4A_0 t + (-2A_1 + 2B_0)]\sin t.$$

Comparing both sides, we have

$$4B_0 = 1$$
$$2A_0 \qquad\qquad +2B_1 = 0$$
$$-4A_0 \qquad\qquad\qquad = 0$$
$$-2A_1 \quad +2B_0 \qquad\qquad = 0$$

which has the solution $A_0 = 0$, $A_1 = 1/4$, $B_0 = 1/4$, $B_1 = 0$. We then have a particular integral

$$y_p = \frac{1}{4} t\cos t + \frac{1}{4} t^2 \sin t.$$

Example 2

Consider the following linear nonhomogeneous differential equation:

$$\frac{dy}{dx} = y + e^x.$$

This is like the first example above, except that the nonhomogeneous part (e^x) is *not* linearly independent to the general solution of the homogeneous part ($c_1 e^x$); as a result, we have to multiply our guess by a sufficiently large power of x to make it linearly independent.

Here our guess becomes:

$$y_p = Axe^x.$$

By substituting this function and its derivative into the differential equation, one can solve for A:

$$\frac{d}{dx}\left(Axe^x\right) = Axe^x + e^x$$

$$Axe^x + Ae^x = Axe^x + e^x$$

$$A = 1.$$

So, the general solution to this differential equation is thus:

$$y = c_1 e^x + xe^x.$$

Example 3

Find the general solution of the equation:

$$\frac{dy}{dt} = t^2 - y$$

f(t), t^2, is a polynomial of degree 2, so we look for a solution using the same form,

$$y_p = At^2 + Bt + C \text{, where}$$

$$\frac{dy_p}{dt} = 2At + B$$

Plugging this particular integral with constants A, B, and C into the original equation yields,

$$2At + B = t^2 - (At^2 + Bt + C) \text{, where}$$

$$t^2 - At^2 = 0 \text{ and } -Bt = 2At \text{ and } -C = B$$

Replacing resulting constants,

$$y_p = t^2 - 2t + 2$$

To solve for the general solution,

$$y = y_p + y_c$$

where y_c is the homogeneous solution $y_c = c_1 e^{-t}$, therefore, the general solution is:

$$y = t^2 - 2t + 2 + c_1 e^{-t}$$

Numerical Methods for Ordinary Differential Equations

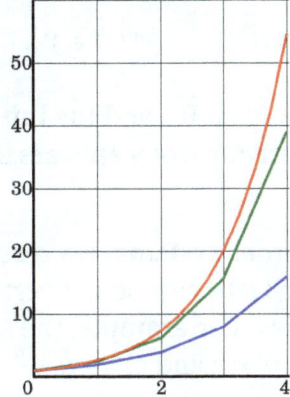

Illustration of numerical integration for the differential equation $y' = y$, $y(0) = 1$. Blue: the Euler method, green: the midpoint method, red: the exact solution, $y = e^t$. The step size is $h = 1.0$.

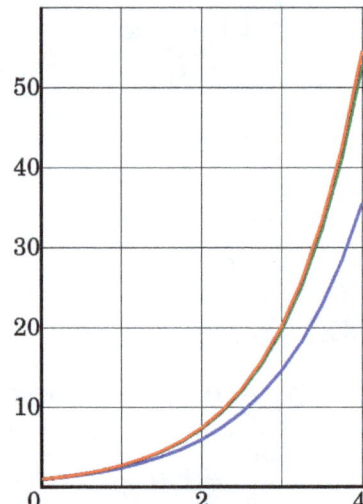

The same illustration for $h = 0.25$. It is seen that the midpoint method converges faster than the Euler method.

Numerical methods for ordinary differential equations are methods used to find numerical approximations to the solutions of ordinary differential equations (ODEs). Their use is also known as "numerical integration", although this term is sometimes taken to mean the computation of integrals.

Many differential equations cannot be solved using symbolic computation ("analysis"). For practical purposes, however – such as in engineering – a numeric approximation to the solution is often sufficient. The algorithms studied here can be used to compute such an approximation. An alternative method is to use techniques from calculus to obtain a series expansion of the solution.

Ordinary differential equations occur in many scientific disciplines, for instance in physics, chemistry, biology, and economics. In addition, some methods in numerical partial differential equations convert the partial differential equation into an ordinary differential equation, which must then be solved.

The Problem

A first-order differential equation is an Initial value problem (IVP) of the form,

$$y'(t) = f(t, y(t)), \qquad y(t_0) = y_0, \qquad (1)$$

where f is a function that maps $[t_0, \infty) \times \mathbf{R}^d$ to $\mathbf{R}^d$, and the initial condition $y_0 \in \mathbf{R}^d$ is a given vector. *First-order* means that only the first derivative of y appears in the equation, and higher derivatives are absent.

Without loss of generality to higher-order systems, we restrict ourselves to *first-order* differential equations, because a higher-order ODE can be converted into a larger system of first-order equations by introducing extra variables. For example, the second-order equation $y'' = -y$ can be rewritten as two first-order equations: $y' = z$ and $z' = -y$.

In this section, we describe numerical methods for IVPs, and remark that *boundary value problems* (BVPs) require a different set of tools. In a BVP, one defines values, or components of the solution y at more than one point. Because of this, different methods need to be used to solve BVPs. For example, the shooting method (and its variants) or global methods like finite differences, Galerkin methods, or collocation methods are appropriate for that class of problems.

The Picard–Lindelöf theorem states that there is a unique solution, provided f is Lipschitz-continuous.

Methods

Numerical methods for solving first-order IVPs often fall into one of two large categories: linear multistep methods, or Runge-Kutta methods. A further division can be realized by dividing methods into those that are explicit and those that are implicit. For example, implicit linear multistep methods include Adams-Moulton methods, and backward differentiation methods (BDF), whereas implicit Runge-Kutta methods include diagonally implicit Runge-Kutta (DIRK), singly diagonally implicit runge kutta (SDIRK), and Gauss-Radau (based on Gaussian quadrature) numerical methods. Explicit examples from the linear multistep family include the Adams-Bashforth methods, and any Runge-Kutta method with a lower diagonal Butcher tableau is explicit. A loose rule of thumb dictates that *stiff* differential equations require the use of implicit schemes, whereas non-stiff problems can be solved more efficiently with explicit schemes.

The so-called general linear methods (GLMs) are a generalization of the above two large classes of methods.

Euler Method

From any point on a curve, you can find an approximation of a nearby point on the curve by moving a short distance along a line tangent to the curve.

Starting with the differential equation (1), we replace the derivative y' by the finite difference approximation

$$y'(t) \approx \frac{y(t+h) - y(t)}{h}, \qquad (2)$$

which when re-arranged yields the following formula

$$y(t+h) \approx y(t) + hy'(t)$$

and using (1) gives:

$$y(t+h) \approx y(t) + hf(t, y(t)). \qquad (3)$$

This formula is usually applied in the following way. We choose a step size h, and we construct the sequence $t_0, t_1 = t_0 + h, t_2 = t_0 + 2h, \ldots$ We denote by y_n a numerical estimate of the exact solution $y(t_n)$. Motivated by (3), we compute these estimates by the following recursive scheme

$$y_{n+1} = y_n + hf(t_n, y_n). \qquad (4)$$

This is the *Euler method* (or *forward Euler method*, in contrast with the *backward Euler method*, to be described below). The method is named after Leonhard Euler who described it in 1768.

The Euler method is an example of an *explicit* method. This means that the new value y_{n+1} is defined in terms of things that are already known, like y_n.

Backward Euler Method

If, instead of (2), we use the approximation

$$y'(t) \approx \frac{y(t) - y(t-h)}{h}, \qquad (5)$$

we get the *backward Euler method*:

$$y_{n+1} = y_n + hf(t_{n+1}, y_{n+1}). \qquad (6)$$

The backward Euler method is an *implicit* method, meaning that we have to solve an equation to find y_{n+1}. One often uses fixed point iteration or (some modification of) the Newton–Raphson method to achieve this.

It costs more time to solve this equation than explicit methods; this cost must be taken into consideration when one selects the method to use. The advantage of implicit methods such as (6) is that they are usually more stable for solving a stiff equation, meaning that a larger step size h can be used.

First-order Exponential Integrator Method

Exponential integrators describe a large class of integrators that have recently seen a lot of development. They date back to at least the 1960s.

In place of (1), we assume the differential equation is either of the form

$$y'(t) = -Ay + \mathcal{N}(y), \qquad (7)$$

or it has been locally linearized about a background state to produce a linear term $-Ay$ and a nonlinear term $\mathcal{N}(y)$.

Exponential integrators are constructed by multiplying (7) by e^{At}, and exactly integrating the result over a time interval $[t_n, t_{n+1} = t_n + h]$:

$$y_{n+1} = e^{-Ah}y_n + \int_0^h e^{-(h-\tau)A}\mathcal{N}\big(y(t_n + \tau)\big)d\tau.$$

This approximation is exact, but it doesn't define the integral.

The first-order exponential integrator can be realized by holding $\mathcal{N}(y(t_n + \tau))$ constant over the full interval:

$$y_{n+1} = e^{-Ah}y_n + A^{-1}(1 - e^{-Ah})\mathcal{N}(y(t_n)). \qquad (8)$$

Generalizations

The Euler method is often not accurate enough. In more precise terms, it only has order one (the concept of *order* is explained below). This caused mathematicians to look for higher-order methods.

One possibility is to use not only the previously computed value y_n to determine y_{n+1}, but to make the solution depend on more past values. This yields a so-called *multistep method*. Perhaps the simplest is the Leapfrog method which is second order and (roughly speaking) relies on two time values.

Almost all practical multistep methods fall within the family of linear multistep methods, which have the form

$$\alpha_k y_{n+k} + \alpha_{k-1} y_{n+k-1} + \cdots + \alpha_0 y_n$$

$$= h\left[\beta_k f(t_{n+k}, y_{n+k}) + \beta_{k-1} f(t_{n+k-1}, y_{n+k-1}) + \cdots + \beta_0 f(t_n, y_n)\right].$$

Another possibility is to use more points in the interval $[t_n, t_{n+1}]$. This leads to the family of Runge–Kutta methods, named after Carl Runge and Martin Kutta. One of their fourth-order methods is especially popular.

Advanced Features

A good implementation of one of these methods for solving an ODE entails more than the time-stepping formula.

It is often inefficient to use the same step size all the time, so *variable step-size methods* have been developed. Usually, the step size is chosen such that the (local) error per step is below some tolerance level. This means that the methods must also compute an *error indicator*, an estimate of the local error.

An extension of this idea is to choose dynamically between different methods of different orders (this is called a *variable order method*). Methods based on Richardson extrapolation, such as the Bulirsch–Stoer algorithm, are often used to construct various methods of different orders.

Other desirable features include:

- *dense output*: cheap numerical approximations for the whole integration interval, and not only at the points t_0, t_1, t_2, ...

- *event location*: finding the times where, say, a particular function vanishes. This typically requires the use of a root-finding algorithm.

- support for parallel computing.

- when used for integrating with respect to time, time reversibility

Alternative Methods

Many methods do not fall within the framework discussed here. Some classes of alternative methods are:

- *multiderivative methods*, which use not only the function f but also its derivatives. This class includes *Hermite–Obreschkoff methods* and *Fehlberg methods*, as well as methods like the Parker–Sochacki method or Bychkov-Scherbakov method, which compute the coefficients of the Taylor series of the solution y recursively.

- *methods for second order ODEs*. We said that all higher-order ODEs can be transformed to first-order ODEs of the form (1). While this is certainly true, it may not be the best way to proceed. In particular, *Nyström methods* work directly with second-order equations.

- *geometric integration methods* are especially designed for special classes of ODEs (e.g., symplectic integrators for the solution of Hamiltonian equations). They take care that the numerical solution respects the underlying structure or geometry of these classes.

- *Quantized State Systems Methods* are a family of ODE integration methods based on the idea of state quantization. They are efficient when simulating sparse systems with frequent discontinuities.

Parallel-in-time Methods

For applications that require parallel computing on supercomputers, the degree of concurrency offered by a numerical method becomes relevant. In view of the challenges from exascale computing systems, numerical methods for initial value problems which can provide concurrency in temporal direction are being studied. Parareal is a relatively well known example of such a *parallel-in-time* integration method, but early ideas go back into the 1960s.

Analysis

Numerical analysis is not only the design of numerical methods, but also their analysis. Three central concepts in this analysis are:

- *convergence*: whether the method approximates the solution,

- *order*: how well it approximates the solution, and

- *stability*: whether errors are damped out.

Convergence

A numerical method is said to be *convergent* if the numerical solution approaches the exact solution as the step size h goes to 0. More precisely, we require that for every ODE (1) with a Lipschitz function f and every $t^* > 0$,

$$\lim_{h \to 0+} \max_{n=0,1,\ldots,\lfloor t^*/h \rfloor} \| y_{n,h} - y(t_n) \| = 0.$$

All the methods mentioned above are convergent. In fact, a numerical scheme has to be convergent to be of any use.

Consistency and Order

Suppose the numerical method is

$$y_{n+k} = \Psi(t_{n+k}; y_n, y_{n+1}, \ldots, y_{n+k-1}; h).$$

The *local (truncation) error* of the method is the error committed by one step of the method. That is, it is the difference between the result given by the method, assuming that no error was made in earlier steps, and the exact solution:

$$\delta_{n+k}^h = \Psi\left(t_{n+k}; y(t_n), y(t_{n+1}), \ldots, y(t_{n+k-1}); h\right) - y(t_{n+k}).$$

The method is said to be *consistent* if

$$\lim_{h \to 0} \frac{\delta_{n+k}^h}{h} = 0.$$

The method has *order* p if

$$\delta_{n+k}^h = O(h^{p+1}) \quad \text{as } h \to 0.$$

Hence a method is consistent if it has an order greater than 0. The (forward) Euler method (4) and the backward Euler method (6) introduced above both have order 1, so they are consistent. Most methods being used in practice attain higher order. Consistency is a necessary condition for convergence, but not sufficient; for a method to be convergent, it must be both consistent and zero-stable.

A related concept is the *global (truncation) error*, the error sustained in all the steps one needs to reach a fixed time t. Explicitly, the global error at time t is $y_N - y(t)$ where $N = (t-t_0)/h$. The global error of a pth order one-step method is $O(h^p)$; in particular, such a method is convergent. This statement is not necessarily true for multi-step methods.

Stability and Stiffness

For some differential equations, application of standard methods —such as the Euler method, explicit Runge–Kutta methods, or multistep methods (e.g., Adams–Bashforth methods)— exhibit instability in the solutions, though other methods may produce stable solutions. This "difficult behaviour" in the equation (which may not necessarily be complex itself) is described as *stiffness*, and is often caused by the presence of different time scales in the underlying problem. For example, a collision in a mechanical system like in an impact oscillator typically occurs at much smaller time scale than the time for the motion of objects; this discrepancy makes for very "sharp turns" in the curves of the state parameters.

Stiff problems are ubiquitous in chemical kinetics, control theory, solid mechanics, weather forecasting, biology, plasma physics, and electronics. One way to overcome stiffness is to extend the notion of differential equation to that of differential inclusion, which allows for and models non-smoothness.

History

Below is a timeline of some important developments in this field.

- 1768 - Leonhard Euler publishes his method.

- 1824 - Augustin Louis Cauchy proves convergence of the Euler method. In this proof, Cauchy uses the implicit Euler method.

- 1855 - First mention of the multistep methods of John Couch Adams in a letter written by F. Bashforth.

- 1895 - Carl Runge publishes the first Runge–Kutta method.

- 1905 - Martin Kutta describes the popular fourth-order Runge–Kutta method.

- 1910 - Lewis Fry Richardson announces his extrapolation method, Richardson extrapolation.

- 1952 - Charles F. Curtiss and Joseph Oakland Hirschfelder coin the term *stiff equations*.

- 1963 - Germund Dahlquist introduces *A-stability* of integration methods.

Numerical Solutions to Second-order One-dimensional Boundary Value Problems

Boundary value problems (BVPs) are usually solved numerically by solving an approximately equivalent matrix problem obtained by discretizing the original BVP. The most commonly used method for numerically solving BVPs in one dimension is called the Finite Difference Method. This method takes advantage of linear combinations of point values to construct finite difference coefficients that describe derivatives of the function. For example, the second-order central difference approximation to the first derivative is given by:

$$\frac{u_{i+1} - u_{i-1}}{2h} = u'(x_i) + \mathcal{O}(h^2),$$

and the second-order central difference for the second derivative is given by:

$$\frac{u_{i+1} - 2u_i + u_{i-1}}{h^2} = u''(x_i) + \mathcal{O}(h^2).$$

In both of these formulae, $h = x_i - x_{i-1}$ is the distance between neighbouring x values on the discretized domain. One then constructs a linear system that can then be solved by standard matrix methods. For instance, suppose the equation to be solved is:

$$\frac{d^2u}{dx^2} - u = 0,$$

$$u(0) = 0,$$

$$u(1) = 1.$$

The next step would be to discretize the problem and use linear derivative approximations such as

$$u_i'' = \frac{u_{i+1} - 2u_i + u_{i-1}}{h^2}$$

and solve the resulting system of linear equations. This would lead to equations such as:

$$\frac{u_{i+1} - 2u_i + u_{i-1}}{h^2} - u_i = 0, \quad \forall i = 1, 2, 3, ..., n-1.$$

On first viewing, this system of equations appears to have difficulty associated with the fact that the equation involves no terms that are not multiplied by variables, but in fact this is false. At $i = 1$ and $n - 1$ there is a term involving the boundary values $u(0) = u_0$ and $u(1) = u_n$ and since these two values are known, one can simply substitute them into this equation and as a result have a non-homogeneous linear system of equations that has non-trivial solutions.

Partial Differential Equation

A visualisation of a solution to the two-dimensional heat equation
with temperature represented by the third dimension

In mathematics, a partial differential equation (PDE) is a differential equation that contains unknown multivariable functions and their partial derivatives. (A special case are ordinary differential equations (ODEs), which deal with functions of a single variable and their derivatives.) PDEs are used to formulate problems involving functions of several variables, and are either solved by hand, or used to create a relevant computer model.

PDEs can be used to describe a wide variety of phenomena such as sound, heat, electrostatics, electrodynamics, fluid dynamics, elasticity, or quantum mechanics. These seemingly distinct physical phenomena can be formalised similarly in terms of PDEs. Just as ordinary differential equations often model one-dimensional dynamical systems, partial differential equations often model multidimensional systems. PDEs find their generalisation in stochastic partial differential equations.

Introduction

Partial differential equations (PDEs) are equations that involve rates of change with respect to continuous variables. The position of a rigid body is specified by six numbers, but the configuration of a fluid is given by the continuous distribution of several parameters, such as the temperature, pressure, and so forth. The dynamics for the rigid body take place in a finite-dimensional configuration space; the dynamics for the fluid occur in an infinite-dimensional configuration space. This distinction usually makes PDEs much harder to solve than ordinary differential equations (ODEs), but here again, there will be simple solutions for linear problems. Classic domains where PDEs are used include acoustics, fluid dynamics, electrodynamics, and heat transfer.

A partial differential equation (PDE) for the function $u(x_1, \cdots, x_n)$ is an equation of the form

$$f\left(x_1, \ldots, x_n, u, \frac{\partial u}{\partial x_1}, \ldots, \frac{\partial u}{\partial x_n}, \frac{\partial^2 u}{\partial x_1 \partial x_1}, \ldots, \frac{\partial^2 u}{\partial x_1 \partial x_n}, \ldots\right) = 0.$$

If f is a linear function of u and its derivatives, then the PDE is called linear. Common examples of linear PDEs include the heat equation, the wave equation, Laplace's equation, Helmholtz equation, Klein–Gordon equation, and Poisson's equation.

A relatively simple PDE is

$$\frac{\partial u}{\partial x}(x, y) = 0.$$

This relation implies that the function $u(x,y)$ is independent of x. However, the equation gives no information on the function's dependence on the variable y. Hence the general solution of this equation is

$$u(x, y) = f(y),$$

where f is an arbitrary function of y. The analogous ordinary differential equation is

$$\frac{du}{dx}(x) = 0,$$

which has the solution

$$u(x) = c,$$

where c is any constant value. These two examples illustrate that general solutions of ordinary differential equations (ODEs) involve arbitrary constants, but solutions of PDEs involve arbitrary functions. A solution of a PDE is generally not unique; additional conditions must generally be specified on the boundary of the region where the solution is defined. For instance, in the simple example above, the function $f(y)$ can be determined if u is specified on the line $x = 0$.

Existence and Uniqueness

Although the issue of existence and uniqueness of solutions of ordinary differential equations has a very satisfactory answer with the Picard–Lindelöf theorem, that is far from the case for partial differential equations. The Cauchy–Kowalevski theorem states that the Cauchy problem for any partial differential equation whose coefficients are analytic in the unknown function and its derivatives, has a locally unique analytic solution. Although this result might appear to settle the existence and uniqueness of solutions, there are examples of linear partial differential equations whose coefficients have derivatives of all orders (which are nevertheless not analytic) but which have no solutions at all. Even if the solution of a partial differential equation exists and is unique, it may nevertheless have undesirable properties. The mathematical study of these questions is usually in the more powerful context of weak solutions.

An example of pathological behavior is the sequence (depending upon n) of Cauchy problems for the Laplace equation

$$\frac{\partial^2 u}{\partial x^2} + \frac{\partial^2 u}{\partial y^2} = 0,$$

with boundary conditions

$$u(x,0) = 0,$$

$$\frac{\partial u}{\partial y}(x,0) = \frac{\sin(nx)}{n},$$

where n is an integer. The derivative of u with respect to y approaches 0 uniformly in x as n increases, but the solution is

$$u(x,y) = \frac{\sinh(ny)\sin(nx)}{n^2}.$$

This solution approaches infinity if nx is not an integer multiple of π for any non-zero value of y. The Cauchy problem for the Laplace equation is called *ill-posed* or *not well-posed*, since the solution does not continuously depend on the data of the problem. Such ill-posed problems are not usually satisfactory for physical applications.

Notation

In PDEs, it is common to denote partial derivatives using subscripts. That is:

$$u_x = \frac{\partial u}{\partial x}$$

$$u_{xx} = \frac{\partial^2 u}{\partial x^2}$$

$$u_{xy} = \frac{\partial^2 u}{\partial y \partial x} = \frac{\partial}{\partial y}\left(\frac{\partial u}{\partial x}\right).$$

Especially in physics, del or Nabla (∇) is often used to denote spatial derivatives, and $\dot{u}\ddot{u}$ for time derivatives. For example, the wave equation (described below) can be written as

$$\ddot{u} = c^2 \nabla^2 u$$

or

$$\ddot{u} = c^2 \Delta u$$

where Δ is the Laplace operator.

Examples

Heat Equation in one Space Dimension

The equation for conduction of heat in one dimension for a homogeneous body has

$$u_t = \alpha u_{xx}$$

where $u(t,x)$ is temperature, and α is a positive constant that describes the rate of diffusion. The Cauchy problem for this equation consists in specifying $u(0, x) = f(x)$, where $f(x)$ is an arbitrary function.

General solutions of the heat equation can be found by the method of separation of variables. Some examples appear in the heat equation article. They are examples of Fourier series for periodic f and Fourier transforms for non-periodic f. Using the Fourier transform, a general solution of the heat equation has the form

$$u(t,x) = \frac{1}{\sqrt{2\pi}}\int_{-\infty}^{\infty} F(\xi)e^{-\alpha\xi^2 t}e^{i\xi x}d\xi,$$

where F is an arbitrary function. To satisfy the initial condition, F is given by the Fourier transform of f, that is

$$F(\xi) = \frac{1}{\sqrt{2\pi}}\int_{-\infty}^{\infty} f(x)e^{-i\xi x}dx.$$

If f represents a very small but intense source of heat, then the preceding integral can be approximated by the delta distribution, multiplied by the strength of the source. For a source whose strength is normalized to 1, the result is

$$F(\xi) = \frac{1}{\sqrt{2\pi}},$$

and the resulting solution of the heat equation is

$$u(t,x) = \frac{1}{2\pi} \int_{-\infty}^{\infty} e^{-\alpha\xi^2 t} e^{i\xi x} d\xi.$$

This is a Gaussian integral. It may be evaluated to obtain

$$u(t,x) = \frac{1}{2\sqrt{\pi\alpha t}} \exp\left(-\frac{x^2}{4\alpha t}\right).$$

This result corresponds to the normal probability density for x with mean 0 and variance $2\alpha t$. The heat equation and similar diffusion equations are useful tools to study random phenomena.

Wave Equation in One Spatial Dimension

The wave equation is an equation for an unknown function $u(k, x)$ of the form

$$u_{kk} = m^2 u_{xx}.$$

Here u might describe the displacement of a stretched string from equilibrium, or the difference in air pressure in a tube, or the magnitude of an electromagnetic field in a tube, and m is a number that corresponds to the velocity of the wave. The Cauchy problem for this equation consists in prescribing the initial displacement and velocity of a string or other medium:

$$u(0,x) = f(x),$$

$$u_k(0,x) = g(x),$$

where f and g are arbitrary given functions. The solution of this problem is given by d'Alembert's formula:

$$u(k,x) = \tfrac{1}{2}\big[f(x-mk) + f(x+mk)\big] + \frac{1}{2m} \int_{x-mk}^{x+mk} g(y)dy.$$

This formula implies that the solution at (k,x) depends on only the data on the segment of the initial line that is cut out by the characteristic curves

$$x - mk = \text{constant}, \quad x + mk = \text{constant},$$

that are drawn backward from that point. These curves correspond to signals that propagate with velocity m forward and backward. Conversely, the influence of the data at any given point on the initial line propagates with the finite velocity m: there is no effect outside a triangle through that point whose sides are characteristic curves. This behavior is very different from the solution for the heat equation, where the effect of a point source appears (with small amplitude) instantaneously at every point in space. The solution given above is also valid if $k < 0$, and the explicit formula shows that the solution depends on smoothly upon the data: both the forward and backward Cauchy problems for the wave equation are well-posed.

Generalised Heat-like Equation in One Space Dimension

Where heat-like equation means equations of the form:

$$\frac{\partial u}{\partial t} = \hat{H}u + f(x,t)u + g(x,t)$$

where $\hat{H}$ is a Sturm–Liouville operator subject to the boundary conditions:

$$u(x,0) = h(x).$$

Then:

If:

$$\hat{H}X_n = \lambda_n X_n$$

$$X_n(a) = X_n(b) = 0$$

$$\dot{a}_n(t) - \lambda_n a_n(t) - \sum_m (X_n f(x,t), X_m)a_m(t) = (g(x,t), X_n)$$

$$a_n(0) = \frac{(h(x), X_n)}{(X_n, X_n)}$$

$$u(x,t) = \sum_n a_n(t)X_n(x)$$

where

$$(f,g) = \int_a^b f(x)g(x)w(x)dx.$$

Spherical Waves

Spherical waves are waves whose amplitude depends on the radial distance r from a central point source only. For such waves, the three-dimensional wave equation takes the form

$$u_{tt} = c^2 \left[u_{rr} + \frac{2}{r}u_r \right].$$

This is equivalent to

$$(ru)_{tt} = c^2 \left[(ru)_{rr} \right],$$

and hence the quantity ru satisfies the one-dimensional wave equation. Therefore, a general solution for spherical waves has the form

$$u(t,r) = \frac{1}{r}\left[F(r-ct) + G(r+ct)\right],$$

where F and G are completely arbitrary functions. Radiation from an antenna corresponds to the case where G is identically zero. Thus, the waveform transmitted from an antenna has no distortion in time: the only distorting factor is $1/r$. This feature of undistorted propagation of waves is not present if there are two spatial dimensions.

Laplace Equation in Two Dimensions

The Laplace equation for an unknown function of two variables φ has the form

$$\varphi_{xx} + \varphi_{yy} = 0.$$

Solutions of Laplace's equation are called harmonic functions.

Connection with Holomorphic Functions

Solutions of the Laplace equation in two dimensions are intimately connected with analytic functions of a complex variable (a.k.a. holomorphic functions): the real and imaginary parts of any analytic function are conjugate harmonic functions: they both satisfy the Laplace equation, and their gradients are orthogonal. If $f=u+iv$, then the Cauchy–Riemann equations state that

$$u_x = v_y, \quad v_x = -u_y,$$

and it follows that

$$u_{xx} + u_{yy} = 0, \quad v_{xx} + v_{yy} = 0.$$

Conversely, given any harmonic function in two dimensions, it is the real part of an analytic function, at least locally. Details are given in Laplace equation.

A Typical Boundary Value Problem

A typical problem for Laplace's equation is to find a solution that satisfies arbitrary values on the boundary of a domain. For example, we may seek a harmonic function that takes on the values $u(\theta)$ on a circle of radius one. The solution was given by Poisson:

$$\varphi(r,\theta) = \frac{1}{2\pi} \int_0^{2\pi} \frac{1-r^2}{1+r^2 - 2r\cos(\theta - \theta')} u(\theta')d\theta'.$$

Petrovsky (1967, p. 248) shows how this formula can be obtained by summing a Fourier series for φ. If $r < 1$, the derivatives of φ may be computed by differentiating under the integral sign, and one can verify that φ is analytic, even if u is continuous but not necessarily differentiable. This behavior is typical for solutions of elliptic partial differential equations: the solutions may be much more smooth than the boundary data. This is in contrast to solutions of the wave equation, and more general hyperbolic partial differential equations, which typically have no more derivatives than the data.

Euler–tricomi Equation

The Euler–Tricomi equation is used in the investigation of transonic flow.

$$u_{xx} = xu_{yy}.$$

Advection Equation

The advection equation describes the transport of a conserved scalar ψ in a velocity field u = (u, v, w). It is:

$$\psi_t + (u\psi)_x + (v\psi)_y + (w\psi)_z = 0.$$

If the velocity field is solenoidal (that is, ∇·u = 0), then the equation may be simplified to

$$\psi_t + u\psi_x + v\psi_y + w\psi_z = 0.$$

In the one-dimensional case where u is not constant and is equal to ψ, the equation is referred to as Burgers' equation.

Ginzburg–landau Equation

The Ginzburg–Landau equation is used in modelling superconductivity. It is

$$iu_t + pu_{xx} + q\,|\,u\,|^2\,u = i\gamma u$$

where $p,q \in$ C and $\gamma \in$ R are constants and i is the imaginary unit.

The Dym Equation

The Dym equation is named for Harry Dym and occurs in the study of solitons. It is

$$u_t = u^3 u_{xxx}.$$

Initial-boundary Value Problems

Many problems of mathematical physics are formulated as initial-boundary value problems.

Vibrating String

If the string is stretched between two points where $x=0$ and $x=L$ and u denotes the amplitude of the displacement of the string, then u satisfies the one-dimensional wave equation in the region where $0 < x < L$ and t is unlimited. Since the string is tied down at the ends, u must also satisfy the boundary conditions

$$u(t,0) = 0, \quad u(t,L) = 0,$$

as well as the initial conditions

$$u(0,x) = f(x), \quad u_t(0,x) = g(x).$$

The method of separation of variables for the wave equation

$$u_{tt} = c^2 u_{xx},$$

leads to solutions of the form

$$u(t,x) = T(t)X(x),$$

where

$$T'' + k^2 c^2 T = 0, \quad X'' + k^2 X = 0,$$

where the constant k must be determined. The boundary conditions then imply that X is a multiple of $\sin kx$, and k must have the form

$$k = \frac{n\pi}{L},$$

where n is an integer. Each term in the sum corresponds to a mode of vibration of the string. The mode with $n = 1$ is called the fundamental mode, and the frequencies of the other modes are all multiples of this frequency. They form the overtone series of the string, and they are the basis for musical acoustics. The initial conditions may then be satisfied by representing f and g as infinite sums of these modes. Wind instruments typically correspond to vibrations of an air column with one end open and one end closed. The corresponding boundary conditions are

$$X(0) = 0, \quad X'(L) = 0.$$

The method of separation of variables can also be applied in this case, and it leads to a series of odd overtones.

The general problem of this type is solved in Sturm–Liouville theory.

Vibrating Membrane

If a membrane is stretched over a curve C that forms the boundary of a domain D in the plane, its vibrations are governed by the wave equation

$$\frac{1}{c^2} u_{tt} = u_{xx} + u_{yy},$$

if $t > 0$ and (x,y) is in D. The boundary condition is $u(t,x,y) = 0$ if (x,y) is on C. The method of separation of variables leads to the form

$$u(t, x, y) = T(t)v(x, y),$$

which in turn must satisfy

$$\frac{1}{c^2}T'' + k^2 T = 0,$$

$$v_{xx} + v_{yy} + k^2 v = 0.$$

The latter equation is called the Helmholtz Equation. The constant k must be determined to allow a non-trivial v to satisfy the boundary condition on C. Such values of k^2 are called the eigenvalues of the Laplacian in D, and the associated solutions are the eigenfunctions of the Laplacian in D. The Sturm–Liouville theory may be extended to this elliptic eigenvalue problem (Jost, 2002).

Other Examples

The Schrödinger equation is a PDE at the heart of non-relativistic quantum mechanics. In the WKB approximation it is the Hamilton–Jacobi equation.

Except for the Dym equation and the Ginzburg–Landau equation, the above equations are linear in the sense that they can be written in the form $Au = f$ for a given linear operator A and a given function f. Other important non-linear equations include the Navier–Stokes equations describing the flow of fluids, and Einstein's field equations of general relativity.

Classification

Some linear, second-order partial differential equations can be classified as parabolic, hyperbolic and elliptic. Others such as the Euler–Tricomi equation have different types in different regions. The classification provides a guide to appropriate initial and boundary conditions, and to the smoothness of the solutions.

Equations of First Order

Linear Equations of Second Order

Assuming $u_{xy} = u_{yx}$, the general second-order PDE in two independent variables has the form

$$Au_{xx} + 2Bu_{xy} + Cu_{yy} + \cdots \text{(lower order terms)} = 0,$$

where the coefficients A, B, C etc. may depend upon x and y. If $A^2 + B^2 + C^2 > 0$ over a region of the xy plane, the PDE is second-order in that region. This form is analogous to the equation for a conic section:

$$Ax^2 + 2Bxy + Cy^2 + \cdots = 0.$$

More precisely, replacing ∂_x by X, and likewise for other variables (formally this is done by a Fourier transform), converts a constant-coefficient PDE into a polynomial of the same degree, with the top degree (a homogeneous polynomial, here a quadratic form) being most significant for the classification.

Just as one classifies conic sections and quadratic forms into parabolic, hyperbolic, and elliptic based on the discriminant $B^2 - 4AC,$, the same can be done for a second-order PDE at a given point. However, the discriminant in a PDE is given by $B^2 - AC,$ due to the convention of the xy term being $2B$ rather than B; formally, the discriminant (of the associated quadratic form) is $(2B)^2 - 4AC = 4(B^2 - AC)$, with the factor of 4 dropped for simplicity.

1. $B^2 - AC < 0$: solutions of elliptic PDEs are as smooth as the coefficients allow, within the interior of the region where the equation and solutions are defined. For example, solutions of Laplace's equation are analytic within the domain where they are defined, but solutions may assume boundary values that are not smooth. The motion of a fluid at subsonic speeds can be approximated with elliptic PDEs, and the Euler–Tricomi equation is elliptic where $x < 0$.

2. $B^2 - AC = 0$: equations that are parabolic at every point can be transformed into a form analogous to the heat equation by a change of independent variables. Solutions smooth out as the transformed time variable increases. The Euler–Tricomi equation has parabolic type on the line where $x = 0$.

3. $B^2 - AC > 0$: hyperbolic equations retain any discontinuities of functions or derivatives in the initial data. An example is the wave equation. The motion of a fluid at supersonic speeds can be approximated with hyperbolic PDEs, and the Euler–Tricomi equation is hyperbolic where $x > 0$.

If there are n independent variables x_1, x_2, ..., x_n, a general linear partial differential equation of second order has the form

$$Lu = \sum_{i=1}^{n}\sum_{j=1}^{n} a_{i,j} \frac{\partial^2 u}{\partial x_i \partial x_j} \quad \text{plus lower-order terms} = 0.$$

The classification depends upon the signature of the eigenvalues of the coefficient matrix $a_{i,j}$.

1. Elliptic: The eigenvalues are all positive or all negative.

2. Parabolic : The eigenvalues are all positive or all negative, save one that is zero.

3. Hyperbolic: There is only one negative eigenvalue and all the rest are positive, or there is only one positive eigenvalue and all the rest are negative.

4. Ultrahyperbolic: There is more than one positive eigenvalue and more than one negative eigenvalue, and there are no zero eigenvalues. There is only a limited theory for ultra-hyperbolic equations (Courant and Hilbert, 1962).

Systems of First-order Equations and Characteristic Surfaces

The classification of partial differential equations can be extended to systems of first-order equations, where the unknown u is now a vector with m components, and the coefficient matrices A_v are m by m matrices for $v = 1, ..., n$. The partial differential equation takes the form

$$Lu = \sum_{v=1}^{n} A_v \frac{\partial u}{\partial x_v} + B = 0,$$

where the coefficient matrices A_v and the vector B may depend upon x and u. If a hypersurface S is given in the implicit form

$$\varphi(x_1, x_2, \ldots, x_n) = 0,$$

where φ has a non-zero gradient, then S is a characteristic surface for the operator L at a given point if the characteristic form vanishes:

$$Q\left(\frac{\partial \varphi}{\partial x_1}, \ldots, \frac{\partial \varphi}{\partial x_n}\right) = \det\left[\sum_{v=1}^{n} A_v \frac{\partial \varphi}{\partial x_v}\right] = 0.$$

The geometric interpretation of this condition is as follows: if data for u are prescribed on the surface S, then it may be possible to determine the normal derivative of u on S from the differential equation. If the data on S and the differential equation determine the normal derivative of u on S, then S is non-characteristic. If the data on S and the differential equation *do not* determine the normal derivative of u on S, then the surface is characteristic, and the differential equation restricts the data on S: the differential equation is *internal* to S.

1. A first-order system $Lu=0$ is *elliptic* if no surface is characteristic for L: the values of u on S and the differential equation always determine the normal derivative of u on S.

2. A first-order system is *hyperbolic* at a point if there is a space-like surface S with normal ξ at that point. This means that, given any non-trivial vector η orthogonal to ξ, and a scalar multiplier λ, the equation $Q(\lambda \xi + \eta) = 0$ has m real roots $\lambda_1, \lambda_2, \ldots, \lambda_m$. The system is strictly hyperbolic if these roots are always distinct. The geometrical interpretation of this condition is as follows: the characteristic form $Q(\zeta) = 0$ defines a cone (the normal cone) with homogeneous coordinates ζ. In the hyperbolic case, this cone has m sheets, and the axis $\zeta = \lambda \xi$ runs inside these sheets: it does not intersect any of them. But when displaced from the origin by η, this axis intersects every sheet. In the elliptic case, the normal cone has no real sheets.

Equations of Mixed Type

If a PDE has coefficients that are not constant, it is possible that it will not belong to any of these categories but rather be of mixed type. A simple but important example is the Euler–Tricomi equation

$$u_{xx} = x u_{yy},$$

which is called elliptic-hyperbolic because it is elliptic in the region $x < 0$, hyperbolic in the region $x > 0$, and degenerate parabolic on the line $x = 0$.

Infinite-order PDEs in Quantum Mechanics

In the phase space formulation of quantum mechanics, one may consider the quantum Hamilton's equations for trajectories of quantum particles. These equations are infinite-order PDEs. However, in the semiclassical expansion, one has a finite system of ODEs at any fixed order of ℏ. The evolution equation of the Wigner function is also an infinite-order PDE. The quantum trajectories are quantum characteristics, with the use of which one could calculate the evolution of the Wigner function.

Analytical Methods to Solve PDEs

Separation of Variables

Linear PDEs can be reduced to systems of ordinary differential equations by the important technique of separation of variables. This technique rests on a characteristic of solutions to differential equations: if one can find any solution that solves the equation and satisfies the boundary conditions, then it is *the* solution (this also applies to ODEs). We assume as an ansatz that the dependence of a solution on the parameters space and time can be written as a product of terms that each depend on a single parameter, and then see if this can be made to solve the problem.

In the method of separation of variables, one reduces a PDE to a PDE in fewer variables, which is an ordinary differential equation if in one variable – these are in turn easier to solve.

This is possible for simple PDEs, which are called separable partial differential equations, and the domain is generally a rectangle (a product of intervals). Separable PDEs correspond to diagonal matrices – thinking of "the value for fixed x" as a coordinate, each coordinate can be understood separately.

This generalizes to the method of characteristics, and is also used in integral transforms.

Method of Characteristics

In special cases, one can find characteristic curves on which the equation reduces to an ODE – changing coordinates in the domain to straighten these curves allows separation of variables, and is called the method of characteristics.

More generally, one may find characteristic surfaces.

Integral Transform

An integral transform may transform the PDE to a simpler one, in particular, a separable PDE. This corresponds to diagonalizing an operator.

An important example of this is Fourier analysis, which diagonalizes the heat equation using the eigenbasis of sinusoidal waves.

If the domain is finite or periodic, an infinite sum of solutions such as a Fourier series is appropriate, but an integral of solutions such as a Fourier integral is generally required for infinite domains. The solution for a point source for the heat equation given above is an example of the use of a Fourier integral.

Change of Variables

Often a PDE can be reduced to a simpler form with a known solution by a suitable change of variables. For example, the Black–Scholes PDE

$$\frac{\partial V}{\partial t} + \frac{1}{2}\sigma^2 S^2 \frac{\partial^2 V}{\partial S^2} + rS\frac{\partial V}{\partial S} - rV = 0$$

is reducible to the heat equation

$$\frac{\partial u}{\partial \tau} = \frac{\partial^2 u}{\partial x^2}$$

by the change of variables

$$V(S,t) = Kv(x,\tau)$$

$$x = \ln\left(\frac{S}{K}\right)$$

$$\tau = \tfrac{1}{2}\sigma^2(T-t)$$

$$v(x,\tau) = \exp(-\alpha x - \beta\tau)u(x,\tau).$$

Fundamental Solution

Inhomogeneous equations can often be solved (for constant coefficient PDEs, always be solved) by finding the fundamental solution (the solution for a point source), then taking the convolution with the boundary conditions to get the solution.

This is analogous in signal processing to understanding afilter by its impulse response.

Superposition Principle

Because any superposition of solutions of a linear, homogeneous PDE is again a solution, the particular solutions may then be combined to obtain more general solutions. if u1 and u2 are solutions of a homogeneous linear pde in same region R, then u= c1u1+c2u2 with any constants c1 and c2 are also a solution of that pde in that same region....

Methods for Non-linear Equations

There are no generally applicable methods to solve nonlinear PDEs. Still, existence and uniqueness results (such as the Cauchy–Kowalevski theorem) are often possible, as are proofs of important qualitative and quantitative properties of solutions (getting these results is a major part of analysis). Computational solution to the nonlinear PDEs, the split-step method, exist for specific equations like nonlinear Schrödinger equation.

Nevertheless, some techniques can be used for several types of equations. The h-principle is the most powerful method to solve underdetermined equations. The Riquier–Janet theory is an effective method for obtaining information about many analytic overdetermined systems.

The method of characteristics (similarity transformation method) can be used in some very special cases to solve partial differential equations.

In some cases, a PDE can be solved via perturbation analysis in which the solution is considered to be a correction to an equation with a known solution. Alternatives are numerical analysis techniques from simple finite difference schemes to the more mature multigrid and finite element methods. Many interesting problems in science and engineering are solved in this way using computers, sometimes high performance supercomputers.

Lie Group Method

From 1870 Sophus Lie's work put the theory of differential equations on a more satisfactory foundation. He showed that the integration theories of the older mathematicians can, by the introduction of what are now called Lie groups, be referred to a common source; and that ordinary differential equations which admit the same infinitesimal transformations present comparable difficulties of integration. He also emphasized the subject of transformations of contact.

A general approach to solving PDE's uses the symmetry property of differential equations, the continuous infinitesimal transformations of solutions to solutions (Lie theory). Continuous group theory, Lie algebras and differential geometry are used to understand the structure of linear and nonlinear partial differential equations for generating integrable equations, to find its Lax pairs, recursion operators, Bäcklund transform and finally finding exact analytic solutions to the PDE.

Symmetry methods have been recognized to study differential equations arising in mathematics, physics, engineering, and many other disciplines.

Semianalytical Methods

The adomian decomposition method, the Lyapunov artificial small parameter method, and He's homotopy perturbation method are all special cases of the more general homotopy analysis method. These are series expansion methods, and except for the Lyapunov method, are independent of small physical parameters as compared to the well known perturbation theory, thus giving these methods greater flexibility and solution generality.

Numerical Methods to Solve PDEs

The three most widely used numerical methods to solve PDEs are the finite element method (FEM), finite volume methods (FVM) and finite difference methods (FDM). The FEM has a prominent position among these methods and especially its exceptionally efficient higher-order version hp-FEM. Other versions of FEM include the generalized finite element method (GFEM), extended finite element method (XFEM), spectral finite element method (SFEM), meshfree finite element method, discontinuous Galerkin finite element method (DGFEM), Element-Free Galerkin Method (EFGM), Interpolating Element-Free Galerkin Method (IEFGM), etc.

Finite Element Method

The finite element method (FEM) (its practical application often known as finite element analysis (FEA)) is a numerical technique for finding approximate solutions of partial differential equations (PDE) as well as of integral equations. The solution approach is based either on eliminating the differential equation completely (steady state problems), or rendering the PDE into an approximating system of ordinary differential equations, which are then numerically integrated using standard techniques such as Euler's method, Runge–Kutta, etc.

Finite Difference Method

Finite-difference methods are numerical methods for approximating the solutions to differential equations using finite difference equations to approximate derivatives.

Finite Volume Method

Similar to the finite difference method or finite element method, values are calculated at discrete places on a meshed geometry. "Finite volume" refers to the small volume surrounding each node point on a mesh. In the finite volume method, surface integrals in a partial differential equation that contain a divergence term are converted to volume integrals, using the divergence theorem. These terms are then evaluated as fluxes at the surfaces of each finite volume. Because the flux entering a given volume is identical to that leaving the adjacent volume, these methods are conservative.

Diffusion Equation

The diffusion equation is a partial differential equation. In physics, it describes the behavior of the collective motion of micro-particles in a material resulting from the random movement of each micro-particle. In mathematics, it is applicable in common to a subject relevant to the Markov process as well as in various other fields, such as the material sciences, information science, life science, social science, and so on. These subjects described by the diffusion equation are generally called Brown problems.

Statement

The equation is usually written as:

$$\frac{\partial \phi(\mathbf{r},t)}{\partial t} = \nabla \cdot \left[D(\phi, \mathbf{r}) \, \nabla \phi(\mathbf{r},t) \right],$$

where $\phi(\mathbf{r}, t)$ is the density of the diffusing material at location r and time t and $D(\phi, \mathbf{r})$ is the collective diffusion coefficient for density ϕ at location r; and ∇ represents the vector differential operator del. If the diffusion coefficient depends on the density then the equation is nonlinear, otherwise it is linear.

More generally, when D is a symmetric positive definite matrix, the equation describes anisotropic diffusion, which is written (for three dimensional diffusion) as:

$$\frac{\partial \phi(\mathbf{r},t)}{\partial t} = \sum_{i=1}^{3}\sum_{j=1}^{3}\frac{\partial}{\partial x_i}\left[D_{ij}(\phi,\mathbf{r})\frac{\partial \phi(\mathbf{r},t)}{\partial x_j}\right]$$

If D is constant, then the equation reduces to the following linear differential equation:

$$\frac{\partial \phi(\mathbf{r},t)}{\partial t} = D\nabla^2 \phi(\mathbf{r},t),$$

also called the heat equation.

History and Development

In mathematics, a great many phenomena in various science fields are expressed by using the well-known evolution equations. The diffusion equation is one of them and mathematically corresponds to the Markov process in relation to the normal distribution rule. In physics, the motion of diffusion particles corresponds to the well-known Brown motion satisfying the parabolic law. It is widely accepted that the Brown problem is a general term of investigating subjects in various science fields relevant to the Markov process, such as the material science, the information science, the life science, the social science, and so on. The extended diffusion equations are used for various sciences fields. In that case, they sometimes have a sink and source of their concerned elements, for example, such as a local equilibrium relation between native defects in silicon crystal in the material science or between predation and prey in the life science. We must then solve a system of diffusion equations. In the following, however, we discuss the fundamental diffusion equation of the so-called Fick's diffusion equation in relation to the material science.

In history, the heat equation proposed by Fourier in 1822 has been applied to investigating a temperature distribution in materials. In 1827, the so-called Brown motion was found, where the self-diffusion of water is visualized by pollen micro particles motion. Nevertheless, the Brown motion had not been recognized as a diffusion problem until the Einstein theory of Brown motion in 1905, although it was a typical diffusion problem. In 1855, Fick applied the heat equation to diffusion phenomena as it had been.

In accordance with the industrial requirement, the solid materials such as alloys, semiconductors, multilayer materials, and so on, have been widely fabricated. The heat treatment is indispensable for their fabrication processes then. The migration of particles in a solid material is caused by the heat treatment. In relation to the migration of their particles, the diffusion problems of various solid materials have been thus widely investigated, although the diffusion equation was mainly applied to problems of liquid material in an early stage after the Fick's proposition.

The Gauss's divergence theorem shows that the diffusion equation is valid in the solid, liquid and gas states in every material as a material conservation law, if there is no sink and source in the given diffusion system. It is also shows that the corresponding Fick's first law to the Fick's second law is mathematically incomplete without a constant diffusion flux relevant to the Brown motion in the localized space. The constant diffusion flux is indispensable for understanding the self-diffusion mechanism. The self-diffusion mechanism itself was not directly investigated, although it had been indirectly investigated by behavior of impurity diffusion in a pure material as shown in the Einstein's Brown theory and the Langevin equation.

We found that the diffusivity of diffusion equation depends generally on the concentration of diffusion particles. In that case, the diffusion equation becomes a nonlinear partial differential equation, and the mathematical solution is almost impossible, even if it is a case of the time and one dimension space coordinate . In accordance with the parabolic law , Boltzmann transformed the diffusion equation of , which is a nonlinear partial differential equation, into a nonlinear ordinary differential equation of in 1894. Since then, however, the Boltzmann transformation equation had not been mathematically solved until recently, although Matano empirically used it for analyzing interdiffusion problems in the metallurgy field.

Here, the analytical method of diffusion equation, which is extremely superior in calculation to the existing analytical method such as the integral transformation method of Fourier or Laplace and/ or the variable separation method, was thus established in the parabolic space.

In 1947, Kirkendall found that an inert marker set at a point in a binary alloy moves from the initial sate point after the diffusion treatment. The phenomena are so called Kirkendall effect and it was considered that we cannot understand it from the existing theory of binary interdiffusion in those days. As a result, a new concept of intrinsic diffusion was then introduced for understanding the Kirkendall effect in the interdiffusion problems. Based on the intrinsic diffusion concept, Darken derived a relation between an interdiffusion coefficient and intrinsic diffusion coefficients in a binary interdiffusion in 1948. At present, however, it is revealed that the so-called Darken equation itself is mathematically wrong in the derivation process. Although the concentration of diffusion particles is a real quantity in physics, the temperature is a thermodynamic state quantity. As far as the shape of heat conduction material is unchangeable during a thermal treatment, the coordinate system of heat equation set in a material is a fixed one, since the coordinate system is not influenced by variations of the material internal structure. On the other hand, strictly speaking, the coordinate system of diffusion equation set in the diffusion field (solvent) is a moving one, since the origin of coordinate system is generally influenced by such variations.

When Fick proposed the diffusion equation, the Gauss divergence theorem had been already reported in 1840. Nevertheless, the problem of coordinate system of diffusion equation had not been mathematically investigated in accordance with the divergence theorem until recently. In general, however, it is indispensable for understanding the diffusion problems to discuss their coordinate systems, since it is, strictly speaking, considered that the diffusion particles, solvent particles and also the diffusion region space simultaneously move against the experimentation system in the diffusion region outside.

Recently, the diffusion equation was thus mathematically investigated in accordance with the divergence theorem and the coordinate transformation theory. As a result, the diffusion flux should be determined by taking account of the concerned coordinate system of diffusion equation. Using the corresponding diffusion flux to the concerned coordinate system of diffusion equation for interdiffusion, one way diffusion, impurity diffusion and self-diffusion, we found that they are uniformly discussed and the foundation of diffusion problems is included in interdiffusion problems. The interdiffusion theory of an elements system applicable to every material was thus reasonably established. In the analysis of interdiffusion problems, the only difference between a binary system and an N elements system is whether the solvent material is one element or elements.

The coordinate transformation theory reveals that the Kirkendall effect is caused by a shift between the coordinate systems of diffusion equation like the Doppler effect relevant to a wave equation is caused by a shift between the fixed coordinate system and the moving one. Further, it was also found that the concept of intrinsic diffusion is an illusion in the diffusion history. All physical information in the given diffusion system is incorporated into the diffusivity. If we can know a diffusivity behavior in the given diffusion equation, the mathematical solution and/or numerical one at least is possible. In the diffusion problems, it is thus extremely dominant to know the diffusivity behavior. The diffusivity is defined by an interaction between a diffusion particle and the diffusion field near the diffusion particle itself. This indicates that the diffusivity should be essentially investigated in the quantum mechanics, since the behavior of a micro particle should be investigated by analyzing the Schrodinger equation.

From applying the diffusion equation to a problem of diffusion elementary process, the equation was reasonably derived. It was revealed that the diffusivity corresponds to the angular momentum operator in the quantum mechanics. As a result, the universal expression of diffusivity, which is applicable to every material in an arbitrary thermodynamic state, was obtained as one with the proportionality constant composed of the product of Planck constant and Avogadro constant . It was also found that the well-known material wave relation proposed by de Broglie in 1923, which is the most fundamental one in materials science, is obtained from a relation between the given diffusivity expressions. This gives evidence for the theory discussed here.

Derivation

The diffusion equation can be trivially derived from the continuity equation, which states that a change in density in any part of the system is due to inflow and outflow of material into and out of that part of the system. Effectively, no material is created or destroyed:

$$\frac{\partial \phi}{\partial t} + \nabla \cdot \mathbf{j} = 0,$$

where j is the flux of the diffusing material. The diffusion equation can be obtained easily from this when combined with the phenomenological Fick's first law, which states that the flux of the diffusing material in any part of the system is proportional to the local density gradient:

$$\mathbf{j} = -D(\phi, \mathbf{r}) \nabla \phi(\mathbf{r}, t).$$

If drift must be taken into account, the Smoluchowski equation provides an appropriate generalization.

Discretization

The diffusion equation is continuous in both space and time. One may discretize space, time, or both space and time, which arise in application. Discretizing time alone just corresponds to taking time slices of the continuous system, and no new phenomena arise. In discretizing space alone, the Green's function becomes the discrete Gaussian kernel, rather than the continuous Gaussian kernel. In discretizing both time and space, one obtains the random walk.

Discretization (Image)

The product rule is used to rewrite the anisotropic tensor diffusion equation, in standard discretization schemes. Because direct discretization of the diffusion equation with only first order spatial central differences leads to checkerboard artifacts. The rewritten diffusion equation used in image filtering:

$$\frac{\partial \phi(\mathbf{r},t)}{\partial t} = \nabla \cdot \left[D(\phi,\mathbf{r}) \right] \nabla \phi(\mathbf{r},t) + \mathrm{tr}\left[D(\phi,\mathbf{r})\left(\nabla \nabla^{T} \phi(\mathbf{r},t) \right) \right]$$

where "tr" denotes the trace of the 2nd rank tensor, and superscript "T" denotes transpose, in which in image filtering $D(\phi, \mathbf{r})$ are symmetric matrices constructed from the eigenvectors of the image structure tensors . The spatial derivatives can then be approximated by two first order and a second order central finite differences. The resulting diffusion algorithm can be written as an image convolution with a varying kernel (stencil) of size 3×3 in 2D and $3 \times 3 \times 3$ in 3D.

Wave Equation

The wave equation is an important second-order linear partial differential equation for the description of waves—as they occur in physics—such as sound waves, light waves and water waves. It arises in fields like acoustics, electromagnetics, and fluid dynamics.

Historically, the problem of a vibrating string such as that of a musical instrument was studied by Jean le Rond d'Alembert, Leonhard Euler, Daniel Bernoulli, and Joseph-Louis Lagrange. In 1746, d'Alembert discovered the one-dimensional wave equation, and within ten years Euler discovered the three-dimensional wave equation.

Introduction

The wave equation is a hyperbolic partial differential equation. It typically concerns a time variable t, one or more spatial variables $x_1, x_2, ..., x_n$, and a scalar function $u = u\,(x_1, x_2, ..., x_n; t)$, whose values could model, for example, the mechanical displacement of a wave. The wave equation for u is

$$\frac{\partial^2 u}{\partial t^2} = c^2 \nabla^2 u$$

where ∇^2 is the (spatial) Laplacian and c is a fixed constant.

Solutions of this equation describe propagation of disturbances out from the region at a fixed speed in one or in all spatial directions, as do physical waves from plane or localized sources; the constant c is identified with the propagation speed of the wave. This equation is linear. Therefore, the sum of any two solutions is again a solution: in physics this property is called the superposition principle.

The wave equation alone does not specify a physical solution; a unique solution is usually obtained by setting a problem with further conditions, such as initial conditions, which prescribe the amplitude and phase of the wave. Another important class of problems occurs in enclosed spaces specified by boundary conditions, for which the solutions represent standing waves, or harmonics, analogous to the harmonics of musical instruments.

The wave equation, and modifications of it, are also found in elasticity, quantum mechanics, plasma physics and general relativity.

Scalar Wave Equation in One Space Dimension

French scientist Jean-Baptiste le Rond d'Alembert (b. 1717) discovered the wave equation in one space dimension.

The wave equation in one space dimension can be written as follows:

$$\frac{\partial^2 u}{\partial t^2} = c^2 \frac{\partial^2 u}{\partial x^2}.$$

This equation is typically described as having only one space dimension "x", because the only other independent variable is the time "t". Nevertheless, the dependent variable "u" may represent a second space dimension, if, for example, the displacement "u" takes place in y-direction, as in the case of a string that is located in the x-y plane.

Derivation of the Wave Equation

The wave equation in one space dimension can be derived in a variety of different physical settings. Most famously, it can be derived for the case of a string that is vibrating in a two-dimensional plane, with each of its elements being pulled in opposite directions by the force of tension.

Another physical setting for derivation of the wave equation in one space dimension utilizes Hooke's Law. In the theory of elasticity, Hooke's Law is an approximation for certain materials, stating that the amount by which a material body is deformed (the strain) is linearly related to the force causing the deformation (the stress).

From Hooke's Law

The wave equation in the one-dimensional case can be derived from Hooke's Law in the following way: Imagine an array of little weights of mass m interconnected with massless springs of length h . The springs have a spring constant of k:

Here the dependent variable $u(x)$ measures the distance from the equilibrium of the mass situated at x, so that $u(x)$ essentially measures the magnitude of a disturbance (i.e. strain) that is traveling in an elastic material. The forces exerted on the mass m at the location $x+h$ are:

$$F_{Newton} = m \cdot a(t) = m \cdot \frac{\partial^2}{\partial t^2} u(x+h,t)$$

$$F_{Hooke} = F_{x+2h} - F_x = k\left[u(x+2h,t) - u(x+h,t)\right] - k[u(x+h,t) - u(x,t)]$$

The equation of motion for the weight at the location $x+h$ is given by equating these two forces:

$$\frac{\partial^2}{\partial t^2} u(x+h,t) = \frac{k}{m}[u(x+2h,t) - u(x+h,t) - u(x+h,t) + u(x,t)]$$

where the time-dependence of $u(x)$ has been made explicit.

If the array of weights consists of N weights spaced evenly over the length $L = Nh$ of total mass $M = Nm$, and the total spring constant of the array $K = k/N$ we can write the above equation as:

$$\frac{\partial^2}{\partial t^2} u(x+h,t) = \frac{KL^2}{M} \frac{u(x+2h,t) - 2u(x+h,t) + u(x,t)}{h^2}$$

Taking the limit $N \to \infty$, $h \to 0$ and assuming smoothness one gets:

$$\frac{\partial^2 u(x,t)}{\partial t^2} = \frac{KL^2}{M} \frac{\partial^2 u(x,t)}{\partial x^2}$$

$(KL^2)/M$ is the square of the propagation speed in this particular case.

Stress Pulse in a Bar

In the case of a stress pulse propagating through a beam the beam acts much like an infinite number of springs in series and can be taken as an extension of the equation derived for Hooke's law. A beam of constant cross section made from a linear elastic material has a stiffness K given by

$$K = \frac{EA}{L}$$

Where A is the cross sectional area and E is the Young's modulus of the material. The wave equation becomes

$$\frac{\partial^2 u(x,t)}{\partial t^2} = \frac{EAL}{M} \frac{\partial^2 u(x,t)}{\partial x^2}$$

AL is equal to the volume of the beam and therefore : $\frac{AL}{M} = \frac{1}{\rho}$ where ρ is the density of the material. The wave equation reduces to

$$\frac{\partial^2 u(x,t)}{\partial t^2} = \frac{E}{\rho} \frac{\partial^2 u(x,t)}{\partial x^2}$$

The speed of a stress wave in a beam is therefore $\sqrt{\dfrac{E}{\rho}}$

General Solution

Algebraic Approach

The one-dimensional wave equation is unusual for a partial differential equation in that a relatively simple general solution may be found. Defining new variables:

$$\xi = x - ct \quad ; \quad \eta = x + ct$$

changes the wave equation into

$$\frac{\partial^2 u}{\partial \xi \partial \eta} = 0$$

which leads to the general solution

$$u(\xi, \eta) = F(\xi) + G(\eta)$$

or equivalently:

$$u(x,t) = F(x - ct) + G(x + ct)$$

In other words, solutions of the 1D wave equation are sums of a right traveling function F and a left traveling function G. "Traveling" means that the shape of these individual arbitrary functions with respect to x stays constant, however the functions are translated left and right with time at the speed c. This was derived by Jean le Rond d'Alembert.

Another way to arrive at this result is to note that the wave equation may be "factored":

$$\left[\frac{\partial}{\partial t} - c \frac{\partial}{\partial x} \right] \left[\frac{\partial}{\partial t} + c \frac{\partial}{\partial x} \right] u = 0$$

and therefore:

$$\text{either} \qquad \frac{\partial u}{\partial t} - c\frac{\partial u}{\partial x} = 0 \qquad \text{or} \qquad \frac{\partial u}{\partial t} + c\frac{\partial u}{\partial x} = 0$$

These last two equations are advection equations, one left traveling and one right, both with constant speed c.

For an initial value problem, the arbitrary functions F and G can be determined to satisfy initial conditions:

$$u(x,0) = f(x)$$

$$u_t(x,0) = g(x)$$

The result is d'Alembert's formula:

$$u(x,t) = \frac{f(x-ct) + f(x+ct)}{2} + \frac{1}{2c}\int_{x-ct}^{x+ct} g(s)ds$$

In the classical sense if $f(x) \in C^k$ and $g(x) \in C^{k-1}$ then $u(t, x) \in C^k$. However, the waveforms F and G may also be generalized functions, such as the delta-function. In that case, the solution may be interpreted as an impulse that travels to the right or the left.

The basic wave equation is a linear differential equation and so it will adhere to the superposition principle. This means that the net displacement caused by two or more waves is the sum of the displacements which would have been caused by each wave individually. In addition, the behavior of a wave can be analyzed by breaking up the wave into components, e.g. the Fourier transform breaks up a wave into sinusoidal components.

Plane Wave Eigenmodes

Another way to solve for the solutions to the one-dimensional wave equation is to first analyze its frequency eigenmodes. A so-called eigenmode is a solution that oscillates in time with a well-defined *constant* angular frequency ω, with which the temporal part of the wave function for such eigenmode takes a specific form $e^{-i\omega t}$. The rest of the wave function is then only dependent on the spatial variable x, hence amounting to separation of variables. Now writing the wave function as

$$u_\omega(x,t) = e^{-i\omega t} f(x),$$

we can obtain an ordinary differential equation for the spatial part $f(x)$

$$\frac{\partial^2 u_\omega}{\partial t^2} = \frac{\partial^2}{\partial t^2}\left(e^{-i\omega t} f(x)\right) = -\omega^2 e^{-i\omega t} f(x) = c^2 \frac{\partial^2}{\partial x^2}\left(e^{-i\omega t} f(x)\right),$$

Therefore:

$$\frac{d^2}{dx^2} f(x) = -\left(\frac{\omega}{c}\right)^2 f(x),$$

which is precisely an eigenvalue equation for $f(x)$, hence the name eigenmode. It has the well-known plane wave solutions

$$f(x) = Ae^{\pm ikx},$$

with wave number $k = \omega / c$.

The total wave function for this eigenmode is then the linear combination

$$u_\omega(x,t) = e^{-i\omega t}\left(Ae^{-ikx} + Be^{ikx}\right) = Ae^{-i(kx+\omega t)} + Be^{i(kx-\omega t)},$$

where complex numbers A, B depend in general on any initial and boundary conditions of the problem.

Eigenmodes are useful in constructing a full solution to the wave equation, because each of them evolves in time trivially with the phase factor $e^{-i\omega t}$. so that a full solution can be decomposed into an eigenmode expansion

$$u(x,t) = \int_{-\infty}^{\infty} s(\omega)u_\omega(x,t)d\omega$$

or in terms of the plane waves,

$$\begin{aligned}
u(x,t) &= \int_{-\infty}^{\infty} s_+(\omega)e^{-i(kx+\omega t)}d\omega + \int_{-\infty}^{\infty} s_-(\omega)e^{i(kx-\omega t)}d\omega \\
&= \int_{-\infty}^{\infty} s_+(\omega)e^{-ik(x+ct)}d\omega + \int_{-\infty}^{\infty} s_-(\omega)e^{ik(x-ct)}d\omega \\
&= F(x-ct) + G(x+ct)
\end{aligned}$$

which is exactly in the same form as in the algebraic approach. Functions $s_\pm(\omega)$ are known as the Fourier component and are determined by initial and boundary conditions. This is a so-called frequency-domain method, alternative to direct time-domain propagations, such as FDTD method, of the wave packet $u(x,t)$, which is complete for representing waves in absence of time dilations. Completeness of the Fourier expansion for representing waves in the presence of time dilations has been challenged by chirp wave solutions allowing for time variation of ω. The chirp wave solutions seem particularly implied by very large but previously inexplicable radar residuals in the flyby anomaly, and differ from the sinusoidal solutions in being receivable at any distance only at proportionally shifted frequencies and time dilations, corresponding to past chirp states of the source.

Scalar Wave Equation in three Space Dimensions

A solution of the initial-value problem for the wave equation in three space dimensions can be obtained from the corresponding solution for a spherical wave. The result can then be also used to obtain the same solution in two space dimensions.

Swiss mathematician and physicist Leonhard Euler (b. 1707) discovered the wave equation in three space dimensions.

Spherical Waves

The wave equation can be solved using the technique of separation of variables. To obtain a solution with constant frequencies, let us first Fourier transform the wave equation in time as

$$\Psi(\vec{r},t) = \int_{-\infty}^{\infty} \Psi(\vec{r},\omega)e^{-i\omega t}d\omega.$$

So we get,

$$\left(\nabla^2 + \frac{\omega^2}{c^2}\right)\Psi(\vec{r},\omega) = 0$$

This is the Helmholtz equation and can be solved using separation of variables. If spherical coordinates are used to describe a problem, then the solution to the angular part of the Helmholtz equation is given by spherical harmonics and the radial equation now becomes

$$\left[\frac{d^2}{dr^2} + \frac{2}{r}\frac{d}{dr} + k^2 - \frac{l(l+1)}{r^2}\right]f_{lm}(r) = 0$$

Here $k \equiv \dfrac{\omega}{c}$ and the complete solution is now given by

$$\Psi(\vec{r},\omega) = \sum_{lm}\left[A_{lm}^{(1)}h_{lm}^{(1)}(kr) + A_{lm}^{(2)}h_{lm}^{(2)}(kr)\right](r)Y_{lm}(\theta,\phi),$$

where $h_{lm}^{(1)}(kr)$ and $h_{lm}^{(2)}(kr)$ are the spherical Hankel functions. To gain a better understanding of the nature of these spherical waves, let us go back and look at the case when $l = 0$. In this case, there is no angular dependence and the amplitude depends only on the radial distance i.e. $\Psi(\vec{r},t) \rightarrow u(r,t)$. In this case, the wave equation reduces to

$$\left(\nabla^2 - \frac{1}{c^2}\frac{\partial^2}{\partial t^2}\right)\Psi(\vec{r},t) = 0 \rightarrow \left(\frac{\partial^2}{\partial r^2} + \frac{2}{r}\frac{\partial}{\partial r} - \frac{1}{c^2}\frac{\partial^2}{\partial t^2}\right)u(r,t) = 0$$

This equation can be rewritten as

$$\frac{\partial^2(ru)}{\partial t^2} - c^2\frac{\partial^2(ru)}{\partial r^2} = 0;$$

where the quantity ru satisfies the one-dimensional wave equation. Therefore, there are solutions in the form

$$u(r,t) = \frac{1}{r}F(r-ct) + \frac{1}{r}G(r+ct),$$

where F and G are general solutions to the one-dimensional wave equation, and can be interpreted as respectively an outgoing or incoming spherical wave. Such waves are generated by a point source, and they make possible sharp signals whose form is altered only by a decrease in amplitude as r increases. Such waves exist only in cases of space with odd dimensions.

For physical examples of non-spherical wave solutions to the 3D wave equation that do possess angular dependence.

Monochromatic Spherical Wave

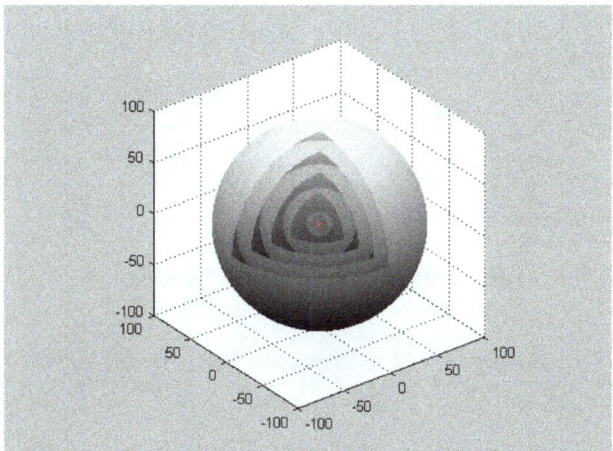

Cut-away of spherical wavefronts, with a wavelength of 10 units, propagating from a point source.

Although the word "monochromatic" is not exactly accurate since it refers to light or electromagnetic radiation with well-defined frequency, the spirit is to discover the eigenmode of the wave equation in three-dimensions. Following the derivation in the previous section on Plane wave eigenmodes, if we again restrict our solutions to spherical waves that oscillate in time with well-defined *constant* angular frequency ω, then the transformed function $ru(r,t)$ has simply plane wave solutions,

$$ru(r,t) = Ae^{i(\omega t \pm kr)},$$

or

$$u(r,t) = \frac{A}{r} e^{i(\omega t \pm kr)}.$$

From this we can observe that the peak intensity of the spherical wave oscillation, characterized as the squared wave amplitude

$$I = |u(r,t)|^2 = \frac{|A|^2}{r^2}.$$

drops at the rate proportional to $1/r^2$, an example of the inverse-square law.

Solution of a General Initial-value Problem

The wave equation is linear in u and it is left unaltered by translations in space and time. Therefore, we can generate a great variety of solutions by translating and summing spherical waves. Let $\varphi(\xi,\eta,\zeta)$ be an arbitrary function of three independent variables, and let the spherical wave form F be a delta-function: that is, let F be a weak limit of continuous functions whose integral is unity, but whose support (the region where the function is non-zero) shrinks to the origin. Let a family of spherical waves have center at (ξ,η,ζ), and let r be the radial distance from that point. Thus

$$r^2 = (x-\xi)^2 + (y-\eta)^2 + (z-\zeta)^2.$$

If u is a superposition of such waves with weighting function φ, then

$$u(t,x,y,z) = \frac{1}{4\pi c} \iiint \varphi(\xi,\eta,\zeta) \frac{\delta(r-ct)}{r} d\xi d\eta d\zeta;$$

the denominator $4\pi c$ is a convenience.

From the definition of the delta-function, u may also be written as

$$u(t,x,y,z) = \frac{t}{4\pi} \iint_S \varphi(x+ct\alpha, y+ct\beta, z+ct\gamma) d\omega,$$

where α, β, and γ are coordinates on the unit sphere S, and ω is the area element on S. This result has the interpretation that $u(t,x)$ is t times the mean value of φ on a sphere of radius ct centered at x:

$$u(t,x,y,z) = tM_{ct}[\phi].$$

It follows that

$$u(0,x,y,z) = 0, \quad u_t(0,x,y,z) = \phi(x,y,z).$$

The mean value is an even function of t, and hence if

$$v(t,x,y,z) = \frac{\partial}{\partial t}\left(tM_{ct}[\psi]\right),$$

then

$$v(0,x,y,z) = \psi(x,y,z), \quad v_t(0,x,y,z) = 0.$$

These formulas provide the solution for the initial-value problem for the wave equation. They show that the solution at a given point P, given (t,x,y,z) depends only on the data on the sphere of radius ct that is intersected by the light cone drawn backwards from P. It does *not* depend upon data on the interior of this sphere. Thus the interior of the sphere is a lacuna for the solution. This phenomenon is called Huygens' principle. It is true for odd numbers of space dimension, where for one dimension the integration is performed over the boundary of an interval with respect to the Dirac measure. It is not satisfied in even space dimensions. The phenomenon of lacunas has been extensively investigated in Atiyah, Bott and Gårding (1970, 1973).

Scalar Wave Equation in Two Space Dimensions

In two space dimensions, the wave equation is

$$u_{tt} = c^2\left(u_{xx} + u_{yy}\right).$$

We can use the three-dimensional theory to solve this problem if we regard u as a function in three dimensions that is independent of the third dimension. If

$$u(0,x,y) = 0, \quad u_t(0,x,y) = \phi(x,y),$$

then the three-dimensional solution formula becomes

$$u(t,x,y) = tM_{ct}[\phi] = \frac{t}{4\pi}\iint_S \phi(x+ct\alpha, y+ct\beta)\,d\omega,$$

where α and β are the first two coordinates on the unit sphere, and $d\omega$ is the area element on the sphere. This integral may be rewritten as a double integral over the disc D with center (x,y) and radius ct:

$$u(t,x,y) = \frac{1}{2\pi c}\iint_D \frac{\phi(x+\xi, y+\eta)}{\sqrt{(ct)^2 - \xi^2 - \eta^2}}\,d\xi\,d\eta.$$

It is apparent that the solution at (t,x,y) depends not only on the data on the light cone where

$$(x-\xi)^2 + (y-\eta)^2 = c^2t^2,$$

but also on data that are interior to that cone.

Scalar Wave Equation in General Dimension and Kirchhoff's Formulae

We want to find solutions to $u_{tt} - \Delta u = 0$ for $u : \mathbb{R}^n \times (0, \infty) \to \mathbb{R}$ with $u(x, 0) = g(x)$ and $u_t(x, 0) = h(x)$.

Odd Dimensions

Assume $n \geq 3$ is an odd integer and $g \in C^{m+1}(\mathbb{R}^n)$, $h \in C^m(\mathbb{R}^n)$ for $m = (n+1)/2$. Let $\gamma_n = 1 \cdot 3 \cdot 5 \cdots (n-2)$ and let

$$u(x,t) = \frac{1}{\gamma_n}\left[\partial_t\left(\frac{1}{t}\partial_t\right)^{\frac{n-3}{2}}\left(t^{n-2}\frac{1}{|\partial B_t(x)|}\int_{\partial B_t(x)} g\,dS\right) + \left(\frac{1}{t}\partial_t\right)^{\frac{n-3}{2}}\left(t^{n-2}\frac{1}{|\partial B_t(x)|}\int_{\partial B_t(x)} h\,dS\right)\right]$$

then

$$u \in C^2(\mathbb{R}^n \times [0, \infty))$$

$$u_{tt} - \Delta u = 0 \text{ in } \mathbb{R}^n \times (0, \infty)$$

$$\lim_{(x,t)\to(x^0,0)} u(x,t) = g(x^0)$$

$$\lim_{(x,t)\to(x^0,0)} u_t(x,t) = h(x^0)$$

Even Dimensions

Assume $n \geq 2$ is an even integer and $g \in C^{m+1}(\mathbb{R}^n)$, $h \in C^m(\mathbb{R}^n)$, for $m = (n+2)/2$. Let $\gamma_n = 2 \cdot 4 \cdots n$ and let

$$u(x,t) = \frac{1}{\gamma_n}\left[\partial_t\left(\frac{1}{t}\partial_t\right)^{\frac{n-2}{2}}\left(t^n\frac{1}{|B_t(x)|}\int_{B_t(x)}\frac{g}{(t^2-|y-x|^2)^{\frac{1}{2}}}dy\right) + \left(\frac{1}{t}\partial_t\right)^{\frac{n-2}{2}}\left(t^n\frac{1}{|B_t(x)|}\int_{B_t(x)}\frac{h}{(t^2-|y-x|^2)^{\frac{1}{2}}}dy\right)\right]$$

then

$$u \in C^2(\mathbb{R}^n \times [0, \infty))$$

$$u_{tt} - \Delta u = 0 \text{ in } \mathbb{R}^n \times (0, \infty)$$

$$\lim_{(x,t)\to(x^0,0)} u(x,t) = g(x^0)$$

$$\lim_{(x,t)\to(x^0,0)} u_t(x,t) = h(x^0)$$

One Space Dimension

The Sturm-liouville Formulation

A flexible string that is stretched between two points $x = 0$ and $x = L$ satisfies the wave equation for $t > 0$ and $0 < x < L$. On the boundary points, u may satisfy a variety of boundary conditions. A general form that is appropriate for applications is

$$-u_x(t,0) + au(t,0) = 0,$$

$$u_x(t,L) + bu(t,L) = 0,$$

where a and b are non-negative. The case where u is required to vanish at an endpoint is the limit of this condition when the respective a or b approaches infinity. The method of separation of variables consists in looking for solutions of this problem in the special form

$$u(t,x) = T(t)v(x).$$

A consequence is that

$$\frac{T''}{c^2 T} = \frac{v''}{v} = -\lambda.$$

The eigenvalue λ must be determined so that there is a non-trivial solution of the boundary-value problem

$$v'' + \lambda v = 0,$$

$$-v'(0) + av(0) = 0, \quad v'(L) + bv(L) = 0.$$

This is a special case of the general problem of Sturm–Liouville theory. If a and b are positive, the eigenvalues are all positive, and the solutions are trigonometric functions. A solution that satisfies square-integrable initial conditions for u and u_t can be obtained from expansion of these functions in the appropriate trigonometric series.

Investigation by Numerical Methods

Approximating the continuous string with a finite number of equidistant mass points one gets the following physical model:

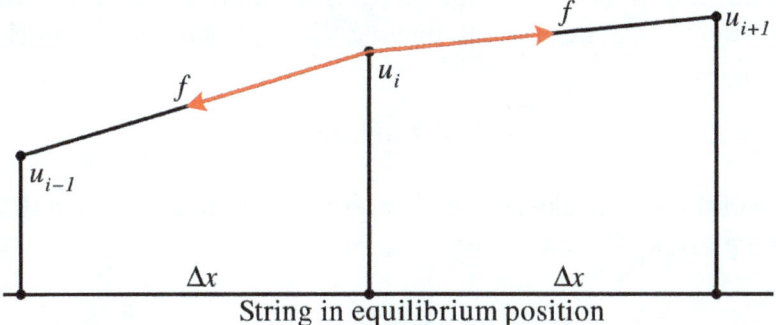

Figure 1: Three consecutive mass points of the discrete model for a string

If each mass point has the mass m, the tension of the string is f, the separation between the mass points is Δx and u_i, $i = 1, ..., n$ are the offset of these n points from their equilibrium points (i.e. their position on a straight line between the two attachment points of the string) the vertical component of the force towards point $i+1$ is

$$\frac{u_{i+1} - u_i}{\Delta x} f \tag{1}$$

and the vertical component of the force towards point $i-1$ is

$$\frac{u_{i-1} - u_i}{\Delta x} f \tag{2}$$

Taking the sum of these two forces and dividing with the mass m one gets for the vertical motion:

$$\ddot{u}_i = \left(\frac{f}{m \, \Delta x}\right)\left(u_{i+1} + u_{i-1} - 2u_i\right) \tag{3}$$

As the mass density is

$$\rho = \frac{m}{\Delta x}$$

this can be written

$$\ddot{u}_i = \left(\frac{f}{\rho \, \Delta x^2}\right)\left(u_{i+1} + u_{i-1} - 2u_i\right) \tag{4}$$

The wave equation is obtained by letting $\Delta x \to 0$ in which case $u_i(t)$ takes the form $u(x, t)$ where $u(x, t)$ is continuous function of two variables, $\ddot{u}_i$ takes the form $\partial^2 u / \partial t^2$ and

$$\frac{u_{i+1} + u_{i-1} - 2u_i}{\Delta x^2} \to \frac{\partial^2 u}{\partial x^2}$$

But the discrete formulation (3) of the equation of state with a finite number of mass point is just the suitable one for a numerical propagation of the string motion. The boundary condition

$$u(0,t) = u(L,t) = 0$$

where L is the length of the string takes in the discrete formulation the form that for the outermost points u_1 and u_n the equation of motion are

$$\ddot{u}_1 = \left(\frac{c}{\Delta x}\right)^2 \left(u_2 - 2u_1\right) \tag{5}$$

and

$$\ddot{u}_n = \left(\frac{c}{\Delta x}\right)^2 \left(u_{n-1} - 2u_n\right) \tag{6}$$

while for $1 < i < n$

$$\ddot{u}_i = \left(\frac{c}{\Delta x}\right)^2 \left(u_{i+1} + u_{i-1} - 2u_i\right) \tag{7}$$

where $c = \sqrt{\dfrac{f}{\rho}}$

If the string is approximated with 100 discrete mass points one gets the 100 coupled second order differential equations (5), (6) and (7) or equivalently 200 coupled first order differential equations.

Propagating these up to the times

$$\frac{L}{c} k\, 0.05 \quad k = 0, \cdots, 5$$

using an 8th order multistep method the 6 states displayed in figure 2 are found:

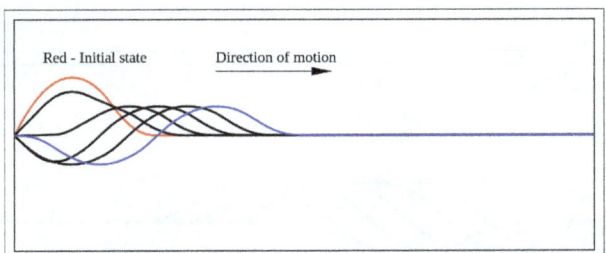

Figure 2: The string at 6 consecutive epochs, the first (red) corresponding to the initial time with the string in rest

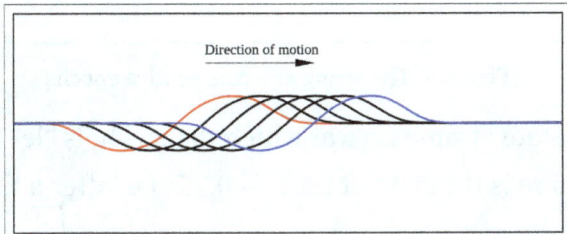

Figure 3: The string at 6 consecutive epochs

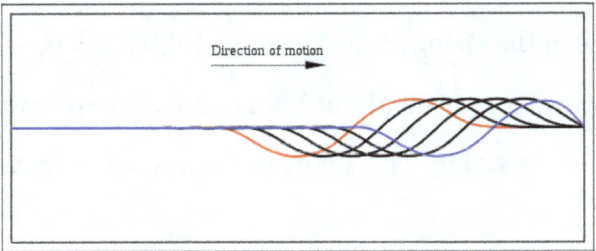

Figure 4: The string at 6 consecutive epochs

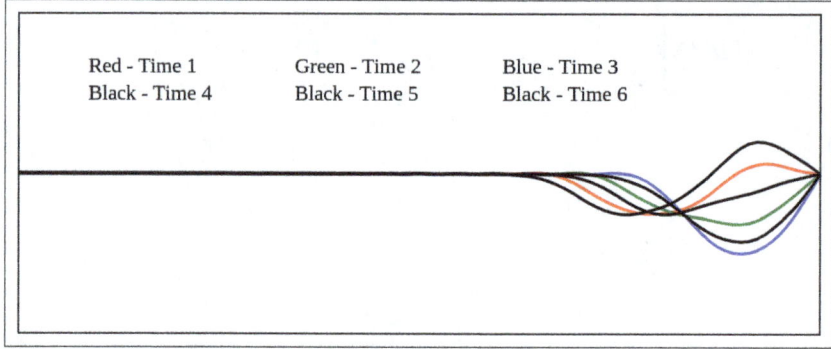

Figure 5: The string at 6 consecutive epochs

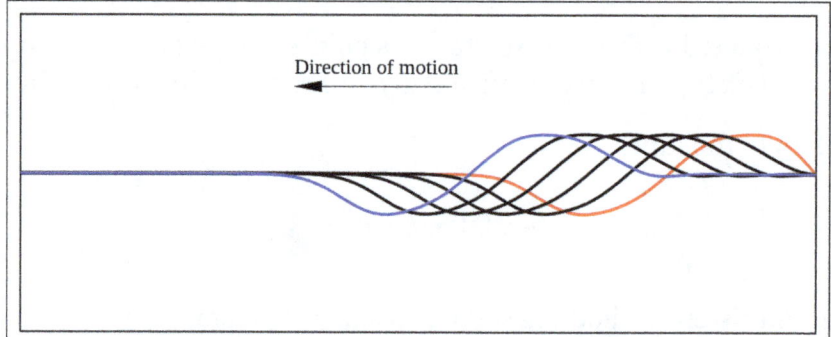

Figure 6: The string at 6 consecutive epochs

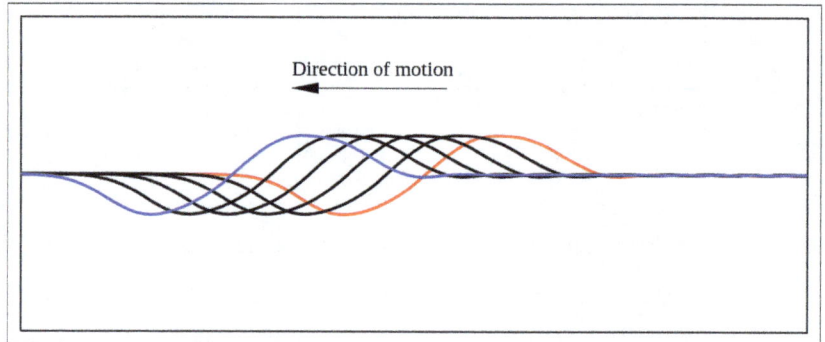

Figure 7: The string at 6 consecutive epochs

The red curve is the initial state at time zero at which the string is "let free" in a predefined shape-with all $\dot{u}_i = 0$. The blue curve is the state at time $\dfrac{L}{c} 0.25$, i.e. after a time that corresponds to the time a wave that is moving with the nominal wave velocity $c = \sqrt{\dfrac{f}{\rho}}$ would need for one fourth of the length of the string.

Figure 3 displays the shape of the string at the times $\dfrac{L}{c} k\, 0.05\ \ k = 6, \cdots, 11$. The wave travels in direction right with the speed $c = \sqrt{\dfrac{f}{\rho}}$ without being actively constraint by the boundary conditions at the two extremes of the string. The shape of the wave is constant, i.e. the curve is indeed of the form $f(x-ct)$.

Figure 4 displays the shape of the string at the times $\dfrac{L}{c} k\, 0.05\ \ k = 12, \cdots, 17$. The constraint on the right extreme starts to interfere with the motion preventing the wave to raise the end of the string.

Figure 5 displays the shape of the string at the times $\dfrac{L}{c} k\, 0.05\ \ k = 18, \cdots, 23$ when the direction of motion is reversed. The red, green and blue curves are the states at the times $\dfrac{L}{c} k\, 0.05\ \ k = 18, \cdots, 20$ while the 3 black curves correspond to the states at times $\dfrac{L}{c} k\, 0.05\ \ k = 21, \cdots, 23$ with the wave starting to move back towards left.

Figure 6 and figure 7 finally display the shape of the string at the times $\dfrac{L}{c} k\, 0.05\ \ k = 24, \cdots, 29$ and $\dfrac{L}{c} k\, 0.05\ \ k = 30, \cdots, 35$. The wave now travels towards left and the constraints at the end points are not active any more. When finally the other extreme of the string the direction will again be reversed in a way similar to what is displayed in figure 6

Several Space Dimensions

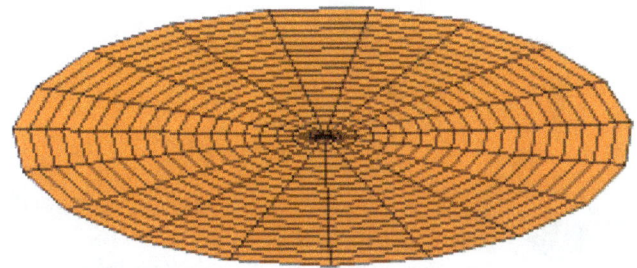

A solution of the wave equation in two dimensions with a zero-displacement boundary condition along the entire outer edge.

The one-dimensional initial-boundary value theory may be extended to an arbitrary number of space dimensions. Consider a domain D in m-dimensional x space, with boundary B. Then the wave equation is to be satisfied if x is in D and $t > 0$. On the boundary of D, the solution u shall satisfy

$$\frac{\partial u}{\partial n} + au = 0,$$

where n is the unit outward normal to B, and a is a non-negative function defined on B. The case where u vanishes on B is a limiting case for a approaching infinity. The initial conditions are

$$u(0, x) = f(x), \quad u_t(0, x) = g(x),$$

where f and g are defined in D. This problem may be solved by expanding f and g in the eigenfunctions of the Laplacian in D, which satisfy the boundary conditions. Thus the eigenfunction v satisfies

$$\nabla \cdot \nabla v + \lambda v = 0,$$

in D, and

$$\frac{\partial v}{\partial n} + av = 0,$$

on B.

In the case of two space dimensions, the eigenfunctions may be interpreted as the modes of vibration of a drumhead stretched over the boundary B. If B is a circle, then these eigenfunctions have an angular component that is a trigonometric function of the polar angle θ, multiplied by a Bessel function (of integer order) of the radial component. Further details are in Helmholtz equation.

If the boundary is a sphere in three space dimensions, the angular components of the eigenfunctions are spherical harmonics, and the radial components are Bessel functions of half-integer order.

Inhomogeneous Wave Equation in one Dimension

The inhomogeneous wave equation in one dimension is the following:

$$c^2 u_{xx}(x,t) - u_{tt}(x,t) = s(x,t)$$

with initial conditions given by

$$u(x,0) = f(x)$$

$$u_t(x,0) = g(x)$$

The function $s(x, t)$ is often called the source function because in practice it describes the effects of the sources of waves on the medium carrying them. Physical examples of source functions include the force driving a wave on a string, or the charge or current density in the Lorenz gauge of electromagnetism.

One method to solve the initial value problem (with the initial values as posed above) is to take advantage of a special property of the wave equation in an odd number of space dimensions, namely that its solutions respect causality. That is, for any point (x_i, t_i), the value of $u(x_i, t_i)$ depends only on the values of $f(x_i+ct_i)$ and $f(x_i-ct_i)$ and the values of the function $g(x)$ between (x_i-ct_i) and (x_i+ct_i). This can be seen in d'Alembert's formula, stated above, where these quantities are the only ones that show up in it. Physically, if the maximum propagation speed is c, then no part of the wave that can't propagate to a given point by a given time can affect the amplitude at the same point and time.

In terms of finding a solution, this causality property means that for any given point on the line being considered, the only area that needs to be considered is the area encompassing all the points that could causally affect the point being considered. Denote the area that casually affects point (x_i, t_i) as R_C. Suppose we integrate the inhomogeneous wave equation over this region.

$$\iint_{R_C} \left(c^2 u_{xx}(x,t) - u_{tt}(x,t) \right) dx\, dt = \iint_{R_C} s(x,t) dx\, dt.$$

To simplify this greatly, we can use Green's theorem to simplify the left side to get the following:

$$\int_{L_0+L_1+L_2}\left(-c^2u_x(x,t)dt-u_t(x,t)dx\right)=\iint_{R_C}s(x,t)dxdt.$$

The left side is now the sum of three line integrals along the bounds of the causality region. These turn out to be fairly easy to compute

$$\int_{x_i-ct_i}^{x_i+ct_i}-u_t(x,0)dx=-\int_{x_i-ct_i}^{x_i+ct_i}g(x)dx.$$

In the above, the term to be integrated with respect to time disappears because the time interval involved is zero, thus $dt = 0$.

For the other two sides of the region, it is worth noting that $x\pm ct$ is a constant, namingly $x_i\pm ct_i$, where the sign is chosen appropriately. Using this, we can get the relation $dx\pm cdt = 0$, again choosing the right sign:

$$\int_{L_1}\left(-c^2u_x(x,t)dt-u_t(x,t)dx\right)\quad=\int_{L_1}\left(cu_x(x,t)dx+cu_t(x,t)dt\right)$$
$$=c\int_{L_1}du(x,t)$$
$$=cu(x_i,t_i)-cf(x_i+ct_i).$$

And similarly for the final boundary segment:

$$\int_{L_2}\left(-c^2u_x(x,t)dt-u_t(x,t)dx\right)\quad=-\int_{L_2}\left(cu_x(x,t)dx+cu_t(x,t)dt\right)$$
$$=-c\int_{L_2}du(x,t)$$
$$=cu(x_i,t_i)-cf(x_i-ct_i).$$

Adding the three results together and putting them back in the original integral:

$$\iint_{R_C}s(x,t)dxdt\quad=-\int_{x_i-ct_i}^{x_i+ct_i}g(x)dx+cu(x_i,t_i)-cf(x_i+ct_i)+cu(x_i,t_i)-cf(x_i-ct_i)$$
$$=2cu(x_i,t_i)-cf(x_i+ct_i)-cf(x_i-ct_i)-\int_{x_i-ct_i}^{x_i+ct_i}g(x)dx$$

Solving for $u(x_i, t_i)$ we arrive at

$$u(x_i,t_i)=\frac{f(x_i+ct_i)+f(x_i-ct_i)}{2}+\frac{1}{2c}\int_{x_i-ct_i}^{x_i+ct_i}g(x)dx+\frac{1}{2c}\int_0^{t_i}\int_{x_i-c(t_i-t)}^{x_i+c(t_i-t)}s(x,t)dxdt.$$

In the last equation of the sequence, the bounds of the integral over the source function have been made explicit. Looking at this solution, which is valid for all choices (x_i, t_i) compatible with the wave equation, it is clear that the first two terms are simply d'Alembert's formula, as stated above as the solution of the homogeneous wave equation in one dimension. The difference is in the third term, the integral over the source.

Other Coordinate Systems

In three dimensions, the wave equation, when written in elliptic cylindrical coordinates, may be solved by separation of variables, leading to the Mathieu differential equation.

Further generalizations

Elastic Waves

The elastic wave equation in three dimensions describes the propagation of waves in an isotropic homogeneous elastic medium. Most solid materials are elastic, so this equation describes such phenomena as seismic waves in the Earth and ultrasonic waves used to detect flaws in materials. While linear, this equation has a more complex form than the equations given above, as it must account for both longitudinal and transverse motion:

$$\rho \ddot{\mathbf{u}} = \mathbf{f} + (\lambda + 2\mu)\nabla(\nabla \cdot \mathbf{u}) - \mu\nabla \times (\nabla \times \mathbf{u})$$

where:

- λ and μ are the so-called Lamé parameters describing the elastic properties of the medium,

- ρ is the density,

- f is the source function (driving force),

- and u is the displacement vector.

Note that in this equation, both force and displacement are vector quantities. Thus, this equation is sometimes known as the vector wave equation. As an aid to understanding, the reader will observe that if f and $\nabla \cdot$ u are set to zero, this becomes (effectively) Maxwell's equation for the propagation of the electric field E, which has only transverse waves.

Dispersion Relation

In dispersive wave phenomena, the speed of wave propagation varies with the wavelength of the wave, which is reflected by a dispersion relation

$$\omega = \omega(\mathbf{k}),$$

where ω is the angular frequency and k is the wavevector describing plane wave solutions. For light waves, the dispersion relation is $\omega = \pm c\,|\mathbf{k}|$, but in general, the constant speed c gets replaced by a variable phase velocity:

$$v_{\mathrm{p}} = \frac{\omega(k)}{k}.$$

Laplace's Equation

In mathematics, Laplace's equation is a second-order partial differential equation named after Pierre-Simon Laplace who first studied its properties. This is often written as:

$$\nabla^2\varphi = 0 \qquad \text{or} \qquad \Delta\varphi = 0$$

where $\Delta = \nabla^2$ is the Laplace operator and φ is a scalar function.

Laplace's equation and Poisson's equation are the simplest examples of elliptic partial differential equations. The general theory of solutions to Laplace's equation is known as potential theory. The solutions of Laplace's equation are the harmonic functions, which are important in many fields of science, notably the fields of electromagnetism, astronomy, and fluid dynamics, because they can be used to accurately describe the behavior of electric, gravitational, and fluid potentials. In the study of heat conduction, the Laplace equation is the steady-state heat equation.

Definition

In three dimensions, the problem is to find twice-differentiable real-valued functions f, of real variables x, y, and z, such that

In Cartesian coordinates

$$\Delta f = \frac{\partial^2 f}{\partial x^2} + \frac{\partial^2 f}{\partial y^2} + \frac{\partial^2 f}{\partial z^2} = 0.$$

In cylindrical coordinates,

$$\Delta f = \frac{1}{r}\frac{\partial}{\partial r}\left(r\frac{\partial f}{\partial r}\right) + \frac{1}{r^2}\frac{\partial^2 f}{\partial \phi^2} + \frac{\partial^2 f}{\partial z^2} = 0$$

In spherical coordinates,

$$\Delta f = \frac{1}{\rho^2}\frac{\partial}{\partial \rho}\left(\rho^2\frac{\partial f}{\partial \rho}\right) + \frac{1}{\rho^2 \sin\theta}\frac{\partial}{\partial \theta}\left(\sin\theta\frac{\partial f}{\partial \theta}\right) + \frac{1}{\rho^2 \sin^2\theta}\frac{\partial^2 f}{\partial \varphi^2} = 0.$$

In curvilinear coordinates,

$$\Delta f = \frac{\partial}{\partial \xi^j}\left(\frac{\partial f}{\partial \xi^k}g^{kj}\right) + \frac{\partial f}{\partial \xi^j}g^{jm}\Gamma^n_{mn} = 0,$$

or

$$\Delta f = \frac{1}{\sqrt{|g|}}\frac{\partial}{\partial \xi^i}\left(\sqrt{|g|}g^{ij}\frac{\partial f}{\partial \xi^j}\right) = 0, \qquad (g = \det\{g_{ij}\}).$$

This is often written as

$$\nabla^2 f = 0$$

or, especially in more general contexts,

$$\Delta f = 0,$$

where $\Delta = \nabla^2$ is the Laplace operator or "Laplacian"

$$\Delta f = \nabla^2 f = \nabla \cdot \nabla f = \operatorname{div} \operatorname{grad} f,$$

where $\nabla \cdot$ is the divergence operator (also symbolized "div") which maps vector functions to scalar functions, and ∇ is the gradient operator (also symbolized "grad") which maps scalar functions to vector functions. Hence the Laplacian $\Delta f \overset{\text{def}}{=} \operatorname{div} \operatorname{grad} f$ maps the scalar function f to a scalar function.

If the right-hand side is specified as a given function, $h(x, y, z)$, i.e., if the whole equation is written as

$$\Delta f = h$$

then it is called "Poisson's equation".

The Laplace equation is also a special case of the Helmholtz equation.

Boundary Conditions

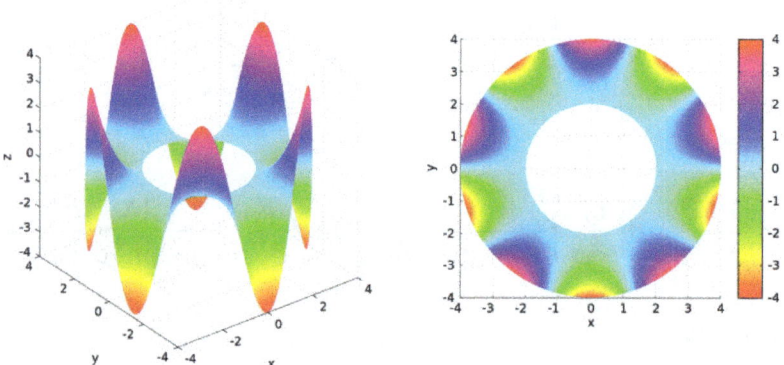

Laplace's equation on an annulus (inner radius r = 2 and outer radius R = 4)
with Dirichlet boundary conditions u(r=2) = 0 and u(R=4) = 4 sin(5 θ)

The Dirichlet problem for Laplace's equation consists of finding a solution φ on some domain D such that φ on the boundary of D is equal to some given function. Since the Laplace operator appears in the heat equation, one physical interpretation of this problem is as follows: fix the temperature on the boundary of the domain according to the given specification of the boundary condition. Allow heat to flow until a stationary state is reached in which the temperature at each point on the domain doesn't change anymore. The temperature distribution in the interior will then be given by the solution to the corresponding Dirichlet problem.

The Neumann boundary conditions for Laplace's equation specify not the function φ itself on the boundary of D, but its normal derivative. Physically, this corresponds to the construction of a potential for a vector field whose effect is known at the boundary of D alone.

Solutions of Laplace's equation are called harmonic functions; they are all analytic within the domain where the equation is satisfied. If any two functions are solutions to Laplace's equation (or any linear homogeneous differential equation), their sum (or any linear combination) is also a solution. This property, called the principle of superposition, is very useful, e.g., solutions to complex problems can be constructed by summing simple solutions.

Laplace Equation in Two Dimensions

The Laplace equation in two independent variables has the form

$$\frac{\partial^2 \psi}{\partial x^2} + \frac{\partial^2 \psi}{\partial y^2} \equiv \psi_{xx} + \psi_{yy} = 0.$$

Analytic Functions

The real and imaginary parts of a complex analytic function both satisfy the Laplace equation. That is, if $z = x + iy$, and if

$$f(z) = u(x, y) + iv(x, y),$$

then the necessary condition that $f(z)$ be analytic is that the Cauchy-Riemann equations be satisfied:

$$u_x = v_y, \quad v_x = -u_y.$$

where u_x is the first partial derivative of u with respect to x. It follows that

$$u_{yy} = (-v_x)_y = -(v_y)_x = -(u_x)_x.$$

Therefore u satisfies the Laplace equation. A similar calculation shows that v also satisfies the Laplace equation. Conversely, given a harmonic function, it is the real part of an analytic function, $f(z)$ (at least locally). If a trial form is

$$f(z) = \varphi(x, y) + i\psi(x, y),$$

then the Cauchy-Riemann equations will be satisfied if we set

$$\psi_x = -\varphi_y, \quad \psi_y = \varphi_x.$$

This relation does not determine ψ, but only its increments:

$$d\psi = -\varphi_y \, dx + \varphi_x \, dy.$$

The Laplace equation for φ implies that the integrability condition for ψ is satisfied:

$$\psi_{xy} = \psi_{yx},$$

and thus ψ may be defined by a line integral. The integrability condition and Stokes' theorem implies that the value of the line integral connecting two points is independent of the path. The resulting pair of solutions of the Laplace equation are called conjugate harmonic functions. This construction is only valid locally, or provided that the path does not loop around a singularity. For example, if r and θ are polar coordinates and

$$\varphi = \log r,$$

then a corresponding analytic function is

$$f(z) = \log z = \log r + i\theta.$$

However, the angle θ is single-valued only in a region that does not enclose the origin.

The close connection between the Laplace equation and analytic functions implies that any solution of the Laplace equation has derivatives of all orders, and can be expanded in a power series, at least inside a circle that does not enclose a singularity. This is in sharp contrast to solutions of the wave equation, which generally have less regularity.

There is an intimate connection between power series and Fourier series. If we expand a function f in a power series inside a circle of radius R, this means that

$$f(z) = \sum_{n=0}^{\infty} c_n z^n,$$

with suitably defined coefficients whose real and imaginary parts are given by

$$c_n = a_n + ib_n.$$

Therefore

$$f(z) = \sum_{n=0}^{\infty} \left[a_n r^n \cos n\theta - b_n r^n \sin n\theta \right] + i \sum_{n=1}^{\infty} \left[a_n r^n \sin n\theta + b_n r^n \cos n\theta \right],$$

which is a Fourier series for f. These trigonometric functions can themselves be expanded, using multiple angle formulae.

Fluid Flow

Let the quantities u and v be the horizontal and vertical components of the velocity field of a steady incompressible, irrotational flow in two dimensions. The condition that the flow be incompressible is that

$$u_x + v_y = 0,$$

and the condition that the flow be irrotational is that

$$\nabla \times \mathbf{V} = v_x - u_y = 0.$$

If we define the differential of a function ψ by

$$d\psi = v\,dx - u\,dy,$$

then the incompressibility condition is the integrability condition for this differential: the resulting function is called the stream function because it is constant along flow lines. The first derivatives of ψ are given by

$$\psi_x = v, \quad \psi_y = -u,$$

and the irrotationality condition implies that ψ satisfies the Laplace equation. The harmonic function φ that is conjugate to ψ is called the velocity potential. The Cauchy-Riemann equations imply that

$$\varphi_x = -u, \quad \varphi_y = -v.$$

Thus every analytic function corresponds to a steady incompressible, irrotational fluid flow in the plane. The real part is the velocity potential, and the imaginary part is the stream function.

Electrostatics

According to Maxwell's equations, an electric field (u,v) in two space dimensions that is independent of time satisfies

$$\nabla \times (u, v, 0) = (v_x - u_y)\hat{\mathbf{k}} = \mathbf{0},$$

and: $\nabla \cdot (u, v) = \rho$, where ρ is the charge density. The first Maxwell equation is the integrability condition for the differential

$$d\varphi = -u\,dx - v\,dy,$$

so the electric potential φ may be constructed to satisfy

$$\varphi_x = -u, \quad \varphi_y = -v.$$

The second of Maxwell's equations then implies that

$$\varphi_{xx} + \varphi_{yy} = -\rho,$$

which is the Poisson equation. It is important to note that the Laplace equation can be used in three-dimensional problems in electrostatics and fluid flow just as in two dimensions.

Laplace Equation in Three Dimensions

Fundamental Solution

A fundamental solution of Laplace's equation satisfies

$$\Delta u = u_{xx} + u_{yy} + u_{zz} = -\delta(x - x', y - y', z - z'),$$

where the Dirac delta function δ denotes a unit source concentrated at the point (x', y', z'). No function has this property: in fact it is a distribution rather than a function; but it can be thought of as a limit of functions whose integrals over space are unity, and whose support (the region where the function is non-zero) shrinks to a point. It is common to take a different sign convention for this equation than one typically does when defining fundamental solutions. This choice of sign is often convenient to work with because $-\Delta$ is a positive operator. The definition of the fundamental solution thus implies that, if the Laplacian of u is integrated over any volume that encloses the source point, then

$$\iiint_V \nabla \cdot \nabla u \, dV = -1.$$

The Laplace equation is unchanged under a rotation of coordinates, and hence we can expect that a fundamental solution may be obtained among solutions that only depend upon the distance r from the source point. If we choose the volume to be a ball of radius a around the source point, then Gauss' divergence theorem implies that

$$-1 = \iiint_V \nabla \cdot \nabla u \, dV = \iint_S \frac{du}{dr} \, dS = 4\pi a^2 \frac{du}{dr}\bigg|_{r=a}.$$

It follows that

$$\frac{du}{dr} = -\frac{1}{4\pi r^2},$$

on a sphere of radius r that is centered on the source point, and hence

$$u = \frac{1}{4\pi r}.$$

Note that, with the opposite sign convention (used in Physics), this is the potential generated by a point particle, for an inverse-square law force, arising in the solution of Poisson equation. A similar argument shows that in two dimensions

$$u = -\frac{\log(r)}{2\pi}.$$

where $\log(r)$ denotes the natural logarithm. Note that, with the opposite sign convention, this is the potential generated by a pointlike sink, which is the solution of the Euler equations in two-dimensional incompressible flow.

Green's Function

A Green's function is a fundamental solution that also satisfies a suitable condition on the boundary S of a volume V. For instance,

$$G(x, y, z; x', y', z')$$

may satisfy

$$\nabla \cdot \nabla G = -\delta(x - x', y - y', z - z') \qquad \text{in } V,$$

$$G = 0 \quad \text{if} \quad (x, y, z) \qquad \text{on } S.$$

Now if u is any solution of the Poisson equation in V:

$$\nabla \cdot \nabla u = -f,$$

and u assumes the boundary values g on S, then we may apply Green's identity, (a consequence of the divergence theorem) which states that

$$\iiint_V [G\nabla \cdot \nabla u - u\nabla \cdot \nabla G]\, dV = \iiint_V \nabla \cdot [G\nabla u - u\nabla G]\, dV = \iint_S [Gu_n - uG_n]\, dS.$$

The notations u_n and G_n denote normal derivatives on S. In view of the conditions satisfied by u and G, this result simplifies to

$$u(x', y', z') = \iiint_V Gf\, dV + \iint_S G_n g\, dS.$$

Thus the Green's function describes the influence at (x', y', z') of the data f and g. For the case of the interior of a sphere of radius a, the Green's function may be obtained by means of a reflection (Sommerfeld 1949): the source point P at distance ρ from the center of the sphere is reflected along its radial line to a point P' that is at a distance

$$\rho' = \frac{a^2}{\rho}.$$

Note that if P is inside the sphere, then P' will be outside the sphere. The Green's function is then given by

$$\frac{1}{4\pi R} - \frac{a}{4\pi \rho R'},$$

where R denotes the distance to the source point P and R' denotes the distance to the reflected point P'. A consequence of this expression for the Green's function is the Poisson integral formula. Let ρ, θ, and φ be spherical coordinates for the source point P. Here θ denotes the angle with the vertical axis, which is contrary to the usual American mathematical notation, but agrees with standard European and physical practice. Then the solution of the Laplace equation with Dirichlet boundary values g inside the sphere is given by

$$u(P) = \frac{1}{4\pi} a^3 \left(1 - \frac{\rho^2}{a^2}\right) \int\limits_0^{2\pi} \int\limits_0^{\pi} \frac{g(\theta', \varphi') \sin \theta'}{(a^2 + \rho^2 - 2a\rho \cos \Theta)^{\frac{3}{2}}} d\theta' d\varphi'$$

where

$$\cos \Theta = \cos \theta \cos \theta' + \sin \theta \sin \theta' \cos(\phi - \phi')$$

is the cosine of the angle between (θ, ϕ) and (θ', φ'). A simple consequence of this formula is that if u is a harmonic function, then the value of u at the center of the sphere is the mean value of its values on the sphere. This mean value property immediately implies that a non-constant harmonic function cannot assume its maximum value at an interior point.

Electrostatics

In free space the Laplace equation of any electrostatic potential must equal zero since ρ (charge density) is zero in free space.

Taking the gradient of the electric potential we get the electrostatic field

$$E = -\nabla V$$

Taking the divergence of the electrostatic field, we obtain Poisson's equation, that relates charge density and electric potential

$$\nabla^2 V = -\frac{\rho}{\varepsilon_0}$$

In the particular case of the empty space (ρ = 0) Poisson's equation reduces to Laplace's equation for the electric potential.

Using a uniqueness theorem and showing that a potential satisfies Laplace's equation (second derivative of V should be zero i.e. in free space) and the potential has the correct values at the boundaries, the potential is then uniquely defined.

A potential that doesn't satisfy Laplace's equation together with the boundary condition is an invalid electrostatic potential.

Linear Differential Equation

In mathematics, linear differential equations are differential equations having solutions which can be added together in particular linear combinations to form further solutions. They equate 0 to a polynomial that is linear in the value and various derivatives of a variable; its linearity means that each term in the polynomial has degree either 0 or 1.

Linear differential equations can be ordinary (ODEs) or partial (PDEs).

The solutions to (homogeneous) linear differential equations form a vector space (unlike non-linear differential equations).

Basic Features

Linear differential equations are of the form

$$Ly = f$$

where the differential operator L is a linear operator, y is the unknown function, and the right hand side f is a given function (called the source term) of the same variable. For a function dependent on time we may write the equation more expressly as

$$Ly(t) = f(t)$$

and, even more precisely by bracketing

$$L[y(t)] = f(t).$$

The linear operator L may be considered to be of the form

$$L_n(y) \equiv \frac{d^n y}{dt^n} + A_1(t)\frac{d^{n-1} y}{dt^{n-1}} + \cdots + A_{n-1}(t)\frac{dy}{dt} + A_n(t)y$$

The linearity condition on L rules out operations such as taking the square of the derivative of y; but permits, for example, taking the second derivative of y. It is convenient to rewrite this equation in an operator form

$$L_n(y) \equiv \left[D^n + A_1(t)D^{n-1} + \cdots + A_{n-1}(t)D + A_n(t) \right]y$$

where D is the differential operator d/dt (i.e. $Dy = y' = dy/dt$, $D^2y = y'' = d^2y/dt^2$,...), and the A_n are given functions.

Such an equation is said to have order n, the index of the highest derivative of y that is involved.

A typical simple example is the linear differential equation used to model radioactive decay. Let $N(t)$ denote the number of radioactive atoms remaining in some sample of material at time t. Then for some constant $k > 0$, the rate at which the radioactive atoms decay can be modelled by

$$\frac{dN}{dt} = -kN$$

If y is assumed to be a function of only one variable, one speaks about an ordinary differential equation, else the derivatives and their coefficients must be understood as (contracted) vectors, matrices or tensors of higher rank, and we have a (linear) partial differential equation.

The case where $f = 0$ is called a homogeneous equation and its solutions are called complementary functions. It is particularly important to the solution of the general case, since any complementary function can be added to a solution of the inhomogeneous equation to give another solution (by a method traditionally called *particular integral and complementary function*). When the A_i are numbers, the equation is said to have *constant coefficients*.

Homogeneous Equations with Constant Coefficients

The first method of solving linear homogeneous ordinary differential equations with constant coefficients is due to Euler, who realized that solutions have the form e^{zx}, for possibly-complex values of z. The exponential function is one of the few functions to keep its shape after differentiation, allowing the sum of its multiple derivatives to cancel out to zero, as required by the equation. Thus, for constant values $A_1,..., A_n$, to solve:

$$y^{(n)} + A_1 y^{(n-1)} + \cdots + A_n y = 0,$$

we set $y = e^{zx}$, leading to

$$z^n e^{zx} + A_1 z^{n-1} e^{zx} + \cdots + A_n e^{zx} = 0.$$

Division by e^{zx} gives the nth-order polynomial:

$$F(z) = z^n + A_1 z^{n-1} + \cdots + A_n = 0.$$

This algebraic equation $F(z) = 0$ is the characteristic equation considered later by Gaspard Monge and Augustin-Louis Cauchy.

Formally, the terms $y^{(k)} (k = 1, 2, \ldots, n)$ of the original differential equation are replaced by z^k. Solving the polynomial gives n values of z, z_1, ..., z_n. Substitution of any of those values for z into e^{zx} gives a solution $e^{z_i x}$. Since homogeneous linear differential equations obey the superposition principle, any linear combination of these functions also satisfies the differential equation.

When these roots are all distinct, we have n distinct solutions to the differential equation. It can be shown that these are linearly independent, by applying the Vandermonde determinant, and together they form a basis of the space of all solutions of the differential equation.

Examples
$$y''' - 2y''' + 2y'' - 2y' + y = 0$$

has the characteristic equation

$$z^4 - 2z^3 + 2z^2 - 2z + 1 = 0.$$

This has zeroes, i, $-i$, and 1 (multiplicity 2). The solution basis is then

$$e^{ix}, e^{-ix}, e^x, xe^x.$$

This corresponds to the real-valued solution basis

$$\cos x, \sin x, e^x, xe^x.$$

The preceding gave a solution for the case when all zeros are distinct, that is, each has multiplicity 1. For the general case, if z is a (possibly complex) zero (or root) of $F(z)$ having multiplicity m, then, for $k \in \{0,1,\ldots,m-1\}$, $y = x^k e^{zx}$ is a solution of the ordinary differential equation. Applying this to all roots gives a collection of n distinct and linearly independent functions, where n is the degree of $F(z)$. As before, these functions make up a basis of the solution space.

If the coefficients A_i of the differential equation are real, then real-valued solutions are generally preferable. Since non-real roots z then come in conjugate pairs, so do their corresponding basis functions $x^k e^{zx}$, and the desired result is obtained by replacing each pair with their real-valued linear combinations $\operatorname{Re}(y)$ and $\operatorname{Im}(y)$, where y is one of the pair.

A case that involves complex roots can be solved with the aid of Euler's formula.

Second-order Case

In the $n=2$ case

$$y'' + (a-1)y' + by = 0,$$

the characteristic equation is of the form

$$z^2 + (a-1)z + b = 0.$$

We then can solve for z. There are three particular cases of interest:

- Case #1: Two distinct roots, z_1 and z_2

- Case #2: One real repeated root, z

- Case #3: Complex roots, $\alpha \pm \beta i$

In case #1, the general solution is given by

$$y = c_1 e^{z_1 x} + c_2 e^{z_2 x}.$$

In case #2, the general solution is given by

$$y = (c_1 + c_2 x)e^{zx}.$$

In case #3, the general solution is given, using Euler's equation, by

$$y = c_1 e^{\alpha x} \cos(\beta x) + c_2 e^{\alpha x} \sin(\beta x),$$

$$\alpha = \operatorname{Re}(z),$$

$$\beta = \operatorname{Im}(z).$$

In each case, the constants c_1, c_2 are functions of the initial conditions $y(0), y'(0)$. They can be found by using the values of the initial conditions in the solution equation for y and in the resulting equation for y', giving two equations in the two unknown parameters.

Examples

Given $y'' - 4y' + 5y = 0$. The characteristic equation is $z^2 - 4z + 5 = 0$ which has roots "$2 \pm i$". Thus the solution basis $\{y_1, y_2\}$ is $\{e^{(2+i)x}, e^{(2-i)x}\}$. Now y is a solution if and only if $y = c_1 y_1 + c_2 y_2$ for $c_1, c_2 \in \mathbf{C}$.

Because the coefficients are real,

- we are likely not interested in the complex solutions

- our basis elements are mutual conjugates

The linear combinations

$$u_1 = \mathrm{Re}(y_1) = \tfrac{1}{2}(y_1 + y_2) = e^{2x} \cos(x),$$

$$u_2 = \mathrm{Im}(y_1) = \tfrac{1}{2i}(y_1 - y_2) = e^{2x} \sin(x),$$

will give us a real basis in $\{u_1, u_2\}$.

Simple Harmonic Oscillator

The second order differential equation

$$D^2 y = -k^2 y,$$

which represents a simple harmonic oscillator, can be restated as

$$(D^2 + k^2)y = 0.$$

The expression in parenthesis can be factored out, yielding

$$((D + ik)(D - ik))y = 0,$$

which has a pair of linearly independent solutions:

$$(D - ik)y = 0$$

$$(D + ik)y = 0.$$

The solutions are, respectively,

$$y_0 = A_0 e^{ikx}$$

and

$$y_1 = A_1 e^{-ikx}.$$

These solutions provide a basis for the two-dimensional solution space of the second order differential equation: meaning that linear combinations of these solutions will also be solutions. In particular, the following solutions can be constructed

$$y_{0'} = \frac{C_0 e^{ikx} + C_0 e^{-ikx}}{2} = C_0 \cos(kx)$$

and

$$y_{1'} = \frac{C_1 e^{ikx} - C_1 e^{-ikx}}{2i} = C_1 \sin(kx).$$

These last two trigonometric solutions are linearly independent, so they can serve as another basis for the solution space, yielding the following general solution:

$$y_H = C_0 \cos(kx) + C_1 \sin(kx).$$

Damped Harmonic Oscillator

Given the equation for the damped harmonic oscillator:

$$\left(D^2 + \frac{b}{m} D + \omega_0^2 \right) y = 0,$$

the expression in parentheses can be factored out: first obtain the characteristic equation by replacing D with z. This equation must be satisfied for all y, thus:

$$z^2 + \frac{b}{m} z + \omega_0^2 = 0.$$

Solve using the quadratic formula:

$$z = \tfrac{1}{2} \left(-\frac{b}{m} \pm \sqrt{\frac{b^2}{m^2} - 4\omega_0^2} \right).$$

Use these characteristic roots to factor the left side of the original differential equation:

$$\left(D + \frac{b}{2m} - \sqrt{\frac{b^2}{4m^2} - \omega_0^2} \right)\left(D + \frac{b}{2m} + \sqrt{\frac{b^2}{4m^2} - \omega_0^2} \right) y = 0.$$

This implies a pair of solutions, one corresponding to

$$\left(D + \frac{b}{2m} - \sqrt{\frac{b^2}{4m^2} - \omega_0^2} \right) y = 0$$

$$\left(D+\frac{b}{2m}+\sqrt{\frac{b^2}{4m^2}-\omega_0^2}\right)y=0$$

The solutions are, respectively,

$$y_0 = A_0 e^{-\omega x+\sqrt{\omega^2-\omega_0^2}\,x} = A_0 e^{-\omega x}e^{\sqrt{\omega^2-\omega_0^2}\,x}$$

$$y_1 = A_1 e^{-\omega x-\sqrt{\omega^2-\omega_0^2}\,x} = A_1 e^{-\omega x}e^{-\sqrt{\omega^2-\omega_0^2}\,x}$$

where $\omega = b/2m$. From this linearly independent pair of solutions can be constructed another linearly independent pair which thus serve as a basis for the two-dimensional solution space:

$$y_H(A_0,A_1)(x)=(A_0\sinh(\sqrt{\omega^2-\omega_0^2}\,x)+A_1\cosh(\sqrt{\omega^2-\omega_0^2}\,x))e^{-\omega x}.$$

However, if $|\omega| < |\omega_0|$ then it is preferable to get rid of the consequential imaginaries, expressing the general solution as

$$y_H(A_0,A_1)(x)=\left(A_0\sin\left(\sqrt{\omega_0^2-\omega^2}\,x\right)+A_1\cos\left(\sqrt{\omega_0^2-\omega^2}\,x\right)\right)e^{-\omega x}.$$

This latter solution corresponds to the underdamped case, whereas the former one corresponds to the overdamped case: the solutions for the underdamped case oscillate whereas the solutions for the overdamped case do not.

Nonhomogeneous Equation with Constant Coefficients

To obtain the solution to the nonhomogeneous equation (sometimes called inhomogeneous equation), find a particular integral $y_p(x)$ by either the method of undetermined coefficients or the method of variation of parameters; the general solution to the linear differential equation is the sum of the general solution of the related homogeneous equation and the particular integral. Or, when the initial conditions are set, use Laplace transform to obtain the particular solution directly.

Suppose we face

$$\frac{d^n y(x)}{dx^n}+A_1\frac{d^{n-1}y(x)}{dx^{n-1}}+\cdots+A_n y(x)=f(x).$$

For later convenience, define the characteristic polynomial

$$P(z)=z^n+A_1 z^{n-1}+\cdots+A_n.$$

We find a solution basis $\{y_1(x),y_2(x),\ldots,y_n(x)\}$ for the homogeneous ($f(x)=0$) case. We now seek a particular integral $y_p(x)$ by the variation of parameters method. Let the coefficients of the linear combination be functions of x:

$$y_p(x) = u_1(x)y_1(x) + u_2(x)y_2(x) + \cdots + u_n(x)y_n(x).$$

For ease of notation we will drop the dependency on x (i.e. the various (x)). Using the operator notation $D = d/dx$, the ODE in question is $P(D)y = f$; so

$$f = P(D)y_p = P(D)(u_1 y_1) + P(D)(u_2 y_2) + \cdots + P(D)(u_n y_n).$$

With the constraints

$$0 = u_1'y_1 + u_2'y_2 + \cdots + u_n'y_n$$

$$0 = u_1'y_1' + u_2'y_2' + \cdots + u_n'y_n'$$

$$\cdots$$

$$0 = u_1'y_1^{(n-2)} + u_2'y_2^{(n-2)} + \cdots + u_n'y_n^{(n-2)}$$

the parameters commute out,

$$f = u_1 P(D)y_1 + u_2 P(D)y_2 + \cdots + u_n P(D)y_n + u_1'y_1^{(n-1)} + u_2'y_2^{(n-1)} + \cdots + u_n'y_n^{(n-1)}.$$

But $P(D)y_j = 0$, therefore

$$f = u_1'y_1^{(n-1)} + u_2'y_2^{(n-1)} + \cdots + u_n'y_n^{(n-1)}.$$

This, with the constraints, gives a linear system in the u_j'. This much can always be solved; in fact, combining Cramer's rule with the Wronskian,

$$u_j' = (-1)^{n+j} \frac{W(y_1, \ldots, y_{j-1}, y_{j+1} \ldots, y_n)\binom{0}{f}}{W(y_1, y_2, \ldots, y_n)}.$$

In the very non-standard notation used above, one should take the i,n-minor of W and multiply it by f. That's why we get a minus-sign. Alternatively, forget about the minus sign and just compute the determinant of the matrix obtained by substituting the j-th W column with (0, 0, ..., f).

The rest is a matter of integrating u_j'.

The particular integral is not unique; $y_p + c_1 y_1 + \cdots + c_n y_n$ also satisfies the ODE for any set of constants c_j.

Example

Suppose $y'' - 4y' + 5y = \sin(kx)$. We take the solution basis found above
$\{e^{(2+i)x} = y_1(x), e^{(2-i)x} = y_2(x)\}$.

$$W = \begin{vmatrix} e^{(2+i)x} & e^{(2-i)x} \\ (2+i)e^{(2+i)x} & (2-i)e^{(2-i)x} \end{vmatrix} = e^{4x} \begin{vmatrix} 1 & 1 \\ 2+i & 2-i \end{vmatrix} = -2ie^{4x}$$

$$u_1' = \frac{1}{W} \begin{vmatrix} 0 & e^{(2-i)x} \\ \sin(kx) & (2-i)e^{(2-i)x} \end{vmatrix} = -\tfrac{i}{2}\sin(kx)e^{(-2-i)x}$$

$$u_2' = \frac{1}{W} \begin{vmatrix} e^{(2+i)x} & 0 \\ (2+i)e^{(2+i)x} & \sin(kx) \end{vmatrix} = \tfrac{i}{2}\sin(kx)e^{(-2+i)x}.$$

Using the list of integrals of exponential functions

$$u_1 = -\tfrac{i}{2}\int \sin(kx)e^{(-2-i)x}\,dx = \frac{ie^{(-2-i)x}}{2(3+4i+k^2)}\big((2+i)\sin(kx) + k\cos(kx)\big)$$

$$u_2 = \tfrac{i}{2}\int \sin(kx)e^{(-2+i)x}\,dx = \frac{ie^{(i-2)x}}{2(3-4i+k^2)}\big((i-2)\sin(kx) - k\cos(kx)\big).$$

And so

$$\begin{aligned} y_p &= u_1(x)y_1(x) + u_2(x)y_2(x) = \frac{i}{2(3+4i+k^2)}\big((2+i)\sin(kx)+k\cos(kx)\big) + \frac{i}{2(3-4i+k^2)}\big((i-2)\sin(kx)-k\cos(kx)\big) \\ &= \frac{(5-k^2)\sin(kx)+4k\cos(kx)}{(3+k^2)^2+16}. \end{aligned}$$

(Notice that u_1 and u_2 had factors that canceled y_1 and y_2; that is typical.)

For interest's sake, this ODE has a physical interpretation as a driven damped harmonic oscillator; y_p represents the steady state, and $c_1y_1 + c_2y_2$ is the transient.

Equation with Variable Coefficients

A linear ODE of order n with variable coefficients has the general form

$$p_n(x)y^{(n)}(x) + p_{n-1}(x)y^{(n-1)}(x) + \cdots + p_0(x)y(x) = r(x).$$

Examples

A simple example is the Cauchy–Euler equation often used in engineering

$$x^n y^{(n)}(x) + a_{n-1}x^{n-1}y^{(n-1)}(x) + \cdots + a_0 y(x) = 0.$$

First-order Equation with Variable Coefficients

Examples
Solve the equation
$$y'(x) + 3y(x) = 2$$
with the initial condition
$$y(0) = 2.$$
Using the general solution method:
$$y = e^{-3x}\left(\int 2e^{3x}\, dx + \kappa\right).$$
The indefinite integral is solved to give:
$$y = e^{-3x}\left((2/3)e^{3x} + \kappa\right).$$
Then we can reduce to:
$$y = 2/3 + \kappa e^{-3x}.$$
where $\kappa = 4/3$ from the initial condition.

A linear ODE of order 1 with variable coefficients has the general form

$$Dy(x) + f(x)y(x) = g(x).$$

Where D is the differential operator. Equations of this form can be solved by multiplying the integrating factor

$$e^{\int f(x)dx}$$

throughout to obtain

$$Dy(x)e^{\int f(x)dx} + f(x)y(x)e^{\int f(x)dx} = g(x)e^{\int f(x)dx},$$

which simplifies due to the product rule (applied backwards) to

$$D\left(y(x)e^{\int f(x)dx}\right) = g(x)e^{\int f(x)dx}$$

which, on integrating both sides and solving for $y(x)$ gives:

$$y(x) = e^{-\int f(x)dx}\left(\int g(x)e^{\int f(x)dx}\, dx + \kappa\right).$$

In other words: The solution of a first-order linear ODE

$$y'(x) + f(x)y(x) = g(x),$$

with coefficients that may or may not vary with x, is:

$$y = e^{-a(x)}\left(\int g(x)e^{a(x)}\,dx + \kappa\right)$$

where κ is the constant of integration, and

$$a(x) = \int f(x)\,dx.$$

A compact form of the general solution based on a Green's function is

$$y(x) = \int_a^x [y(a)\delta(t-a) + g(t)]e^{-\int_t^x f(u)\,du}\,dt.$$

where $\delta(x)$ is the generalized Dirac delta function.

Examples

Consider a first order differential equation with constant coefficients:

$$\frac{dy}{dx} + by = 1.$$

This equation is particularly relevant to first order systems such as RC circuits and mass-damper systems.

In this case, $f(x) = b$, $g(x) = 1$.

Hence its solution is

$$y(x) = e^{-bx}\left(\frac{e^{bx}}{b} + C\right) = \frac{1}{b} + Ce^{-bx}.$$

Systems of Linear Differential Equations

An arbitrary linear ordinary differential equation or even a system of such equations can be converted into a first order system of linear differential equations by adding variables for all but the highest order derivatives. A linear system can be viewed as a single equation with a vector-valued variable. The general treatment is analogous to the treatment above of ordinary first order linear differential equations, but with complications stemming from noncommutativity of matrix multiplication.

To solve

$$\begin{cases} \mathbf{y}'(x) & = A(x)\mathbf{y}(x) + \mathbf{b}(x) \\ \mathbf{y}(x_0) & = \mathbf{y}_0 \end{cases}$$

(here $\mathbf{y}(x)$ is a vector or matrix, and $A(x)$ is a matrix), let $U(x)$ be the solution of $\mathbf{y}'(x) = A(x)\mathbf{y}(x)$ with $U(x_0) = I$ (the identity matrix). U is a fundamental matrix for the equation — the columns of

U form a complete linearly independent set of solutions for the homogeneous equation. After substituting $\mathbf{y}(x) = U(x)\mathbf{z}(x)$, the equation $\mathbf{y}'(x) = A(x)\mathbf{y}(x) + \mathbf{b}(x)$ simplifies to $U(x)\mathbf{z}'(x) = \mathbf{b}(x)$. Thus,

$$\mathbf{y}(x) = U(x)\mathbf{y_0} + U(x)\int_{x_0}^{x} U^{-1}(t)\mathbf{b}(t)dt$$

If $A(x_1)$ commutes with $A(x_2)$ for all x_1 and x_2, then

$$U(x) = e^{\int_{x_0}^{x} A(x)dx}$$

and thus

$$U^{-1}(x) = e^{-\int_{x_0}^{x} A(x)dx},$$

but in the general case there is no closed form solution, and an approximation method such as Magnus expansion may have to be used. Note that the exponentials are matrix exponentials.

Cauchy–Euler Equation

In mathematics, a Cauchy–Euler equation (also known as the Euler–Cauchy equation, or simply Euler's equation) is a linear homogeneous ordinary differential equation with variable coefficients. It is sometimes referred to as an *equidimensional* equation. Because of the particularly simple equidimensional structure the equation can be replaced with an equivalent equation with constant coefficients which can then be solved explicitly.

The Equation

Let $y^{(n)}(x)$ be the nth derivative of the unknown function $y(x)$. Then a Cauchy–Euler equation of order n has the form

$$a_n x^n y^{(n)}(x) + a_{n-1}x^{n-1}y^{(n-1)}(x) + \cdots + a_0 y(x) = 0.$$

The substitution $x = e^u$ reduces this equation to a linear differential equation with constant coefficients. Alternatively a trial solution $y = x^m$ may be used to solve for the basis solutions.

Second Order – solving Through Trial Solution

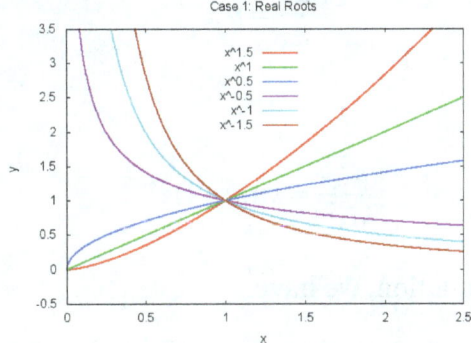

Typical solution curves for a second-order Euler–Cauchy equation for the case of two real roots

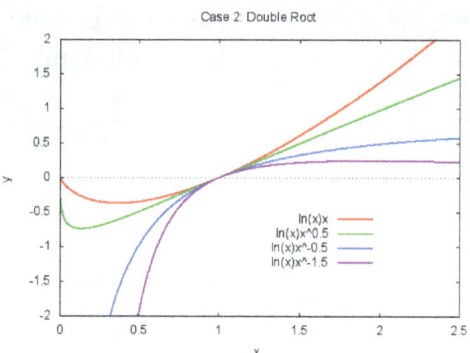

Typical solution curves for a second-order Euler–Cauchy equation for the case of a double root

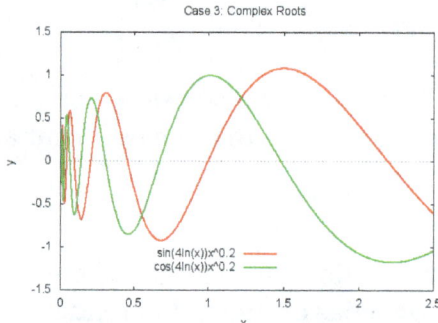

Typical solution curves for a second-order Euler–Cauchy equation for the case of complex roots

The most common Cauchy–Euler equation is the second-order equation, appearing in a number of physics and engineering applications, such as when solving Laplace's equation in polar coordinates. It is given by the equation:

$$x^2 \frac{d^2 y}{dx^2} + ax \frac{dy}{dx} + by = 0.$$

We assume a trial solution given by

$$y = x^m.$$

Differentiating, we have:

$$\frac{dy}{dx} = mx^{m-1}$$

and

$$\frac{d^2 y}{dx^2} = m(m-1)x^{m-2}.$$

Substituting into the original equation, we have:

$$x^2 \left(m(m-1)x^{m-2} \right) + ax(mx^{m-1}) + b(x^m) = 0$$

Or rearranging gives:

$$m^2 + (a-1)m + b = 0.$$

We then can solve for m. There are three particular cases of interest:

- Case #1: Two distinct roots, m_1 and m_2

- Case #2: One real repeated root, m

- Case #3: Complex roots, $\alpha \pm \beta i$

In case #1, the solution is given by:

$$y = c_1 x^{m_1} + c_2 x^{m_2}$$

In case #2, the solution is given by

$$y = c_1 x^m \ln(x) + c_2 x^m$$

To get to this solution, the method of reduction of order must be applied after having found one solution $y = x^m$.

In case #3, the solution is given by:

$$y = c_1 x^{\alpha} \cos(\beta \ln(x)) + c_2 x^{\alpha} \sin(\beta \ln(x))$$

$$\alpha = \mathrm{Re}(m)$$

$$\beta = \mathrm{Im}(m)$$

For c_1 and c_2 in the real plane

This form of the solution is derived by setting $x = e^t$ and using Euler's formula

Second Order – solution Through Change of Variables

$$x^2 \frac{d^2 y}{dx^2} + ax \frac{dy}{dx} + by = 0$$

We operate the variable substitution defined by

$$t = \ln(x).$$

$$y(x) = \phi(\ln(x)) = \phi(t).$$

Differentiating:

$$\frac{dy}{dx} = \frac{1}{x} \frac{d\phi}{dt}$$

$$\frac{d^2 y}{dx^2} = \frac{1}{x^2}\left(\frac{d^2\phi}{dt^2} - \frac{d\phi}{dt}\right).$$

Substituting $\phi(t)$, we have

$$\frac{d^2\phi}{dt^2} + (a-1)\frac{d\phi}{dt} + b\phi = 0.$$

This equation in $\phi(t)$ can be easily solved using its characteristic polynomial

$$\lambda^2 + (a-1)\lambda + b = 0.$$

Now, if λ_1 and λ_2 are the roots of this polynomial, we analyze the two main cases: distinct roots and double roots:

If the roots are distinct, the general solution is given by

$$\phi(t) = c_1 e^{\lambda_1 t} + c_2 e^{\lambda_2 t},$$ where the exponentials may be complex.

If the roots are equal, the general solution is given by

$$\phi(t) = c_1 e^{\lambda_1 t} + c_2 t e^{\lambda_1 t}.$$

In both cases, the solution $y(x)$ may be found by setting $t = \ln(x)$, hence $\phi(\ln(x)) = y(x)$.

Hence, in the first case,

$$y(x) = c_1 x^{\lambda_1} + c_2 x^{\lambda_2},$$

and in the second case,

$$y(x) = c_1 x^{\lambda_1} + c_2 \ln(x) x^{\lambda_1}.$$

Example

Given

$$x^2 u'' - 3xu' + 3u = 0,$$

we substitute the simple solution x^α:

$$x^2(\alpha(\alpha-1)x^{\alpha-2}) - 3x(\alpha x^{\alpha-1}) + 3x^\alpha = \alpha(\alpha-1)x^\alpha - 3\alpha x^\alpha + 3x^\alpha = (\alpha^2 - 4\alpha + 3)x^\alpha = 0.$$

For x^α to be a solution, either $x = 0$, which gives the trivial solution, or the coefficient of x^α is zero. Solving the quadratic equation, we get $\alpha = 1, 3$. The general solution is therefore

$$u = c_1 x + c_2 x^3.$$

Difference Equation Analogue

There is a difference equation analogue to the Cauchy–Euler equation. For a fixed $m > 0$, define the sequence $f_m(n)$ as

$$f_m(n) := n(n+1)\cdots(n+m-1).$$

Applying the difference operator to f_m, we find that

$$Df_m(n) \qquad\qquad = f_m(n+1) - f_m(n)$$

$$= m(n+1)(n+2)\cdots(n+m-1) = \frac{m}{n}f_m(n).$$

If we do this k times, we will find that

$$f_m^{(k)}(n) = \frac{m(m-1)\cdots(m-k+1)}{n(n+1)\cdots(n+k-1)}f_m(n)$$

$$= m(m-1)\cdots(m-k+1)\frac{f_m(n)}{f_k(n)},$$

where the superscript $^{(k)}$ denotes applying the difference operator k times. Comparing this to the fact that the k-th derivative of x^m equals

$$m(m-1)\cdots(m-k+1)\frac{x^m}{x^k}$$

suggests that we can solve the N-th order difference equation

$$f_N(n)y^{(N)}(n) + a_{N-1}f_{N-1}(n)y^{(N-1)}(n) + \cdots + a_0 y(n) = 0,$$

in a similar manner to the differential equation case. Indeed, substituting the trial solution

$$y(n) = f_m(n)$$

brings us to the same situation as the differential equation case,

$$m(m-1)\cdots(m-N+1) + a_{N-1}m(m-1)\cdots(m-N+2) + \cdots + a_1 m + a_0 = 0.$$

One may now proceed as in the differential equation case, since the general solution of an N-th order linear difference equation is also the linear combination of N linearly independent solutions. Applying reduction of order in case of a multiple root m_1 will yield expressions involving a discrete version of ln,

$$\varphi(n) = \sum_{k=1}^{n}\frac{1}{k-m_1}.$$

(Compare with: $\ln(x - m_1) = \int\limits_{1+m_1}^{x} \dfrac{1}{t - m_1} dt.$)

In cases where fractions become involved, one may use

$$f_m(n) := \frac{\Gamma(n+m)}{\Gamma(n)}$$

instead (or simply use it in all cases), which coincides with the definition before for integer m.

Power Series Solution of Differential Equations

In mathematics, the power series method is used to seek a power series solution to certain differential equations. In general, such a solution assumes a power series with unknown coefficients, then substitutes that solution into the differential equation to find a recurrence relation for the coefficients.

Method

Consider the second-order linear differential equation

$$a_2(z)f''(z) + a_1(z)f'(z) + a_0(z)f(z) = 0.$$

Suppose a_2 is nonzero for all z. Then we can divide throughout to obtain

$$f'' + \frac{a_1(z)}{a_2(z)} f' + \frac{a_0(z)}{a_2(z)} f = 0.$$

Suppose further that a_1/a_2 and a_0/a_2 are analytic functions.

The power series method calls for the construction of a power series solution

$$f = \sum_{k=0}^{\infty} A_k z^k.$$

If a_2 is zero for some z, then the Frobenius method, a variation on this method, is suited to deal with so called *singular points*. The method works analogously for higher order equations as well as for systems.

Example Usage

Let us look at the Hermite differential equation,

$$f'' - 2zf' + \lambda f = 0; \; \lambda = 1$$

We can try to construct a series solution

$$f = \sum_{k=0}^{\infty} A_k z^k$$

$$f' = \sum_{k=1}^{\infty} k A_k z^{k-1}$$

$$f'' = \sum_{k=2}^{\infty} k(k-1) A_k z^{k-2}$$

Substituting these in the differential equation

$$\sum_{k=2}^{\infty} k(k-1) A_k z^{k-2} - 2z \sum_{k=1}^{\infty} k A_k z^{k-1} + \sum_{k=0}^{\infty} A_k z^k = 0$$

$$= \sum_{k=2}^{\infty} k(k-1) A_k z^{k-2} - \sum_{k=1}^{\infty} 2k A_k z^k + \sum_{k=0}^{\infty} A_k z^k$$

Making a shift on the first sum

$$= \sum_{k=0}^{\infty} (k+2)(k+1) A_{k+2} z^k - \sum_{k=1}^{\infty} 2k A_k z^k + \sum_{k=0}^{\infty} A_k z^k$$

$$= 2A_2 + \sum_{k=1}^{\infty} (k+2)(k+1) A_{k+2} z^k - \sum_{k=1}^{\infty} 2k A_k z^k + A_0 + \sum_{k=1}^{\infty} A_k z^k$$

$$= 2A_2 + A_0 + \sum_{k=1}^{\infty} \left((k+2)(k+1) A_{k+2} + (-2k+1) A_k \right) z^k$$

If this series is a solution, then all these coefficients must be zero, so for both k=0 and k>0:

$$(k+2)(k+1) A_{k+2} + (-2k+1) A_k = 0$$

We can rearrange this to get a recurrence relation for A_{k+2}.

$$(k+2)(k+1) A_{k+2} = -(-2k+1) A_k$$

$$A_{k+2} = \frac{(2k-1)}{(k+2)(k+1)} A_k$$

Now, we have

$$A_2 = \frac{-1}{(2)(1)} A_0 = \frac{-1}{2} A_0, A_3 = \frac{1}{(3)(2)} A_1 = \frac{1}{6} A_1$$

We can determine A_0 and A_1 if there are initial conditions, i.e. if we have an initial value problem.

So we have

$$A_4 = \frac{1}{4}A_2 = \left(\frac{1}{4}\right)\left(\frac{-1}{2}\right)A_0 = \frac{-1}{8}A_0$$

$$A_5 = \frac{1}{4}A_3 = \left(\frac{1}{4}\right)\left(\frac{1}{6}\right)A_1 = \frac{1}{24}A_1$$

$$A_6 = \frac{7}{30}A_4 = \left(\frac{7}{30}\right)\left(\frac{-1}{8}\right)A_0 = \frac{-7}{240}A_0$$

$$A_7 = \frac{3}{14}A_5 = \left(\frac{3}{14}\right)\left(\frac{1}{24}\right)A_1 = \frac{1}{112}A_1$$

and the series solution is

$$f = A_0 z^0 + A_1 z^1 + A_2 z^2 + A_3 z^3 + A_4 z^4 + A_5 z^5 + A_6 z^6 + A_7 z^7 + \cdots$$

$$= A_0 z^0 + A_1 z^1 + \frac{-1}{2}A_0 z^2 + \frac{1}{6}A_1 z^3 + \frac{-1}{8}A_0 z^4 + \frac{1}{24}A_1 z^5 + \frac{-7}{240}A_0 z^6 + \frac{1}{112}A_1 z^7 + \cdots$$

$$= A_0 z^0 + \frac{-1}{2}A_0 z^2 + \frac{-1}{8}A_0 z^4 + \frac{-7}{240}A_0 z^6 + A_1 z + \frac{1}{6}A_1 z^3 + \frac{1}{24}A_1 z^5 + \frac{1}{112}A_1 z^7 + \cdots$$

which we can break up into the sum of two linearly independent series solutions:

$$f = A_0\left(1 + \frac{-1}{2}z^2 + \frac{-1}{8}z^4 + \frac{-7}{240}z^6 + \cdots\right) + A_1\left(z + \frac{1}{6}z^3 + \frac{1}{24}z^5 + \frac{1}{112}z^7 + \cdots\right)$$

which can be further simplified by the use of hypergeometric series.

A simpler way Using Taylor Series

A much simpler way of solving this equation (and power series solution in general) using the Taylor series form of the expansion. Here we assume the answer is of the form

$$f = \sum_{k=0}^{\infty} \frac{A_k z^k}{k!}$$

If we do this, the general rule for obtaining the recurrence relationship for the coefficients is

$$y^{[n]} \rightarrow A_{k+n}$$

and

$$x^m y^{[n]} \rightarrow (k)(k-1)\cdots(k-m+1)A_{k+n-m}$$

In this case we can solve the Hermite equation in fewer steps:

$$f'' - 2zf' + \lambda f = 0; \lambda = 1$$

becomes

$$A_{k+2} - 2kA_k + \lambda A_k = 0$$

or

$$A_{k+2} = (2k - \lambda)A_k$$

in the series

$$f = \sum_{k=0}^{\infty} \frac{A_k z^k}{k!}$$

Nonlinear Equations

The power series method can be applied to certain nonlinear differential equations, though with less flexibility. A very large class of nonlinear equations can be solved analytically by using the Parker–Sochacki method. Since the Parker–Sochacki method involves an expansion of the original system of ordinary differential equations through auxiliary equations, it is not simply referred to as the power series method. The Parker–Sochacki method is done before the power series method to make the power series method possible on many nonlinear problems. An ODE problem can be expanded with the auxiliary variables which make the power series method trivial for an equivalent, larger system. Expanding the ODE problem with auxiliary variables produces the same coefficients (since the power series for a function is unique) at the cost of also calculating the coefficients of auxiliary equations. Many times, without using auxiliary variables, there is no known way to get the power series for the solution to a system, hence the power series method alone is difficult to apply to most nonlinear equations.

The power series method will give solutions only to initial value problems (opposed to boundary value problems), this is not an issue when dealing with linear equations since the solution may turn up multiple linearly independent solutions which may be combined (by superposition) to solve boundary value problems as well. A further restriction is that the series coefficients will be specified by a nonlinear recurrence (the nonlinearities are inherited from the differential equation).

In order for the solution method to work, as in linear equations, it is necessary to express every term in the nonlinear equation as a power series so that all of the terms may be combined into one power series.

As an example, consider the initial value problem

$$FF'' + 2F'^2 + \eta F' = 0 \quad ; \quad F(1) = 0 \,, F'(1) = -\frac{1}{2}$$

which describes a solution to capillary-driven flow in a groove. Note the two nonlinearities: the first and second terms involve products. Note also that the initial values are given at $\eta = 1,$, which hints that the power series must be set up as:

$$F(\eta) = \sum_{i=0}^{\infty} c_i (\eta - 1)^i$$

since in this way

$$\frac{d^n F}{d\eta^n}\Big|_{\eta=1} = n! \, c_n$$

which makes the initial values very easy to evaluate. It is necessary to rewrite the equation slightly in light of the definition of the power series,

$$FF'' + 2F'^2 + (\eta - 1)F' + F' = 0 \quad ; \quad F(1) = 0 \,, \, F'(1) = -\frac{1}{2}$$

so that the third term contains the same form $\eta - 1$ that shows in the power series.

The last consideration is what to do with the products; substituting the power series in would result in products of power series when it's necessary that each term be its own power series. This is where the Cauchy product

$$\left(\sum_{i=0}^{\infty} a_i x^i\right)\left(\sum_{i=0}^{\infty} b_i x^i\right) = \sum_{i=0}^{\infty} x^i \sum_{j=0}^{i} a_{i-j} b_j$$

is useful; substituting the power series into the differential equation and applying this identity leads to an equation where every term is a power series. After much rearrangement, the recurrence

$$\sum_{j=0}^{i} \left((j+1)(j+2)c_{i-j}c_{j+2} + 2(i-j+1)(j+1)c_{i-j+1}c_{j+1}\right) + ic_i + (i+1)c_{i+1} = 0$$

is obtained, specifying exact values of the series coefficients. From the initial values, $c_0 = 0$ and $c_1 = -1/2$, thereafter the above recurrence is used. For example, the next few coefficients:

$$c_2 = -\frac{1}{6} \quad ; \quad c_3 = -\frac{1}{108} \quad ; \quad c_4 = \frac{7}{3240} \quad ; \quad c_5 = -\frac{19}{48600} \cdots$$

A limitation of the power series solution shows itself in this example. A numeric solution of the problem shows that the function is smooth and always decreasing to the left of $\eta = 1$, and zero to the right. At $\eta = 1$, a slope discontinuity exists, a feature which the power series is incapable of rendering, for this reason the series solution continues decreasing to the right of $\eta = 1$ instead of suddenly becoming zero.

Laplace Transform

In mathematics the Laplace transform is an integral transform named after its discoverer Pierre-Simon Laplace. It takes a function of a positive real variable t (often time) to a function of a complex variable s (frequency).

The Laplace transform is very similar to the Fourier transform. While the Fourier transform of a function is a complex function of a *real* variable (frequency), the Laplace transform of a function is a complex function of a *complex variable*. Laplace transforms are usually restricted to functions of t with $t > 0$. A consequence of this restriction is that the Laplace transform of a function is a holomorphic function of the variable s. Unlike the Fourier transform, the Laplace transform of a distribution is generally a well-behaved function. Also techniques of complex variables can be used directly to study Laplace transforms. As a holomorphic function, the Laplace transform has a power series representation. This power series expresses a function as a linear superposition of moments of the function. This perspective has applications in probability theory.

The Laplace transform is invertible on a large class of functions. The inverse Laplace transform takes a function of a complex variable s (often frequency) and yields a function of a real variable t (time). Given a simple mathematical or functional description of an input or output to a system, the Laplace transform provides an alternative functional description that often simplifies the process of analyzing the behavior of the system, or in synthesizing a new system based on a set of specifications. So, for example, Laplace transformation from the time domain to the frequency domain transforms differential equations into algebraic equations and convolution into multiplication. It has many applications in the sciences and technology.

History

The Laplace transform is named after mathematician and astronomer Pierre-Simon Laplace, who used a similar transform (now called the z-transform) in his work on probability theory. The current widespread use of the transform (mainly in engineering) came about during and soon after World War II although it had been used in the 19th century by Abel, Lerch, Heaviside, and Bromwich.

The early history of methods having some similarity to Laplace transform is as follows. From 1744, Leonhard Euler investigated integrals of the form

$$z = \int X(x)e^{ax}\,dx \quad \text{and} \quad z = \int X(x)x^A\,dx$$

as solutions of differential equations but did not pursue the matter very far.

Joseph Louis Lagrange was an admirer of Euler and, in his work on integrating probability density functions, investigated expressions of the form

$$\int X(x)e^{-ax}a^x\,dx,$$

which some modern historians have interpreted within modern Laplace transform theory.

These types of integrals seem first to have attracted Laplace's attention in 1782 where he was following in the spirit of Euler in using the integrals themselves as solutions of equations. However, in 1785, Laplace took the critical step forward when, rather than just looking for a solution in the form of an integral, he started to apply the transforms in the sense that was later to become popular. He used an integral of the form

$$\int x^s \phi(x) dx,$$

akin to a Mellin transform, to transform the whole of a difference equation, in order to look for solutions of the transformed equation. He then went on to apply the Laplace transform in the same way and started to derive some of its properties, beginning to appreciate its potential power.

Laplace also recognised that Joseph Fourier's method of Fourier series for solving the diffusion equation could only apply to a limited region of space because those solutions were periodic. In 1809, Laplace applied his transform to find solutions that diffused indefinitely in space.

Formal Definition

The Laplace transform is a frequency-domain approach for continuous time signals irrespective of whether the system is stable or unstable. The Laplace transform of a function $f(t)$, defined for all real numbers $t \geq 0$, is the function $F(s)$, which is a unilateral transform defined by

$$F(s) = \int_0^\infty e^{-st} f(t) dt$$

where s is a complex number frequency parameter

$$s = \sigma + i\omega, \text{ with real numbers } \sigma \text{ and } \omega.$$

Other notations for the Laplace transform include $L\{f\}$, or alternatively $L\{f(t)\}$ instead of F.

The meaning of the integral depends on types of functions of interest. A necessary condition for existence of the integral is that f must be locally integrable on $[0, \infty)$. For locally integrable functions that decay at infinity or are of exponential type, the integral can be understood to be a (proper) Lebesgue integral. However, for many applications it is necessary to regard it to be a conditionally convergent improper integral at ∞. Still more generally, the integral can be understood in a weak sense, and this is dealt with below.

One can define the Laplace transform of a finite Borel measure μ by the Lebesgue integral

$$\mathcal{L}\{\mu\}(s) = \int_{[0,\infty)} e^{-st} d\mu(t).$$

An important special case is where μ is a probability measure, for example, the Dirac delta function. In operational calculus, the Laplace transform of a measure is often treated as though the measure came from a probability density function f. In that case, to avoid potential confusion, one often writes

$$\mathcal{L}\{f\}(s) = \int_{0^-}^\infty e^{-st} f(t) dt,$$

where the lower limit of 0^- is shorthand notation for

$$\lim_{\varepsilon \downarrow 0} \int_{-\varepsilon}^{\infty} .$$

This limit emphasizes that any point mass located at 0 is entirely captured by the Laplace transform. Although with the Lebesgue integral, it is not necessary to take such a limit, it does appear more naturally in connection with the Laplace–Stieltjes transform.

Probability Theory

In pure and applied probability, the Laplace transform is defined as an expected value. If X is a random variable with probability density function f, then the Laplace transform of f is given by the expectation

$$\mathcal{L}\{f\}(s) = E\left[e^{-sX}\right].$$

By abuse of language, this is referred to as the Laplace transform of the random variable X itself. Replacing s by $-t$ gives the moment generating function of X. The Laplace transform has applications throughout probability theory, including first passage times of stochastic processes such as Markov chains, and renewal theory.

Of particular use is the ability to recover the cumulative distribution function of a continuous random variable X by means of the Laplace transform as follows

$$F_X(x) = \mathcal{L}^{-1}\left\{\frac{1}{s}E\left[e^{-sX}\right]\right\}(x) = \mathcal{L}^{-1}\left\{\frac{1}{s}\mathcal{L}\{f\}(s)\right\}(x).$$

Bilateral Laplace Transform

When one says "the Laplace transform" without qualification, the unilateral or one-sided transform is normally intended. The Laplace transform can be alternatively defined as the *bilateral Laplace transform* or two-sided Laplace transform by extending the limits of integration to be the entire real axis. If that is done the common unilateral transform simply becomes a special case of the bilateral transform where the definition of the function being transformed is multiplied by the Heaviside step function.

The bilateral Laplace transform is defined as follows,

$$\mathcal{B}\{f\}(s) = \int_{-\infty}^{\infty} e^{-st} f(t)dt.$$

Inverse Laplace Transform

Two integrable functions have the same Laplace transform only if they differ on a set of Lebesgue measure zero. This means that, on the range of the transform, there is an inverse transform. In fact, besides integrable functions, the Laplace transform is a one-to-one mapping from one function space into another in many other function spaces as well, although there is usually no easy characterization of the range. Typical function spaces in which this is true include the spaces of bounded continuous functions, the space $L^\infty(0, \infty)$, or more generally tempered functions (that is, functions of at worst polynomial growth) on $(0, \infty)$. The Laplace transform is also defined and injective for suitable spaces of tempered distributions.

In these cases, the image of the Laplace transform lives in a space of analytic functions in the region of convergence. The inverse Laplace transform is given by the following complex integral, which is known by various names (the Bromwich integral, the Fourier–Mellin integral, and Mellin's inverse formula):

$$f(t) = \mathcal{L}^{-1}\{F\}(t) = \frac{1}{2\pi i} \lim_{T \to \infty} \int_{\gamma-iT}^{\gamma+iT} e^{st} F(s)\,ds,$$

where γ is a real number so that the contour path of integration is in the region of convergence of $F(s)$. An alternative formula for the inverse Laplace transform is given by Post's inversion formula. The limit here is interpreted in the weak-* topology.

In practice, it is typically more convenient to decompose a Laplace transform into known transforms of functions obtained from a table, and construct the inverse by inspection.

Region of Convergence

If f is a locally integrable function (or more generally a Borel measure locally of bounded variation), then the Laplace transform $F(s)$ of f converges provided that the limit

$$\lim_{R \to \infty} \int_0^R f(t)e^{-st}\,dt$$

exists.

The Laplace transform converges absolutely if the integral

$$\int_0^\infty \left| f(t)e^{-st} \right| dt$$

exists (as a proper Lebesgue integral). The Laplace transform is usually understood as conditionally convergent, meaning that it converges in the former instead of the latter sense.

The set of values for which $F(s)$ converges absolutely is either of the form $\mathrm{Re}(s) > a$ or else $\mathrm{Re}(s) \geq a$, where a is an extended real constant, $-\infty \leq a \leq \infty$. (This follows from the dominated convergence theorem.) The constant a is known as the abscissa of absolute convergence, and depends on the growth behavior of $f(t)$. Analogously, the two-sided transform converges absolutely in a strip of the form $a < \mathrm{Re}(s) < b$, and possibly including the lines $\mathrm{Re}(s) = a$ or $\mathrm{Re}(s) = b$. The subset of values of s for which the Laplace transform converges absolutely is called the region of absolute convergence or the domain of absolute convergence. In the two-sided case, it is sometimes called the strip of absolute convergence. The Laplace transform is analytic in the region of absolute convergence.

Similarly, the set of values for which $F(s)$ converges (conditionally or absolutely) is known as the region of conditional convergence, or simply the region of convergence (ROC). If the Laplace transform converges (conditionally) at $s = s_0$, then it automatically converges for all s with $\text{Re}(s) > \text{Re}(s_0)$. Therefore, the region of convergence is a half-plane of the form $\text{Re}(s) > a$, possibly including some points of the boundary line $\text{Re}(s) = a$.

In the region of convergence $\text{Re}(s) > \text{Re}(s_0)$, the Laplace transform of f can be expressed by integrating by parts as the integral

$$F(s) = (s - s_0)\int_0^\infty e^{-(s-s_0)t}\beta(t)dt, \quad \beta(u) = \int_0^u e^{-s_0 t}f(t)dt.$$

That is, in the region of convergence $F(s)$ can effectively be expressed as the absolutely convergent Laplace transform of some other function. In particular, it is analytic.

There are several Paley–Wiener theorems concerning the relationship between the decay properties of f and the properties of the Laplace transform within the region of convergence.

In engineering applications, a function corresponding to a linear time-invariant (LTI) system is *stable* if every bounded input produces a bounded output. This is equivalent to the absolute convergence of the Laplace transform of the impulse response function in the region $\text{Re}(s) \geq 0$. As a result, LTI systems are stable provided the poles of the Laplace transform of the impulse response function have negative real part.

This ROC is used in knowing about the causality and stability of a system.

Properties and Theorems

The Laplace transform has a number of properties that make it useful for analyzing linear dynamical systems. The most significant advantage is that differentiation and integration become multiplication and division, respectively, by s (similarly to logarithms changing multiplication of numbers to addition of their logarithms).

Because of this property, the Laplace variable s is also known as *operator variable* in the L domain: either *derivative operator* or (for s^{-1}) *integration operator*. The transform turns integral equations and differential equations to polynomial equations, which are much easier to solve. Once solved, use of the inverse Laplace transform reverts to the time domain.

Given the functions $f(t)$ and $g(t)$, and their respective Laplace transforms $F(s)$ and $G(s)$,

$$f(t) = \mathcal{L}^{-1}\{F(s)\},$$
$$g(t) = \mathcal{L}^{-1}\{G(s)\},$$

- Initial value theorem:

$$f(0^+) = \lim_{s \to \infty} sF(s).$$

- Final value theorem:

$$f(\infty) = \lim_{s \to 0} sF(s), \text{ if all poles of } sF(s) \text{ are in the left half-plane.}$$

The final value theorem is useful because it gives the long-term behaviour without having to perform partial fraction decompositions or other difficult algebra. If $F(s)$ has a pole in the right-hand plane or poles on the imaginary axis (e.g., if $f(t) = e^t$ or $f(t) = \sin(t)$), the behaviour of this formula is undefined.

Relation to Power Series

The Laplace transform can be viewed as a continuous analogue of a power series. If $a(n)$ is a discrete function of a positive integer n, then the power series associated to $a(n)$ is the series

$$\sum_{n=0}^{\infty} a(n)x^n$$

where x is a real variable. Replacing summation over n with integration over t, a continuous version of the power series becomes

$$\int_0^{\infty} f(t)x^t \, dt$$

where the discrete function $a(n)$ is replaced by the continuous one $f(t)$.

Changing the base of the power from x to e gives

$$\int_0^{\infty} f(t)\left(e^{\ln x}\right)^t \, dt$$

For this to converge for, say, all bounded functions f, it is necessary to require that $\ln x < 0$. Making the substitution $-s = \ln x$ gives just the Laplace transform:

$$\int_0^{\infty} f(t)e^{-st} \, dt$$

In other words, the Laplace transform is a continuous analog of a power series in which the discrete parameter n is replaced by the continuous parameter t, and x is replaced by e^{-s}.

Relation to Moments

The quantities

$$\mu_n = \int_0^{\infty} t^n f(t)dt$$

are the *moments* of the function f. If the first n moments of f converge absolutely, then by repeated differentiation under the integral,

$$(-1)^n (\mathcal{L}f)^{(n)}(0) = \mu_n.$$

This is of special significance in probability theory, where the moments of a random variable X are given by the expectation values $\mu_n = E[X^n]$. Then, the relation holds

$$\mu_n = (-1)^n \frac{d^n}{ds^n} E\left[e^{-sX}\right](0).$$

Proof of the Laplace Transform of a Function's Derivative

It is often convenient to use the differentiation property of the Laplace transform to find the transform of a function's derivative. This can be derived from the basic expression for a Laplace transform as follows:

$$\mathcal{L}\{f(t)\} = \int_{0^-}^{\infty} e^{-st} f(t) dt$$

$$= \left[\frac{f(t)e^{-st}}{-s}\right]_{0^-}^{\infty} - \int_{0^-}^{\infty} \frac{e^{-st}}{-s} f'(t) dt \quad \text{(by parts)}$$

$$= \left[-\frac{f(0^-)}{-s}\right] + \frac{1}{s}\mathcal{L}\{f'(t)\},$$

yielding

$$\mathcal{L}\{f'(t)\} = s \cdot \mathcal{L}\{f(t)\} - f(0^-),$$

and in the bilateral case,

$$\mathcal{L}\{f'(t)\} = s \int_{-\infty}^{\infty} e^{-st} f(t) dt = s \cdot \mathcal{L}\{f(t)\}.$$

The general result

$$\mathcal{L}\{f^{(n)}(t)\} = s^n \cdot \mathcal{L}\{f(t)\} - s^{n-1} f(0^-) - \cdots - f^{(n-1)}(0^-),$$

where $f^{(n)}$ denotes the nth derivative of f, can then be established with an inductive argument.

Evaluating Integrals Over the Positive Real Axis

A really useful property of the Laplace transform is the following:

$$\int_0^{+\infty} f(x)g(x)dx = \int_0^{+\infty} (\mathcal{L}f)(s)\cdot(\mathcal{L}^{-1}g)(s)ds$$

under suitable assumptions on the behaviour of f, g in a right neighbourhood of 0 and on the decay rate of f, g in a left neighbourhood of $+\infty$. The above formula is a variation of integration by parts, with the operators $\dfrac{d}{dx}$ and $\int dx$ being replaced by $\mathcal{L}$ and $\mathcal{L}^{-1}$. Let us prove the equivalent formulation:

$$\int_0^{+\infty} (\mathcal{L}f)(x)g(x)dx = \int_0^{+\infty} f(s)(\mathcal{L}g)(s)ds.$$

By plugging in $(\mathcal{L}f)(x) = \int_0^{+\infty} f(s)e^{-sx}ds$ the left-hand side turns into:

$$\int_0^{+\infty}\int_0^{+\infty} f(s)g(x)e^{-sx}dsdx,$$

but assuming Fubini's theorem holds, by reversing the order of integration we get the wanted right-hand side.

Evaluating Improper Integrals

Let $\mathcal{L}\{f(t)\} = F(s)$, then

$$\mathcal{L}\left\{\frac{f(t)}{t}\right\} = \int_s^\infty F(p)dp,$$

or

$$\int_0^\infty \frac{f(t)}{t}e^{-st}dt = \int_s^\infty F(p)dp.$$

Letting $s \to 0$, gives one the identity

$$\int_0^\infty \frac{f(t)}{t}dt = \int_0^\infty F(p)dp.$$

provided that the interchange of limits can be justified. Even when the interchange cannot be justified the calculation can be suggestive. For example, proceeding formally one has

$$\int_0^\infty \frac{1}{t}\big(\cos(at)-\cos(bt)\big)dt = \int_0^\infty \left(\frac{p}{p^2+a^2} - \frac{p}{p^2+b^2}\right)dp = \frac{1}{2}\ln\frac{p^2+a^2}{p^2+b^2}\bigg|_0^\infty = \ln b - \ln a.$$

The validity of this identity can be proved by other means. It is an example of a Frullani integral.

Another example is Dirichlet integral.

Relationship to other Transforms

Laplace–Stieltjes Transform

The (unilateral) Laplace–Stieltjes transform of a function $g : \mathrm{R} \to \mathrm{R}$ is defined by the Lebesgue–Stieltjes integral

$$\{\mathcal{L}^*g\}(s) = \int_0^\infty e^{-st} dg(t).$$

The function g is assumed to be of bounded variation. If g is the antiderivative of f:

$$g(x) = \int_0^x f(t) dt$$

then the Laplace–Stieltjes transform of g and the Laplace transform of f coincide. In general, the Laplace–Stieltjes transform is the Laplace transform of the Stieltjes measure associated to g. So in practice, the only distinction between the two transforms is that the Laplace transform is thought of as operating on the density function of the measure, whereas the Laplace–Stieltjes transform is thought of as operating on its cumulative distribution function.

Fourier Transform

The continuous Fourier transform is equivalent to evaluating the bilateral Laplace transform with imaginary argument $s = i\omega$ or $s = 2\pi f i$,

$$\hat{f}(\omega) = \mathcal{F}\{f(t)\}$$
$$= \mathcal{L}\{f(t)\}|_{s=i\omega} = F(s)|_{s=i\omega}$$
$$= \int_{-\infty}^{\infty} e^{-i\omega t} f(t) dt .$$

This definition of the Fourier transform requires a prefactor of $1/2\pi$ on the reverse Fourier transform. This relationship between the Laplace and Fourier transforms is often used to determine the frequency spectrum of a signal or dynamical system.

The above relation is valid as stated if and only if the region of convergence (ROC) of $F(s)$ contains the imaginary axis, $\sigma = 0$.

For example, the function $f(t) = \cos(\omega_0 t)$ has a Laplace transform $F(s) = s/(s^2 + \omega_0^2)$ whose ROC is $\mathrm{Re}(s) > 0$. As $s = i\omega$ is a pole of $F(s)$, substituting $s = i\omega$ in $F(s)$ does not yield the Fourier transform of $f(t)u(t)$, which is proportional to the Dirac delta-function $\delta(\omega - \omega_0)$.

However, a relation of the form

$$\lim_{\sigma \to 0^+} F(\sigma + i\omega) = \hat{f}(\omega)$$

holds under much weaker conditions. For instance, this holds for the above example provided that the limit is understood as a weak limit of measures. General conditions relating the limit of the Laplace transform of a function on the boundary to the Fourier transform take the form of Paley-Wiener theorems.

Mellin Transform

The Mellin transform and its inverse are related to the two-sided Laplace transform by a simple change of variables.

If in the Mellin transform

$$G(s) = \mathcal{M}\{g(\theta)\} = \int_0^\infty \theta^s g(\theta) \frac{d\theta}{\theta}$$

we set $\theta = e^{-t}$ we get a two-sided Laplace transform.

Z-transform

The unilateral or one-sided Z-transform is simply the Laplace transform of an ideally sampled signal with the substitution of

$$z \overset{\text{def}}{=} e^{sT},$$

where $T = 1/f_s$ is the sampling period (in units of time e.g., seconds) and f_s is the sampling rate (in samples per second or hertz).

Let

$$\Delta_T(t) \overset{\text{def}}{=} \sum_{n=0}^\infty \delta(t - nT)$$

be a sampling impulse train (also called a Dirac comb) and

$$x_q(t) \overset{\text{def}}{=} x(t)\Delta_T(t) = x(t)\sum_{n=0}^\infty \delta(t - nT)$$

$$= \sum_{n=0}^\infty x(nT)\delta(t - nT) = \sum_{n=0}^\infty x[n]\delta(t - nT)$$

be the sampled representation of the continuous-time $x(t)$

$$x[n] \overset{\text{def}}{=} x(nT).$$

The Laplace transform of the sampled signal $x_{q(t)}$ is

$$X_q(s) = \int_{0^-}^{\infty} x_q(t)e^{-st}\, dt$$

$$= \int_{0^-}^{\infty} \sum_{n=0}^{\infty} x[n]\delta(t - nT)e^{-st}\, dt$$

$$= \sum_{n=0}^{\infty} x[n]\int_{0^-}^{\infty} \delta(t - nT)e^{-st}\, dt$$

$$= \sum_{n=0}^{\infty} x[n]e^{-nsT} \ .$$

This is the precise definition of the unilateral Z-transform of the discrete function $x[n]$

$$X(z) = \sum_{n=0}^{\infty} x[n]z^{-n}$$

with the substitution of $z \to e^{sT}$.

Comparing the last two equations, we find the relationship between the unilateral Z-transform and the Laplace transform of the sampled signal,

$$X_q(s) = X(z)\Big|_{z=e^{sT}} \ .$$

The similarity between the Z and Laplace transforms is expanded upon in the theory of time scale calculus.

Borel Transform

The integral form of the Borel transform

$$F(s) = \int_0^{\infty} f(z)e^{-sz}\, dz$$

is a special case of the Laplace transform for f an entire function of exponential type, meaning that

$$|f(z)| \leq Ae^{B|z|}$$

for some constants A and B. The generalized Borel transform allows a different weighting function to be used, rather than the exponential function, to transform functions not of exponential type. Nachbin's theorem gives necessary and sufficient conditions for the Borel transform to be well defined.

Fundamental Relationships

Since an ordinary Laplace transform can be written as a special case of a two-sided transform, and since the two-sided transform can be written as the sum of two one-sided transforms, the theory of the Laplace-, Fourier-, Mellin-, and Z-transforms are at bottom the same subject. However, a different point of view and different characteristic problems are associated with each of these four major integral transforms.

s-domain Equivalent Circuits and Impedances

The Laplace transform is often used in circuit analysis, and simple conversions to the s-domain of circuit elements can be made. Circuit elements can be transformed into impedances, very similar to phasor impedances.

Here is a summary of equivalents:

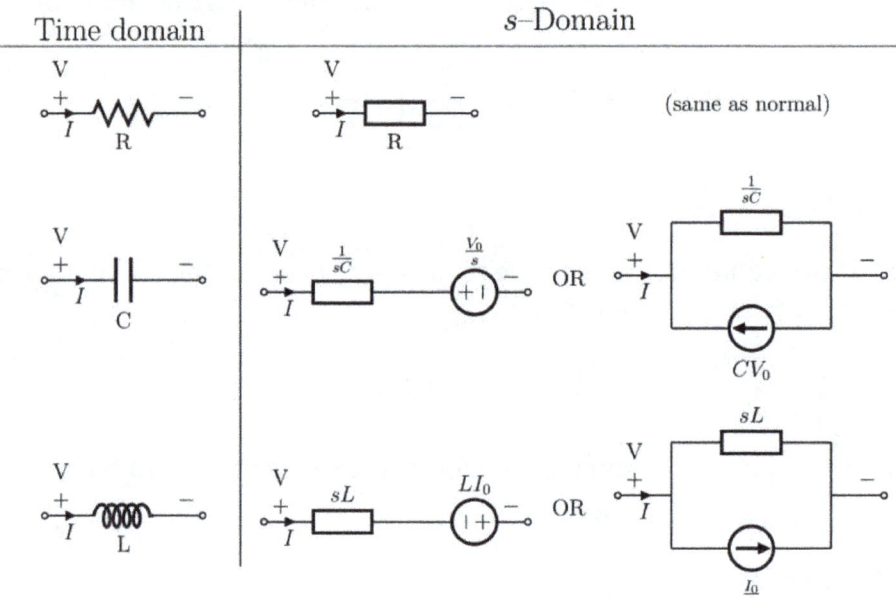

Note that the resistor is exactly the same in the time domain and the s-domain. The sources are put in if there are initial conditions on the circuit elements. For example, if a capacitor has an initial voltage across it, or if the inductor has an initial current through it, the sources inserted in the s-domain account for that.

The equivalents for current and voltage sources are simply derived from the transformations in the table above.

Examples: How to Apply the Properties and Theorems

The Laplace transform is used frequently in engineering and physics; the output of a linear time-invariant system can be calculated by convolving its unit impulse response with the input signal. Performing this calculation in Laplace space turns the convolution into a multiplication; the latter being easier to solve because of its algebraic form.

The Laplace transform can also be used to solve differential equations and is used extensively in electrical engineering. The Laplace transform reduces a linear differential equation to an algebraic equation, which can then be solved by the formal rules of algebra. The original differential equation can then be solved by applying the inverse Laplace transform. The English electrical engineer Oliver Heaviside first proposed a similar scheme, although without using the Laplace transform; and the resulting operational calculus is credited as the Heaviside calculus.

Example 1: Solving a Differential Equation

In nuclear physics, the following fundamental relationship governs radioactive decay: the number of radioactive atoms N in a sample of a radioactive isotope decays at a rate proportional to N. This leads to the first order linear differential equation

$$\frac{dN}{dt} = -\lambda N,$$

where λ is the decay constant. The Laplace transform can be used to solve this equation.

Rearranging the equation to one side, we have

$$\frac{dN}{dt} + \lambda N = 0.$$

Next, we take the Laplace transform of both sides of the equation:

$$\left(s\tilde{N}(s) - N_0 \right) + \lambda \tilde{N}(s) = 0,$$

where

$$\tilde{N}(s) = \mathcal{L}\{N(t)\}$$

and

$$N_0 = N(0).$$

Solving, we find

$$\tilde{N}(s) = \frac{N_0}{s+\lambda}.$$

Finally, we take the inverse Laplace transform to find the general solution

$$N(t) = \mathcal{L}^{-1}\{\tilde{N}(s)\} = \mathcal{L}^{-1}\left\{ \frac{N_0}{s+\lambda} \right\} = N_0 e^{-\lambda t},$$

which is indeed the correct form for radioactive decay.

Example 2: Deriving the Complex Impedance for a Capacitor

In the theory of electrical circuits, the current flow in a capacitor is proportional to the capacitance and rate of change in the electrical potential (in SI units). Symbolically, this is expressed by the differential equation

$$i = C\frac{dv}{dt},$$

where C is the capacitance (in farads) of the capacitor, $i = i(t)$ is the electric current (in amperes) through the capacitor as a function of time, and $v = v(t)$ is the voltage (in volts) across the terminals of the capacitor, also as a function of time.

Taking the Laplace transform of this equation, we obtain

$$I(s) = C(sV(s) - V_0),$$

where

$$I(s) = \mathcal{L}\{i(t)\}, V(s) = \mathcal{L}\{v(t)\},$$

and

$$V_0 = v(t)\big|_{t=0}.$$

Solving for $V(s)$ we have

$$V(s) = \frac{I(s)}{sC} + \frac{V_0}{s}.$$

The definition of the complex impedance Z (in ohms) is the ratio of the complex voltage V divided by the complex current I while holding the initial state V_0 at zero:

$$Z(s) = \frac{V(s)}{I(s)}\bigg|_{V_0=0}.$$

Using this definition and the previous equation, we find:

$$Z(s) \quad \frac{1}{sC},$$

which is the correct expression for the complex impedance of a capacitor.

Example 3: Method of Partial Fraction Expansion

Consider a linear time-invariant system with transfer function

$$H(s) = \frac{1}{(s+\alpha)(s+\beta)}.$$

The impulse response is simply the inverse Laplace transform of this transfer function:

$$h(t) = \mathcal{L}^{-1}\{H(s)\}.$$

To evaluate this inverse transform, we begin by expanding $H(s)$ using the method of partial fraction expansion,

$$\frac{1}{(s+\alpha)(s+\beta)} = \frac{P}{s+\alpha} + \frac{R}{s+\beta}.$$

The unknown constants P and R are the residues located at the corresponding poles of the transfer function. Each residue represents the relative contribution of that singularity to the transfer function's overall shape.

By the residue theorem, the inverse Laplace transform depends only upon the poles and their residues. To find the residue P, we multiply both sides of the equation by $s + \alpha$ to get

$$\frac{1}{s+\beta} = P + \frac{R(s+\alpha)}{s+\beta}.$$

Then by letting $s = -\alpha$, the contribution from R vanishes and all that is left is

$$P = \left.\frac{1}{s+\beta}\right|_{s=-\alpha} = \frac{1}{\beta-\alpha}.$$

Similarly, the residue R is given by

$$R = \left.\frac{1}{s+\alpha}\right|_{s=-\beta} = \frac{1}{\alpha-\beta}.$$

Note that

$$R = \frac{-1}{\beta-\alpha} = -P$$

and so the substitution of R and P into the expanded expression for $H(s)$ gives

$$H(s) = \left(\frac{1}{\beta-\alpha}\right)\left(\frac{1}{s+\alpha} - \frac{1}{s+\beta}\right).$$

Finally, using the linearity property and the known transform for exponential decay, we can take the inverse Laplace transform of $H(s)$ to obtain

$$h(t) = \mathcal{L}^{-1}\{H(s)\} = \frac{1}{\beta-\alpha}\left(e^{-\alpha t} - e^{-\beta t}\right),$$

which is the impulse response of the system.

Example 3.2: Convolution

The same result can be achieved using the convolution property as if the system is a series of filters with transfer functions of $1/(s + a)$ and $1/(s + b)$. That is, the inverse of

$$H(s) = \frac{1}{(s+a)(s+b)} = \frac{1}{s+a} \cdot \frac{1}{s+b}$$

is

$$\mathcal{L}^{-1}\left\{\frac{1}{s+a}\right\} * \mathcal{L}^{-1}\left\{\frac{1}{s+b}\right\} = e^{-at} * e^{-bt} = \int_0^t e^{-ax}e^{-b(t-x)}\,dx = \frac{e^{-at} - e^{-bt}}{b-a}.$$

Example 4: Mixing Sines, Cosines, and Exponentials

Time function	Laplace transform
$e^{-\alpha t}\left[\cos(\omega t) + \left(\dfrac{\beta - \alpha}{\omega}\right)\sin(\omega t)\right]u(t)$	$\dfrac{s+\beta}{(s+\alpha)^2 + \omega^2}$

Starting with the Laplace transform

$$X(s) = \frac{s+\beta}{(s+\alpha)^2 + \omega^2},$$

we find the inverse transform by first adding and subtracting the same constant α to the numerator:

$$X(s) = \frac{s+\alpha}{(s+\alpha)^2 + \omega^2} + \frac{\beta - \alpha}{(s+\alpha)^2 + \omega^2}.$$

By the shift-in-frequency property, we have

$$x(t) = e^{-\alpha t}\mathcal{L}^{-1}\left\{\frac{s}{s^2 + \omega^2} + \frac{\beta - \alpha}{s^2 + \omega^2}\right\}$$

$$= e^{-\alpha t}\mathcal{L}^{-1}\left\{\frac{s}{s^2 + \omega^2} + \left(\frac{\beta - \alpha}{\omega}\right)\left(\frac{\omega}{s^2 + \omega^2}\right)\right\}$$

$$= e^{-\alpha t}\left[\mathcal{L}^{-1}\left\{\frac{s}{s^2 + \omega^2}\right\} + \left(\frac{\beta - \alpha}{\omega}\right)\mathcal{L}^{-1}\left\{\frac{\omega}{s^2 + \omega^2}\right\}\right].$$

Finally, using the Laplace transforms for sine and cosine, we have

$$x(t) = e^{-at}\left[\cos(\omega t)u(t) + \left(\frac{\beta - \alpha}{\omega}\right)\sin(\omega t)u(t)\right].$$

$$x(t) = e^{-at}\left[\cos(\omega t) + \left(\frac{\beta - \alpha}{\omega}\right)\sin(\omega t)\right]u(t).$$

Example 5: Phase Delay

Time function	Laplace transform
$\sin(\omega t + \phi)$	$\dfrac{s\sin(\phi) + \omega\cos(\phi)}{s^2 + \omega^2}$
$\cos(\omega t + \phi)$	$\dfrac{s\cos(\phi) - \omega\sin(\phi)}{s^2 + \omega^2}.$

Starting with the Laplace transform,

$$X(s) = \frac{s\sin(\phi) + \omega\cos(\phi)}{s^2 + \omega^2}$$

we find the inverse by first rearranging terms in the fraction:

$$X(s) = \frac{s\sin(\phi)}{s^2 + \omega^2} + \frac{\omega\cos(\phi)}{s^2 + \omega^2}$$

$$= \sin(\phi)\left(\frac{s}{s^2 + \omega^2}\right) + \cos(\phi)\left(\frac{\omega}{s^2 + \omega^2}\right).$$

We are now able to take the inverse Laplace transform of our terms:

$$x(t) = \sin(\phi)\mathcal{L}^{-1}\left\{\frac{s}{s^2 + \omega^2}\right\} + \cos(\phi)\mathcal{L}^{-1}\left\{\frac{\omega}{s^2 + \omega^2}\right\}$$

$$= \sin(\phi)\cos(\omega t) + \sin(\omega t)\cos(\phi).$$

This is just the sine of the sum of the arguments, yielding:

$$x(t) = \sin(\omega t + \phi).$$

We can apply similar logic to find that

$$\left\{\frac{\cos\phi}{2} \quad \frac{\omega\sin\phi}{2}\right\} = \cos(\omega + \phi).$$

Example 6: Determining Structure of Astronomical Object from Spectrum

The wide and general applicability of the Laplace transform and its inverse is illustrated by an application in astronomy which provides some information on the *spatial distribution* of matter of an astronomical source of radiofrequency thermal radiation too distant to resolve as more than a point, given its flux density spectrum, rather than relating the *time* domain with the spectrum (frequency domain).

Assuming certain properties of the object, e.g. spherical shape and constant temperature, calculations based on carrying out an inverse Laplace transformation on the spectrum of the object can produce the only possible model of the distribution of matter in it (density as a function of distance from the center) consistent with the spectrum. When independent information on the structure of an object is available, the inverse Laplace transform method has been found to be in good agreement.

Separation of Variables

In mathematics, separation of variables (also known as the Fourier method) is any of several methods for solving ordinary and partial differential equations, in which algebra allows one to rewrite an equation so that each of two variables occurs on a different side of the equation.

Ordinary Differential Equations (ODE)

Suppose a differential equation can be written in the form

$$\frac{d}{dx} f(x) = g(x)h(f(x))$$

which we can write more simply by letting $y = f(x)$:

$$\frac{dy}{dx} = g(x)h(y).$$

As long as $h(y) \neq 0$, we can rearrange terms to obtain:

$$\frac{dy}{h(y)} = g(x)dx,$$

so that the two variables x and y have been separated. dx (and dy) can be viewed, at a simple level, as just a convenient notation, which provides a handy mnemonic aid for assisting with manipulations. A formal definition of dx as a differential (infinitesimal) is somewhat advanced.

Alternative Notation

Some who dislike Leibniz's notation may prefer to write this as

$$\frac{1}{h(y)}\frac{dy}{dx} = g(x),$$

but that fails to make it quite as obvious why this is called "separation of variables". Integrating both sides of the equation with respect to x, we have

$$\int \frac{1}{h(y)}\frac{dy}{dx}dx = \int g(x)dx, \qquad (1)$$

or equivalently,

$$\int \frac{1}{h(y)}dy = \int g(x)dx$$

because of the substitution rule for integrals.

If one can evaluate the two integrals, one can find a solution to the differential equation. Observe that this process effectively allows us to treat the derivative $\frac{dy}{dx}$ as a fraction which can be separated. This allows us to solve separable differential equations more conveniently, as demonstrated in the example below.

(Note that we do not need to use two constants of integration, in equation (1) as in

$$\int \frac{1}{h(y)}dy + C_1 = \int g(x)dx + C_2,$$

because a single constant $C = C_2 - C_1$ is equivalent.)

Example

Population growth is often modeled by the differential equation

$$\frac{dP}{dt} = kP\left(1 - \frac{P}{K}\right)$$

where P is the population with respect to time t, k is the rate of growth, and K is the carrying capacity of the environment.

Separation of variables may be used to solve this differential equation.

$$\frac{dP}{dt} = kP\left(1 - \frac{P}{K}\right)$$

$$\int \frac{dP}{P\left(1 - \frac{P}{K}\right)} = \int k\,dt$$

To evaluate the integral on the left side, we simplify the fraction

$$\frac{1}{P\left(1-\dfrac{P}{K}\right)} = \frac{K}{P(K-P)}$$

and then, we decompose the fraction into partial fractions

$$\frac{K}{P(K-P)} = \frac{1}{P} + \frac{1}{K-P}$$

Thus we have

$$\int \left(\frac{1}{P} + \frac{1}{K-P}\right) dP = \int k\,dt$$

$$\ln|P| - \ln|K-P| = kt + C$$

$$\ln|K-P| - \ln|P| = -kt - C$$

$$\ln\left|\frac{K-P}{P}\right| = -kt - C$$

$$\left|\frac{K-P}{P}\right| = e^{-kt-C}$$

$$\left|\frac{K-P}{P}\right| = e^{-C}e^{-kt}$$

$$\frac{K-P}{P} = \pm e^{-C}e^{-kt}$$

Let $A = \pm e^{-C}$.

$$\frac{K-P}{P} = Ae^{-kt}$$

$$\frac{K}{P} - 1 = Ae^{-kt}$$

$$\frac{K}{P} = 1 + Ae^{-kt}$$

$$\frac{P}{K} = \frac{1}{1 + Ae^{-kt}}$$

$$P = \frac{K}{1 + Ae^{-kt}}$$

Therefore, the solution to the logistic equation is

$$P(t) = \frac{K}{1 + Ae^{-kt}}$$

To find A, let $t = 0$ and $P(0) = P_0$ Then we have

$$P_0 = \frac{K}{1 + Ae^0}$$

Noting that $e^0 = 1$, and solving for A we get

$$A = \frac{K - P_0}{P_0}.$$

Partial Differential Equations

The method of separation of variables is also used to solve a wide range of linear partial differential equations with boundary and initial conditions, such as heat equation, wave equation, Laplace equation and Helmholtz equation.

Homogeneous Case

Consider the one-dimensional heat equation. The equation is

$$\frac{\partial u}{\partial t} - \alpha \frac{\partial^2 u}{\partial x^2} = 0 \tag{1}$$

The boundary condition is homogeneous, that is

$$u\big|_{x=0} = u\big|_{x=L} = 0 \tag{2}$$

Let us attempt to find a solution which is not identically zero satisfying the boundary conditions but with the following property: u is a product in which the dependence of u on x, t is separated, that is:

$$u(x,t) = X(x)T(t). \tag{3}$$

Substituting u back into equation (1) and using the product rule,

$$\frac{T'(t)}{\alpha T(t)} = \frac{X''(x)}{X(x)}. \tag{4}$$

Since the right hand side depends only on x and the left hand side only on t, both sides are equal to some constant value $-\lambda$. Thus:

$$T'(t) = -\lambda \alpha T(t), \tag{5}$$

and

$$X''(x) = -\lambda X(x).$$

(6)

$-\lambda$ here is the eigenvalue for both differential operators, and $T(t)$ and $X(x)$ are corresponding eigenfunctions.

We will now show that solutions for $X(x)$ for values of $\lambda \leq 0$ cannot occur:

Suppose that $\lambda < 0$. Then there exist real numbers B, C such that

$$X(x) = Be^{\sqrt{-\lambda}x} + Ce^{-\sqrt{-\lambda}x}.$$

From (2) we get

$$X(0) = 0 = X(L),$$

(7)

and therefore $B = 0 = C$ which implies u is identically 0.

Suppose that $\lambda = 0$. Then there exist real numbers B, C such that

$$X(x) = Bx + C.$$

From (7) we conclude in the same manner as in 1 that u is identically 0.

Therefore, it must be the case that $\lambda > 0$. Then there exist real numbers A, B, C such that

$$T(t) = Ae^{-\lambda \alpha t},$$

and

$$X(x) = B\sin(\sqrt{\lambda}x) + C\cos(\sqrt{\lambda}x).$$

From (7) we get $C = 0$ and that for some positive integer n,

$$\sqrt{\lambda} = n\frac{\pi}{L}.$$

This solves the heat equation in the special case that the dependence of u has the special form of (3).

In general, the sum of solutions to (1) which satisfy the boundary conditions (2) also satisfies (1) and (3). Hence a complete solution can be given as

$$u(x,t) = \sum_{n=1}^{\infty} D_n \sin\frac{n\pi x}{L} \exp\left(-\frac{n^2\pi^2\alpha t}{L^2}\right),$$

where D_n are coefficients determined by initial condition.

Given the initial condition

$$u\big|_{t=0} = f(x),$$

we can get

$$f(x) = \sum_{n=1}^{\infty} D_n \sin \frac{n\pi x}{L}.$$

This is the sine series expansion of $f(x)$. Multiplying both sides with $\sin \dfrac{n\pi x}{L}$ and integrating over $[0,L]$ result in

$$D_n = \frac{2}{L} \int_0^L f(x) \sin \frac{n\pi x}{L} dx.$$

This method requires that the eigenfunctions of x, here $\left\{ \sin \dfrac{n\pi x}{L} \right\}_{n=1}^{\infty}$, are orthogonal and complete. In general this is guaranteed by Sturm-Liouville theory.

Nonhomogeneous Case

Suppose the equation is nonhomogeneous,

$$\frac{\partial u}{\partial t} - \alpha \frac{\partial^2 u}{\partial x^2} = h(x,t) \tag{8}$$

with the boundary condition the same as (2).

Expand $h(x,t)$, $u(x,t)$ and $f(x)$ into

$$h(x,t) = \sum_{n=1}^{\infty} h_n(t) \sin \frac{n\pi x}{L}, \tag{9}$$

$$u(x,t) = \sum_{n=1}^{\infty} u_n(t) \sin \frac{n\pi x}{L}, \tag{10}$$

$$f(x) = \sum_{n=1}^{\infty} b_n \sin \frac{n\pi x}{L}, \tag{11}$$

where $h_n(t)$ and b_n can be calculated by integration, while $u_n(t)$ is to be determined.

Substitute (9) and (10) back to (8) and considering the orthogonality of sine functions we get

$$u_n'(t) + \alpha \frac{n^2 \pi^2}{L^2} u_n(t) = h_n(t),$$

which are a sequence of linear differential equations that can be readily solved with, for instance, Laplace transform, or Integrating factor. Finally, we can get

$$u_n(t) = e^{-\alpha \frac{n^2\pi^2}{L^2}t}\left(b_n + \int_0^t h_n(s)e^{\alpha \frac{n^2\pi^2}{L^2}s}\, ds \right).$$

If the boundary condition is nonhomogeneous, then the expansion of (9) and (10) is no longer valid. One has to find a function v that satisfies the boundary condition only, and subtract it from u. The function u-v then satisfies homogeneous boundary condition, and can be solved with the above method.

In orthogonal curvilinear coordinates, separation of variables can still be used, but in some details different from that in Cartesian coordinates. For instance, regularity or periodic condition may determine the eigenvalues in place of boundary conditions.

Matrices

The matrix form of the separation of variables is the Kronecker sum.

As an example we consider the 2D discrete Laplacian on a regular grid:

$$L = \mathbf{D}_{xx} \oplus \mathbf{D}_{yy} = \mathbf{D}_{xx} \otimes \mathbf{I} + \mathbf{I} \otimes \mathbf{D}_{yy},$$

where $\mathbf{D}_{xx}$ and $\mathbf{D}_{yy}$ are 1D discrete Laplacians in the x- and y-directions, correspondingly, and $\mathbf{I}$ are the identities of appropriate sizes. Kronecker sum of discrete Laplacians for details.

References

- Hairer, Ernst; Nørsett, Syvert Paul; Wanner, Gerhard (1993), Solving ordinary differential equations I: Nonstiff problems, Berlin, New York: Springer-Verlag, ISBN 978-3-540-56670-0

- Polyanin, A. D. and V. F. Zaitsev, Handbook of Exact Solutions for Ordinary Differential Equations (2nd edition)", Chapman & Hall/CRC Press, Boca Raton, 2003. ISBN 1-58488-297-2

- Tipler, Paul A. (1991), Physics for Scientists and Engineers: Extended version (3rd ed.), New York: Worth Publishers, ISBN 0-87901-432-6

- Boyce, W. E.; DiPrima, R. C. (1986). Elementary Differential Equations and Boundary Value Problems (4th ed.). John Wiley & Sons. ISBN 0-471-83824-1.

- Riley, K. F.; Bence, S. J. (2010). Mathematical Methods for Physics and Engineering. Cambridge University Press. ISBN 978-0-521-86153-3.

- Ernst Hairer, Syvert Paul Nørsett and Gerhard Wanner, Solving ordinary differential equations I: Nonstiff problems, second edition, Springer Verlag, Berlin, 1993. ISBN 3-540-56670-8.

- Bradie, Brian (2006). A Friendly Introduction to Numerical Analysis. Upper Saddle River, New Jersey: Pearson Prentice Hall. ISBN 0-13-013054-0.

- Holubová, Pavel Drábek ; Gabriela (2007). Elements of partial differential equations ([Online-Ausg.]. ed.). Berlin: de Gruyter. ISBN 9783110191240.

- Ibragimov, Nail H (1993), CRC Handbook of Lie Group Analysis of Differential Equations Vol. 1-3, Providence:

CRC-Press, ISBN 0-8493-4488-3 .

- Liao, S.J. (2003), Beyond Perturbation: Introduction to the Homotopy Analysis Method, Boca Raton: Chapman & Hall/ CRC Press, ISBN 1-58488-407-X

- Pinchover, Y. & Rubinstein, J. (2005), An Introduction to Partial Differential Equations, New York: Cambridge University Press, ISBN 0-521-84886-5 .

- Polyanin, A. D.; Zaitsev, V. F. & Moussiaux, A. (2002), Handbook of First Order Partial Differential Equations, London: Taylor & Francis, ISBN 0-415-27267-X .

- Roubíček, T. (2013), Nonlinear Partial Differential Equations with Applications (2nd ed.), Basel, Boston, Berlin: Birkhäuser, ISBN 978-3-0348-0512-4, MR MR3014456

- Solin, P.; Segeth, K. & Dolezel, I. (2003), Higher-Order Finite Element Methods, Boca Raton: Chapman & Hall/ CRC Press, ISBN 1-58488-438-X .

- Wazwaz, Abdul-Majid (2009). Partial Differential Equations and Solitary Waves Theory. Higher Education Press. ISBN 978-3-642-00251-9.

- Gershenfeld, N. (1999), The Nature of Mathematical Modeling (1st ed.), New York: Cambridge University Press, New York, NY, USA, ISBN 0-521-57095-6 .

- Vinogradov, A.M. (2001), Cohomological Analysis of Partial Differential Equations and Secondary Calculus, American Mathematical Society, Providence, Rhode Island,USA, ISBN 0-8218-2922-X .

- Polyanin, A. D. (2002). Handbook of Linear Partial Differential Equations for Engineers and Scientists. Boca Raton: Chapman & Hall/CRC Press. ISBN 1-58488-299-9.

- Robinson, James C. (2004), An Introduction to Ordinary Differential Equations, Cambridge, UK.: Cambridge University Press, ISBN 0-521-82650-0

3

Linear Algebra: An Overview

Linear algebra is the study of the organization of information and this is done through vectors and linear equations related to vectors. System of linear equations, Cramer's rule, Gaussian elimination, vector space and linear maps are the topics that have been elucidated in the following text.

Linear Algebra

Linear algebra is the branch of mathematics concerning vector spaces and linear mappings between such spaces. It includes the study of lines, planes, and subspaces, but is also concerned with properties common to all vector spaces.

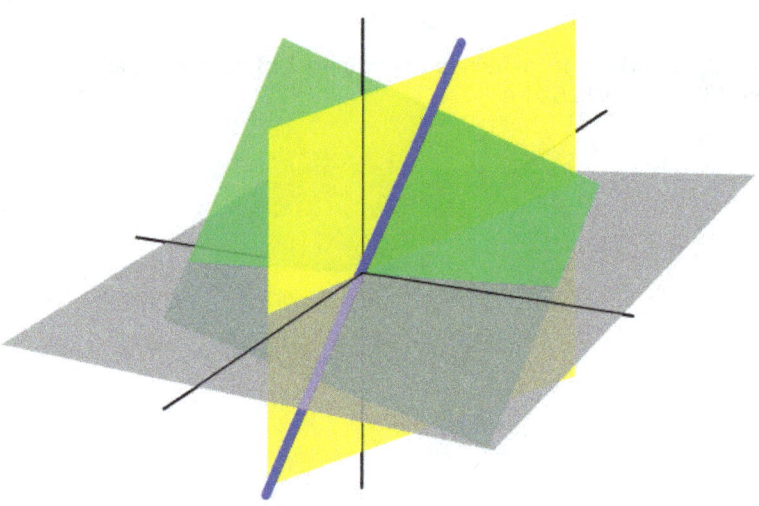

The three-dimensional Euclidean space R³ is a vector space, and lines and planes passing through the origin are vector subspaces in R³.

The set of points with coordinates that satisfy a linear equation forms a hyperplane in an n-dimensional space. The conditions under which a set of n hyperplanes intersect in a single point is an important focus of study in linear algebra. Such an investigation is initially motivated by a system of linear equations containing several unknowns. Such equations are naturally represented using the formalism of matrices and vectors.

Linear algebra is central to both pure and applied mathematics. For instance, abstract algebra arises by relaxing the axioms of a vector space, leading to a number of generalizations. Functional analysis studies the infinite-dimensional version of the theory of vector spaces. Combined with calculus, linear algebra facilitates the solution of linear systems of differential equations.

Techniques from linear algebra are also used in analytic geometry, engineering, physics, natural sciences, computer science, computer animation, advanced facial recognition algorithms and the social sciences (particularly in economics). Because linear algebra is such a well-developed theory, nonlinear mathematical models are sometimes approximated by linear models.

History

The study of linear algebra first emerged from the study of determinants, which were used to solve systems of linear equations. Determinants were used by Leibniz in 1693, and subsequently, Gabriel Cramer devised Cramer's Rule for solving linear systems in 1750. Later, Gauss further developed the theory of solving linear systems by using Gaussian elimination, which was initially listed as an advancement in geodesy.

The study of matrix algebra first emerged in England in the mid-1800s. In 1844 Hermann Grassmann published his "Theory of Extension" which included foundational new topics of what is today called linear algebra. In 1848, James Joseph Sylvester introduced the term matrix, which is Latin for "womb". While studying compositions of linear transformations, Arthur Cayley was led to define matrix multiplication and inverses. Crucially, Cayley used a single letter to denote a matrix, thus treating a matrix as an aggregate object. He also realized the connection between matrices and determinants, and wrote "There would be many things to say about this theory of matrices which should, it seems to me, precede the theory of determinants".

In 1882, Hüseyin Tevfik Pasha wrote the book titled "Linear Algebra". The first modern and more precise definition of a vector space was introduced by Peano in 1888; by 1900, a theory of linear transformations of finite-dimensional vector spaces had emerged. Linear algebra took its modern form in the first half of the twentieth century, when many ideas and methods of previous centuries were generalized as abstract algebra. The use of matrices in quantum mechanics, special relativity, and statistics helped spread the subject of linear algebra beyond pure mathematics. The development of computers led to increased research in efficient algorithms for Gaussian elimination and matrix decompositions, and linear algebra became an essential tool for modelling and simulations.

Educational History

Linear algebra first appeared in graduate textbooks in the 1940s and in undergraduate textbooks in the 1950s. Following work by the School Mathematics Study Group, U.S. high schools asked 12th grade students to do "matrix algebra, formerly reserved for college" in the 1960s. In France during the 1960s, educators attempted to teach linear algebra through affine dimensional vector spaces in the first year of secondary school. This was met with a backlash in the 1980s that removed linear algebra from the curriculum. In 1993, the U.S.-based Linear Algebra Curriculum Study Group recommended that undergraduate linear algebra courses be given an application-based "matrix orientation" as opposed to a theoretical orientation.

Scope of Study

Vector Spaces

The main structures of linear algebra are vector spaces. A vector space over a field F is a set V together with two binary operations. Elements of V are called *vectors* and elements of F are called *scalars*. The first operation, *vector addition*, takes any two vectors v and w and outputs a third vector $v + w$. The second operation, *scalar multiplication*, takes any scalar a and any vector v and outputs a new vector av. The operations of addition and multiplication in a vector space must satisfy the following axioms. In the list below, let u, v and w be arbitrary vectors in V, and a and b scalars in F.

Axiom	Signification
Associativity of addition	$u + (v + w) = (u + v) + w$
Commutativity of addition	$u + v = v + u$
Identity element of addition	There exists an element $0 \in V$, called the *zero vector*, such that $v + 0 = v$ for all $v \in V$.
Inverse elements of addition	For every $v \in V$, there exists an element $-v \in V$, called the *additive inverse* of v, such that $v + (-v) = 0$
Distributivity of scalar multiplication with respect to vector addition	$a(u + v) = au + av$
Distributivity of scalar multiplication with respect to field addition	$(a + b)v = av + bv$
Compatibility of scalar multiplication with field multiplication	$a(bv) = (ab)v$ [nb 1]
Identity element of scalar multiplication	$1v = v$, where 1 denotes the multiplicative identity in F.

The first four axioms are those of V being an abelian group under vector addition. Vector spaces may be diverse in nature, for example, containing functions, polynomials or matrices. Linear algebra is concerned with properties common to all vector spaces.

Linear Transformations

Similarly as in the theory of other algebraic structures, linear algebra studies mappings between vector spaces that preserve the vector-space structure. Given two vector spaces V and W over a field F, a linear transformation (also called linear map, linear mapping or linear operator) is a map

$$T : V \rightarrow W$$

that is compatible with addition and scalar multiplication:

$$T(u + v) = T(u) + T(v), \quad T(av) = aT(v)$$

for any vectors $u, v \in V$ and a scalar $a \in F$.

Additionally for any vectors $u, v \in V$ and scalars $a, b \in F$:

$$T(au+bv) = T(au)+T(bv) = aT(u)+bT(v)$$

When a bijective linear mapping exists between two vector spaces (that is, every vector from the second space is associated with exactly one in the first), we say that the two spaces are isomorphic. Because an isomorphism preserves linear structure, two isomorphic vector spaces are "essentially the same" from the linear algebra point of view. One essential question in linear algebra is whether a mapping is an isomorphism or not, and this question can be answered by checking if the determinant is nonzero. If a mapping is not an isomorphism, linear algebra is interested in finding its range (or image) and the set of elements that get mapped to zero, called the kernel of the mapping.

Linear transformations have geometric significance. For example, 2 × 2 real matrices denote standard planar mappings that preserve the origin.

Subspaces, Span, and Basis

Again, in analogue with theories of other algebraic objects, linear algebra is interested in subsets of vector spaces that are themselves vector spaces; these subsets are called linear subspaces. For example, both the range and kernel of a linear mapping are subspaces, and are thus often called the range space and the nullspace; these are important examples of subspaces. Another important way of forming a subspace is to take a linear combination of a set of vectors $v_1, v_2, ..., v_k$:

$$a_1 v_1 + a_2 v_2 + \cdots + a_k v_k,$$

where $a_1, a_2, ..., a_k$ are scalars. The set of all linear combinations of vectors $v_1, v_2, ..., v_k$ is called their span, which forms a subspace.

A linear combination of any system of vectors with all zero coefficients is the zero vector of V. If this is the only way to express the zero vector as a linear combination of $v_1, v_2, ..., v_k$ then these vectors are linearly independent. Given a set of vectors that span a space, if any vector w is a linear combination of other vectors (and so the set is not linearly independent), then the span would remain the same if we remove w from the set. Thus, a set of linearly dependent vectors is redundant in the sense that there will be a linearly independent subset which will span the same subspace. Therefore, we are mostly interested in a linearly independent set of vectors that spans a vector space V, which we call a basis of V. Any set of vectors that spans V contains a basis, and any linearly independent set of vectors in V can be extended to a basis. It turns out that if we accept the axiom of choice, every vector space has a basis; nevertheless, this basis may be unnatural, and indeed, may not even be constructible. For instance, there exists a basis for the real numbers, considered as a vector space over the rationals, but no explicit basis has been constructed.

Any two bases of a vector space V have the same cardinality, which is called the dimension of V. The dimension of a vector space is well-defined by the dimension theorem for vector spaces. If a basis of V has finite number of elements, V is called a finite-dimensional vector space. If V is finite-dimensional and U is a subspace of V, then $\dim U \le \dim V$. If U_1 and U_2 are subspaces of V, then

$$\dim(U_1 + U_2) = \dim U_1 + \dim U_2 - \dim(U_1 \cap U_2).$$

One often restricts consideration to finite-dimensional vector spaces. A fundamental theorem of linear algebra states that all vector spaces of the same dimension are isomorphic, giving an easy way of characterizing isomorphism.

Matrix Theory

A particular basis $\{v_1, v_2, ..., v_n\}$ of V allows one to construct a coordinate system in V: the vector with coordinates $(a_1, a_2, ..., a_n)$ is the linear combination

$$a_1 v_1 + a_2 v_2 + \cdots + a_n v_n.$$

The condition that $v_1, v_2, ..., v_n$ span V guarantees that each vector v can be assigned coordinates, whereas the linear independence of $v_1, v_2, ..., v_n$ assures that these coordinates are unique (i.e. there is only one linear combination of the basis vectors that is equal to v). In this way, once a basis of a vector space V over F has been chosen, V may be identified with the coordinate n-space F^n. Under this identification, addition and scalar multiplication of vectors in V correspond to addition and scalar multiplication of their coordinate vectors in F^n. Furthermore, if V and W are an n-dimensional and m-dimensional vector space over F, and a basis of V and a basis of W have been fixed, then any linear transformation $T: V \rightarrow W$ may be encoded by an $m \times n$ matrix A with entries in the field F, called the matrix of T with respect to these bases. Two matrices that encode the same linear transformation in different bases are called similar. Matrix theory replaces the study of linear transformations, which were defined axiomatically, by the study of matrices, which are concrete objects. This major technique distinguishes linear algebra from theories of other algebraic structures, which usually cannot be parameterized so concretely.

There is an important distinction between the coordinate n-space R^n and a general finite-dimensional vector space V. While R^n has a standard basis $\{e_1, e_2, ..., e_n\}$, a vector space V typically does not come equipped with such a basis and many different bases exist (although they all consist of the same number of elements equal to the dimension of V).

One major application of the matrix theory is calculation of determinants, a central concept in linear algebra. While determinants could be defined in a basis-free manner, they are usually introduced via a specific representation of the mapping; the value of the determinant does not depend on the specific basis. It turns out that a mapping has an inverse if and only if the determinant has an inverse (every non-zero real or complex number has an inverse). If the determinant is zero, then the nullspace is nontrivial. Determinants have other applications, including a systematic way of seeing if a set of vectors is linearly independent (we write the vectors as the columns of a matrix, and if the determinant of that matrix is zero, the vectors are linearly dependent). Determinants could also be used to solve systems of linear equations, but in real applications, Gaussian elimination is a faster method.

Eigenvalues and Eigenvectors

In general, the action of a linear transformation may be quite complex. Attention to low-dimensional examples gives an indication of the variety of their types. One strategy for a general n-dimensional transformation T is to find "characteristic lines" that are invariant sets under T. If v is a non-zero vector such that Tv is a scalar multiple of v, then the line through 0 and v is an invariant set under T and v is called a characteristic vector or eigenvector. The scalar λ such that $Tv = \lambda v$ is called a characteristic value or eigenvalue of T.

To find an eigenvector or an eigenvalue, we note that

$$Tv - \lambda v = (T - \lambda \mathrm{I})v = 0,$$

where I is the identity matrix. For there to be nontrivial solutions to that equation, $\det(T - \lambda \mathrm{I}) = 0$. The determinant is a polynomial, and so the eigenvalues are not guaranteed to exist if the field is R. Thus, we often work with an algebraically closed field such as the complex numbers when dealing with eigenvectors and eigenvalues so that an eigenvalue will always exist. It would be particularly nice if given a transformation T taking a vector space V into itself we can find a basis for V consisting of eigenvectors. If such a basis exists, we can easily compute the action of the transformation on any vector: if v_1, v_2, ..., v_n are linearly independent eigenvectors of a mapping of n-dimensional spaces T with (not necessarily distinct) eigenvalues λ_1, λ_2, ..., λ_n, and if $v = a_1 v_1 + ... + a_n v_n$, then,

$$T(v) = T(a_1 v_1) + \cdots + T(a_n v_n) = a_1 T(v_1) + \cdots + a_n T(v_n) = a_1 \lambda_1 v_1 + \cdots + a_n \lambda_n v_n.$$

Such a transformation is called a diagonalizable matrix since in the eigenbasis, the transformation is represented by a diagonal matrix. Because operations like matrix multiplication, matrix inversion, and determinant calculation are simple on diagonal matrices, computations involving matrices are much simpler if we can bring the matrix to a diagonal form. Not all matrices are diagonalizable (even over an algebraically closed field).

Inner-product Spaces

Besides these basic concepts, linear algebra also studies vector spaces with additional structure, such as an inner product. The inner product is an example of a bilinear form, and it gives the vector space a geometric structure by allowing for the definition of length and angles. Formally, an *inner product* is a map

$$\langle \cdot, \cdot \rangle : V \times V \to F$$

that satisfies the following three axioms for all vectors u, v, w in V and all scalars a in F:

- Conjugate symmetry:

$$\langle u, v \rangle = \overline{\langle v, u \rangle}.$$

Note that in R, it is symmetric.

- Linearity in the first argument:

$$\langle au, v \rangle = a \langle u, v \rangle.$$

$$\langle u + v, w \rangle = \langle u, w \rangle + \langle v, w \rangle.$$

- Positive-definiteness:

$$\langle v, v \rangle \geq 0 \text{ with equality only for } v = 0.$$

We can define the length of a vector v in V by

$$\| v \|^2 = \langle v, v \rangle,$$

and we can prove the Cauchy–Schwarz inequality:

$$| \langle u, v \rangle | \leq \| u \| \cdot \| v \|.$$

In particular, the quantity

$$\frac{| \langle u, v \rangle |}{\| u \| \cdot \| v \|} \leq 1,$$

and so we can call this quantity the cosine of the angle between the two vectors.

Two vectors are orthogonal if $\langle u, v \rangle = 0$. An orthonormal basis is a basis where all basis vectors have length 1 and are orthogonal to each other. Given any finite-dimensional vector space, an orthonormal basis could be found by the Gram–Schmidt procedure. Orthonormal bases are particularly nice to deal with, since if $v = a_1 v_1 + \ldots + a_n v_n$, then $a_i = \langle v, v_i \rangle$.

The inner product facilitates the construction of many useful concepts. For instance, given a transform T, we can define its Hermitian conjugate T^* as the linear transform satisfying

$$\langle Tu, v \rangle = \langle u, T^* v \rangle.$$

If T satisfies $TT^* = T^*T$, we call T normal. It turns out that normal matrices are precisely the matrices that have an orthonormal system of eigenvectors that span V.

Some Main Useful Theorems

- A matrix is invertible, or non-singular, if and only if the linear map represented by the matrix is an isomorphism.

- Any vector space over a field F of dimension n is isomorphic to F^n as a vector space over F.

- Corollary: Any two vector spaces over F of the same finite dimension are isomorphic to each other.

- A linear map is an isomorphism if and only if the determinant is nonzero.

Applications

Because of the ubiquity of vector spaces, linear algebra is used in many fields of mathematics, natural sciences, computer science, and social science. Below are just some examples of applications of linear algebra.

Solution of Linear Systems

Linear algebra provides the formal setting for the linear combination of equations used in the Gaussian method. Suppose the goal is to find and describe the solution(s), if any, of the following system of linear equations:

$$2x + y - z = 8 \quad (L_1)$$
$$-3x - y + 2z = -11 \quad (L_2)$$
$$-2x + y + 2z = -3 \quad (L_3)$$

The Gaussian-elimination algorithm is as follows: eliminate x from all equations below L_1, and then eliminate y from all equations below L_2. This will put the system into triangular form. Then, using back-substitution, each unknown can be solved for.

In the example, x is eliminated from L_2 by adding $(3/2)L_1$ to L_2. x is then eliminated from L_3 by adding L_1 to L_3. Formally:

$$L_2 + \tfrac{3}{2}L_1 \rightarrow L_2$$

$$L_3 + L_1 \rightarrow L_3$$

The result is:

$$2x + y - z = 8$$
$$\frac{1}{2}y + \frac{1}{2}z = 1$$
$$2y + z = 5$$

Now y is eliminated from L_3 by adding $-4L_2$ to L_3:

$$L_3 + -4L_2 \rightarrow L_3$$

The result is:

$$2x + y - z = 8$$
$$\frac{1}{2}y + \frac{1}{2}z = 1$$
$$-z = 1$$

This result is a system of linear equations in triangular form, and so the first part of the algorithm is complete.

The last part, back-substitution, consists of solving for the known in reverse order. It can thus be seen that

$$z = -1 \quad (L_3)$$

Then, z can be substituted into L_2, which can then be solved to obtain

$$y = 3 \quad (L_2)$$

Next, z and y can be substituted into L_1, which can be solved to obtain

$$x = 2 \quad (L_1)$$

The system is solved.

We can, in general, write any system of linear equations as a matrix equation:

$$Ax = b.$$

The solution of this system is characterized as follows: first, we find a particular solution x_0 of this equation using Gaussian elimination. Then, we compute the solutions of $Ax = 0$; that is, we find the null space N of A. The solution set of this equation is given by $x_0 + N = \{x_0 + n : n \in N\}$. If the number of variables is equal to the number of equations, then we can characterize when the system has a unique solution: since N is trivial if and only if $\det A \neq 0$, the equation has a unique solution if and only if $\det A \neq 0$.

Least-squares Best Fit Line

The least squares method is used to determine the best fit line for a set of data. This line will minimize the sum of the squares of the residuals.

Fourier Series Expansion

Fourier series are a representation of a function $f: [-\pi, \pi] \to \mathbb{R}$ as a trigonometric series:

$$f(x) = \frac{a_0}{2} + \sum_{n=1}^{\infty} [a_n \cos(nx) + b_n \sin(nx)].$$

This series expansion is extremely useful in solving partial differential equations. In this section, we will not be concerned with convergence issues; it is nice to note that all Lipschitz-continuous functions have a converging Fourier series expansion, and nice enough discontinuous functions have a Fourier series that converges to the function value at most points.

The space of all functions that can be represented by a Fourier series form a vector space (technically speaking, we call functions that have the same Fourier series expansion the "same" function, since two different discontinuous functions might have the same Fourier series). Moreover, this space is also an inner product space with the inner product

$$\langle f, g \rangle = \frac{1}{\pi} \int_{-\pi}^{\pi} f(x)g(x)dx.$$

The functions $g_n(x) = \sin(nx)$ for $n > 0$ and $h_n(x) = \cos(nx)$ for $n \geq 0$ are an orthonormal basis for the space of Fourier-expandable functions. We can thus use the tools of linear algebra to find the expansion of any function in this space in terms of these basis functions. For instance, to find the coefficient a_k, we take the inner product with h_k:

$$\langle f, h_k \rangle = \frac{a_0}{2} \langle h_0, h_k \rangle + \sum_{n=1}^{\infty} [a_n \langle h_n, h_k \rangle + b_n \langle g_n, h_k \rangle],$$

and by orthonormality, $\langle f, h_k \rangle = a_k$; that is,

$$a_k = \frac{1}{\pi} \int_{-\pi}^{\pi} f(x) \cos(kx) dx.$$

Quantum Mechanics

Quantum mechanics is highly inspired by notions in linear algebra. In quantum mechanics, the physical state of a particle is represented by a vector, and observables (such as momentum, energy, and angular momentum) are represented by linear operators on the underlying vector space. More concretely, the wave function of a particle describes its physical state and lies in the vector space

L^2 (the functions $\varphi \colon \mathbb{R}^3 \to \mathbb{C}$ such that $\int_{-\infty}^{\infty} \int_{-\infty}^{\infty} \int_{-\infty}^{\infty} |\phi|^2 \, dx dy dz$ is finite), and it evolves according to the Schrödinger equation. Energy is represented as the operator $H = -\frac{\hbar^2}{2m} \nabla^2 + V(x, y, z)$, where V is the potential energy. H is also known as the Hamiltonian operator. The eigenvalues of H represents the possible energies that can be observed. Given a particle in some state φ, we can expand φ into a linear combination of eigenstates of H. The component of H in each eigenstate determines the probability of measuring the corresponding eigenvalue, and the measurement forces the particle to assume that eigenstate (wave function collapse).

Geometric Introduction

Many of the principles and techniques of linear algebra can be seen in the geometry of lines in a real two dimensional plane E. When formulated using vectors and matrices the geometry of points and lines in the plane can be extended to the geometry of points and hyperplanes in high-dimensional spaces.

Point coordinates in the plane E are ordered pairs of real numbers, (x,y), and a line is defined as the set of points (x,y) that satisfy the linear equation

$$\lambda : ax + by + c = 0,,$$

where a, b and c are not all zero. Then,

$$\lambda : \begin{bmatrix} a & b & c \end{bmatrix} \begin{Bmatrix} x \\ y \\ 1 \end{Bmatrix} = 0,$$

or

$$A\mathbf{x} = 0,$$

where x = $(x, y, 1)$ is the 3×1 set of homogeneous coordinates associated with the point (x, y).

Homogeneous coordinates identify the plane E with the $z = 1$ plane in three dimensional space. The x–y coordinates in E are obtained from homogeneous coordinates y = (y_1, y_2, y_3) by dividing by the third component (if it is nonzero) to obtain y = $(y_1/y_3, y_2/y_3, 1)$.

The linear equation, λ, has the important property, that if x_1 and x_2 are homogeneous coordinates of points on the line, then the point $\alpha x_1 + \beta x_2$ is also on the line, for any real α and β.

Now consider the equations of the two lines λ_1 and λ_2,

$$\lambda_1 : a_1 x + b_1 y + c_1 = 0, \quad \lambda_2 : a_2 x + b_2 y + c_2 = 0,$$

which forms a system of linear equations. The intersection of these two lines is defined by x = $(x, y, 1)$ that satisfy the matrix equation,

$$\lambda_{1,2} : \begin{bmatrix} a_1 & b_1 & c_1 \\ a_2 & b_2 & c_2 \end{bmatrix} \begin{Bmatrix} x \\ y \\ 1 \end{Bmatrix} = \begin{Bmatrix} 0 \\ 0 \end{Bmatrix},$$

or using homogeneous coordinates,

$$B\mathbf{x} = 0.$$

The point of intersection of these two lines is the unique non-zero solution of these equations. In homogeneous coordinates, the solutions are multiples of the following solution:

$$x_1 = \begin{vmatrix} b_1 & c_1 \\ b_2 & c_2 \end{vmatrix}, x_2 = -\begin{vmatrix} a_1 & c_1 \\ a_2 & c_2 \end{vmatrix}, x_3 = \begin{vmatrix} a_1 & b_1 \\ a_2 & b_2 \end{vmatrix}$$

if the rows of B are linearly independent (i.e., λ_1 and λ_2 represent distinct lines). Divide through by x_3 to get Cramer's rule for the solution of a set of two linear equations in two unknowns. Notice that this yields a point in the $z = 1$ plane only when the 2×2 submatrix associated with x_3 has a non-zero determinant.

It is interesting to consider the case of three lines, λ_1, λ_2 and λ_3, which yield the matrix equation,

$$\lambda_{1,2,3} : \begin{bmatrix} a_1 & b_1 & c_1 \\ a_2 & b_2 & c_2 \\ a_3 & b_3 & c_3 \end{bmatrix} \begin{bmatrix} x \\ y \\ 1 \end{bmatrix} = \begin{Bmatrix} 0 \\ 0 \\ 0 \end{Bmatrix}.$$

which in homogeneous form yields,

$$C\mathbf{x} = 0.$$

Clearly, this equation has the solution x = $(0,0,0)$, which is not a point on the $z = 1$ plane E. For a solution to exist in the plane E, the coefficient matrix C must have rank 2, which means its determinant must be zero. Another way to say this is that the columns of the matrix must be linearly dependent.

Introduction to Linear Transformations

Another way to approach linear algebra is to consider linear functions on the two dimensional real plane $E=R^2$. Here R denotes the set of real numbers. Let x=(x, y) be an arbitrary vector in E and consider the linear function $\lambda: E \rightarrow R$, given by

$$\lambda : \begin{bmatrix} a & b \end{bmatrix} \begin{Bmatrix} x \\ y \end{Bmatrix} = c,$$

or

$$A\mathbf{x} = c.$$

This transformation has the important property that if Ay=d, then

$$A(\alpha \mathbf{x} + \beta \mathbf{y}) = \alpha A\mathbf{x} + \beta A\mathbf{y} = \alpha c + \beta d.$$

This shows that the sum of vectors in E map to the sum of their images in R. This is the defining characteristic of a linear map, or linear transformation. For this case, where the image space is a real number the map is called a linear functional.

Consider the linear functional a little more carefully. Let i=(1,0) and j =(0,1) be the natural basis vectors on E, so that x=xi+yj. It is now possible to see that

$$A\mathbf{x} = A(x\mathbf{i} + y\mathbf{j}) = xA\mathbf{i} + yA\mathbf{j} = \begin{bmatrix} A\mathbf{i} & A\mathbf{j} \end{bmatrix} \begin{Bmatrix} x \\ y \end{Bmatrix} = \begin{bmatrix} a & b \end{bmatrix} \begin{Bmatrix} x \\ y \end{Bmatrix} = c.$$

Thus, the columns of the matrix A are the image of the basis vectors of E in R.

This is true for any pair of vectors used to define coordinates in E. Suppose we select a non-orthogonal non-unit vector basis v and w to define coordinates of vectors in E. This means a vector x has coordinates (α,β), such that x=αv+βw. Then, we have the linear functional

$$\lambda : A\mathbf{x} = \begin{bmatrix} A\mathbf{v} & A\mathbf{w} \end{bmatrix} \begin{Bmatrix} \alpha \\ \beta \end{Bmatrix} = \begin{bmatrix} d & e \end{bmatrix} \begin{Bmatrix} \alpha \\ \beta \end{Bmatrix} = c,$$

where Av=d and Aw=e are the images of the basis vectors v and w. This is written in matrix form as

$$\begin{bmatrix} a & b \end{bmatrix} \begin{bmatrix} v_1 & w_1 \\ v_2 & w_2 \end{bmatrix} = \begin{bmatrix} d & e \end{bmatrix}.$$

Coordinates Relative to a Basis

This leads to the question of how to determine the coordinates of a vector x relative to a general basis v and w in E. Assume that we know the coordinates of the vectors, x, v and w in the natural basis i=(1,0) and j =(0,1). Our goal is two find the real numbers α, β, so that x=αv+βw, that is

$$\begin{Bmatrix} x \\ y \end{Bmatrix} = \begin{bmatrix} v_1 & w_1 \\ v_2 & w_2 \end{bmatrix} \begin{Bmatrix} \alpha \\ \beta \end{Bmatrix}.$$

To solve this equation for α, β, we compute the linear coordinate functionals σ and τ for the basis v, w, which are given by,

$$\sigma = \begin{bmatrix} \sigma_1 & \sigma_2 \end{bmatrix} = \frac{1}{v_1 w_2 - v_2 w_1} \begin{bmatrix} w_2 & -w_1 \end{bmatrix}, \tau = \begin{bmatrix} \tau_1 & \tau_2 \end{bmatrix} = \frac{1}{v_1 w_2 - v_2 w_1} \begin{bmatrix} -v_2 & v_1 \end{bmatrix},$$

The functionals σ and τ compute the components of x along the basis vectors v and w, respectively, that is,

$$\sigma \mathbf{x} = \alpha, \tau \mathbf{x} = \beta,$$

which can be written in matrix form as

$$\begin{bmatrix} \sigma_1 & \sigma_2 \\ \tau_1 & \tau_2 \end{bmatrix} \begin{Bmatrix} x \\ y \end{Bmatrix} = \begin{Bmatrix} \alpha \\ \beta \end{Bmatrix}.$$

These coordinate functionals have the properties,

$$\sigma \mathbf{v} = 1, \sigma \mathbf{w} = 0, \tau \mathbf{w} = 1, \tau \mathbf{v} = 0.$$

These equations can be assembled into the single matrix equation,

$$\begin{bmatrix} \sigma_1 & \sigma_2 \\ \tau_1 & \tau_2 \end{bmatrix} \begin{bmatrix} v_1 & w_1 \\ v_2 & w_2 \end{bmatrix} = \begin{bmatrix} 1 & 0 \\ 0 & 1 \end{bmatrix}.$$

Thus, the matrix formed by the coordinate linear functionals is the inverse of the matrix formed by the basis vectors.

Inverse Image

The set of points in the plane E that map to the same image in R under the linear functional λ define a line in E. This line is the image of the inverse map, $\lambda^{-1}: R \rightarrow E$. This inverse image is the set of the points x=(x, y) that solve the equation,

$$A\mathbf{x} = \begin{bmatrix} a & b \end{bmatrix} \begin{Bmatrix} x \\ y \end{Bmatrix} = c.$$

Notice that a linear functional operates on known values for x=(x, y) to compute a value c in R, while the inverse image seeks the values for x=(x, y) that yield a specific value c.

In order to solve the equation, we first recognize that only one of the two unknowns (x,y) can be determined, so we select y to be determined, and rearrange the equation

$$by = c - ax.$$

Solve for y and obtain the inverse image as the set of points,

$$\mathbf{x}(t) = \begin{Bmatrix} 0 \\ c/b \end{Bmatrix} + t \begin{Bmatrix} 1 \\ -a/b \end{Bmatrix} = \mathbf{p} + t\mathbf{h}.$$

For convenience the free parameter x has been relabeled t.

The vector p defines the intersection of the line with the y-axis, known as the y-intercept. The vector h satisfies the homogeneous equation,

$$A\mathbf{h} = \begin{bmatrix} a & b \end{bmatrix} \begin{Bmatrix} 1 \\ -a/b \end{Bmatrix} = 0.$$

Notice that if h is a solution to this homogeneous equation, then t h is also a solution.

The set of points of a linear functional that map to zero define the *kernel* of the linear functional. The line can be considered to be the set of points h in the kernel translated by the vector p.

Generalizations and Related Topics

Since linear algebra is a successful theory, its methods have been developed and generalized in other parts of mathematics. In module theory, one replaces the field of scalars by a ring. The concepts of linear independence, span, basis, and dimension (which is called rank in module theory) still make sense. Nevertheless, many theorems from linear algebra become false in module theory. For instance, not all modules have a basis (those that do are called free modules), the rank of a free module is not necessarily unique, not every linearly independent subset of a module can be extended to form a basis, and not every subset of a module that spans the space contains a basis.

In multilinear algebra, one considers multivariable linear transformations, that is, mappings that are linear in each of a number of different variables. This line of inquiry naturally leads to the idea of the dual space, the vector space V^* consisting of linear maps $f: V \to F$ where F is the field of scalars. Multilinear maps $T: V^n \to F$ can be described via tensor products of elements of V^*.

If, in addition to vector addition and scalar multiplication, there is a bilinear vector product $V \times V \to V$, the vector space is called an algebra; for instance, associative algebras are algebras with an associate vector product (like the algebra of square matrices, or the algebra of polynomials).

Functional analysis mixes the methods of linear algebra with those of mathematical analysis and studies various function spaces, such as L^p spaces.

Representation theory studies the actions of algebraic objects on vector spaces by representing these objects as matrices. It is interested in all the ways that this is possible, and it does so by finding subspaces invariant under all transformations of the algebra. The concept of eigenvalues and eigenvectors is especially important.

Algebraic geometry considers the solutions of systems of polynomial equations.

There are several related topics in the field of Computer Programming that utilizes much of the techniques and theorems Linear Algebra encompasses and refers to.

System of Linear Equations

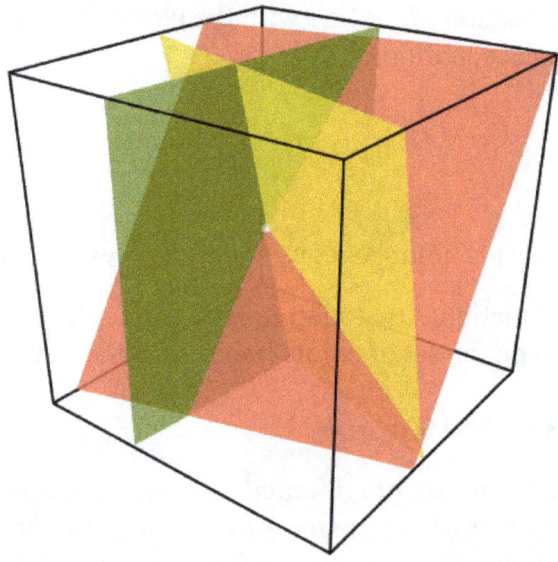

A linear system in three variables determines a collection of planes. The intersection point is the solution.

In mathematics, a system of linear equations (or linear system) is a collection of two or more linear equations involving the same set of variables. For example,

$$3x + 2y - z = 1$$
$$2x - 2y + 4z = -2$$
$$-x + \tfrac{1}{2}y - z = 0$$

is a system of three equations in the three variables x, y, z. A solution to a linear system is an assignment of values to the variables such that all the equations are simultaneously satisfied. A solution to the system above is given by

$$x = 1$$
$$y = -2$$
$$z = -2$$

since it makes all three equations valid. The word "*system*" indicates that the equations are to be considered collectively, rather than individually.

In mathematics, the theory of linear systems is the basis and a fundamental part of linear algebra, a subject which is used in most parts of modern mathematics. Computational algorithms for finding the solutions are an important part of numerical linear algebra, and play a prominent role in engineering, physics, chemistry, computer science, and economics. A system of non-linear equations can often be approximated by a linear system, a helpful technique when making a mathematical model or computer simulation of a relatively complex system.

Very often, the coefficients of the equations are real or complex numbers and the solutions are searched in the same set of numbers, but the theory and the algorithms apply for coefficients and solutions in any field. For solutions in an integral domain like the ring of the integers, or in other algebraic structures, other theories have been developed. Integer linear programming is a collection of methods for finding the "best" integer solu-tion (when there are many). Gröbner basis theory provides algorithms when coefficients and unknowns are polynomials. Also tropical geometry is an example of linear algebra in a more exotic structure.

Elementary Example

The simplest kind of linear system involves two equations and two variables:

$$2x + 3y = 6$$
$$4x + 9y = 15$$

One method for solving such a system is as follows. First, solve the top equation for x in terms of y:

$$x = 3 - \frac{3}{2}y.$$

Now substitute this expression for x into the bottom equation:

$$4\left(3 - \frac{3}{2}y\right) + 9y = 15.$$

This results in a single equation involving only the variable y. Solving gives $y = 1$,, and substituting this back into the equation for x yields $x = 3/2$. This method generalizes to systems with additional variables.

General Form

A general system of m linear equations with n unknowns can be written as

$$a_{11}x_1 + a_{12}x_2 + \cdots + a_{1n}x_n = b_1$$
$$a_{21}x_1 + a_{22}x_2 + \cdots + a_{2n}x_n = b_2$$
$$a_{m1}x_1 + a_{m2}x_2 + \cdots + a_{mn}x_n = b_m.$$

Here $x_1, x_2, \ldots, x_n$ are the unknowns, $a_{11}, a_{12}, \ldots, a_{mn}$ are the coefficients of the system, and $b_1, b_2, \ldots, b_m$ are the constant terms.

Often the coefficients and unknowns are real or complex numbers, but integers and rational numbers are also seen, as are polynomials and elements of an abstract algebraic structure.

Vector Equation

One extremely helpful view is that each unknown is a weight for a column vector in a linear combination.

$$x_1 \begin{bmatrix} a_{11} \\ a_{21} \\ \vdots \\ a_{m1} \end{bmatrix} + x_2 \begin{bmatrix} a_{12} \\ a_{22} \\ \vdots \\ a_{m2} \end{bmatrix} + \cdots + x_n \begin{bmatrix} a_{1n} \\ a_{2n} \\ \vdots \\ a_{mn} \end{bmatrix} = \begin{bmatrix} b_1 \\ b_2 \\ \vdots \\ b_m \end{bmatrix}$$

This allows all the language and theory of *vector spaces* (or more generally, *modules*) to be brought to bear. For example, the collection of all possible linear combinations of the vectors on the left-hand side is called their *span*, and the equations have a solution just when the right-hand vector is within that span. If every vector within that span has exactly one expression as a linear combination of the given left-hand vectors, then any solution is unique. In any event, the span has a *basis* of linearly independent vectors that do guarantee exactly one expression; and the number of vectors in that basis (its *dimension*) cannot be larger than m or n, but it can be smaller. This is important because if we have m independent vectors a solution is guaranteed regardless of the right-hand side, and otherwise not guaranteed.

Matrix Equation

The vector equation is equivalent to a matrix equation of the form

$$A\mathbf{x} = \mathbf{b}$$

where A is an $m \times n$ matrix, x is a column vector with n entries, and b is a column vector with m entries.

$$A = \begin{bmatrix} a_{11} & a_{12} & \cdots & a_{1n} \\ a_{21} & a_{22} & \cdots & a_{2n} \\ \vdots & \vdots & \ddots & \vdots \\ a_{m1} & a_{m2} & \cdots & a_{mn} \end{bmatrix}, \quad \mathbf{x} = \begin{bmatrix} x_1 \\ x_2 \\ \vdots \\ x_n \end{bmatrix}, \quad \mathbf{b} = \begin{bmatrix} b_1 \\ b_2 \\ \vdots \\ b_m \end{bmatrix}$$

The number of vectors in a basis for the span is now expressed as the *rank* of the matrix.

Solution Set

A solution of a linear system is an assignment of values to the variables $x_1, x_2, ..., x_n$ such that each of the equations is satisfied. The set of all possible solutions is called the solution set.

A linear system may behave in any one of three possible ways:

1. The system has *infinitely many solutions*.

2. The system has a single *unique solution*.

3. The system has *no solution*.

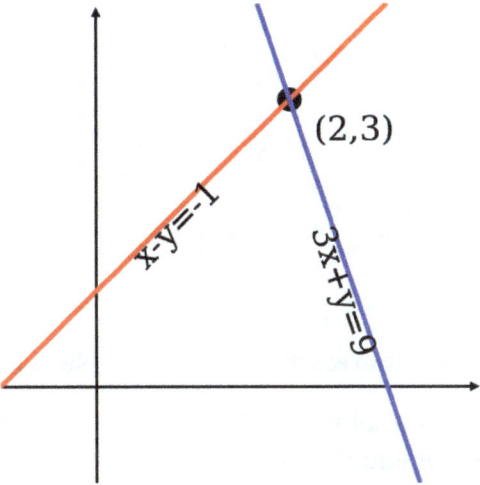

The solution set for the equations $x - y = -1$ and $3x + y = 9$ is the single point $(2, 3)$.

Geometric Interpretation

For a system involving two variables (x and y), each linear equation determines a line on the xy-plane. Because a solution to a linear system must satisfy all of the equations, the solution set is the intersection of these lines, and is hence either a line, a single point, or the empty set.

For three variables, each linear equation determines a plane in three-dimensional space, and the solution set is the intersection of these planes. Thus the solution set may be a plane, a line, a single point, or the empty set. For example, as three parallel planes do not have a common point, the solution set of their equations is empty; the solution set of the equations of three planes intersecting at a point is single point; if three planes pass through two points, their equations have at least two common solutions; in fact the solution set is infinite and consists in all the line passing through these points.

For n variables, each linear equation determines a hyperplane in n-dimensional space. The solution set is the intersection of these hyperplanes, which may be a flat of any dimension.

General Behavior

In general, the behavior of a linear system is determined by the relationship between the number of equations and the number of unknowns:

*   Usually, a system with fewer equations than unknowns has infinitely many solutions, but it may have no solution. Such a system is known as an underdetermined system.

*   Usually, a system with the same number of equations and unknowns has a single unique solution.

*   Usually, a system with more equations than unknowns has no solution. Such a system is also known as an overdetermined system.

The solution set for two equations in three variables is usually a line.

In the first case, the dimension of the solution set is usually equal to $n - m$, where n is the number of variables and m is the number of equations.

The following pictures illustrate this trichotomy in the case of two variables:

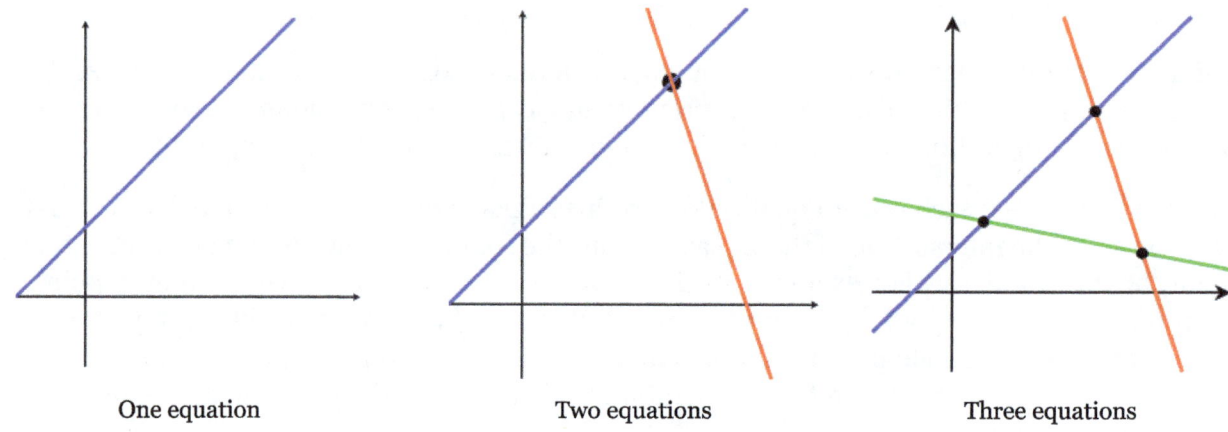

| One equation | Two equations | Three equations |

The first system has infinitely many solutions, namely all of the points on the blue line. The second system has a single unique solution, namely the intersection of the two lines. The third system has no solutions, since the three lines share no common point.

Keep in mind that the pictures above show only the most common case. It is possible for a system of two equations and two unknowns to have no solution (if the two lines are parallel), or for a system of three equations and two unknowns to be solvable (if the three lines intersect at a single point). In general, a system of linear equations may behave differently from expected if the equations are linearly dependent, or if two or more of the equations are inconsistent.

Properties

Independence

The equations of a linear system are independent if none of the equations can be derived algebraically from the others. When the equations are independent, each equation contains new information about the variables, and removing any of the equations increases the size of the solution set. For linear equations, logical independence is the same as linear independence.

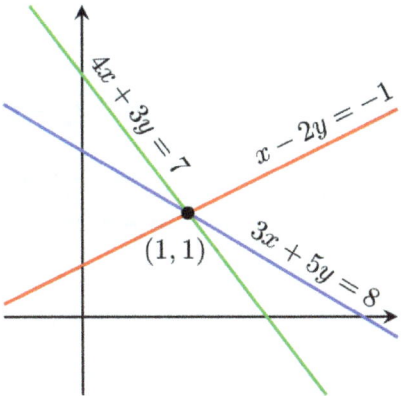

The equations $x - 2y = -1$, $3x + 5y = 8$, and $4x + 3y = 7$ are linearly dependent.

For example, the equations

$$3x + 2y = 6 \quad \text{and} \quad 6x + 4y = 12$$

are not independent — they are the same equation when scaled by a factor of two, and they would produce identical graphs. This is an example of equivalence in a system of linear equations.

For a more complicated example, the equations

$$\begin{aligned} x - 2y &= -1 \\ 3x + 5y &= 8 \\ 4x + 3y &= 7 \end{aligned}$$

are not independent, because the third equation is the sum of the other two. Indeed, any one of these equations can be derived from the other two, and any one of the equations can be removed without affecting the solution set. The graphs of these equations are three lines that intersect at a single point.

Consistency

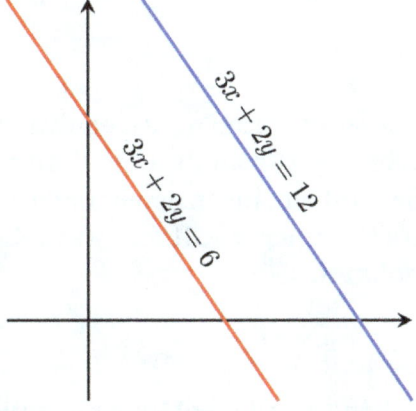

The equations $3x + 2y = 6$ and $3x + 2y = 12$ are inconsistent.

A linear system is inconsistent if it has no solution, and otherwise it is said to be consistent. When the system is inconsistent, it is possible to derive a contradiction from the equations, that may always be rewritten as the statement 0 = 1.

For example, the equations

$$3x + 2y = 6 \quad \text{and} \quad 3x + 2y = 12$$

are inconsistent. In fact, by subtracting the first equation from the second one and multiplying both sides of the result by 1/6, we get 0 = 1. The graphs of these equations on the xy-plane are a pair of parallel lines.

It is possible for three linear equations to be inconsistent, even though any two of them are consistent together. For example, the equations

$$x + y = 1$$
$$2x + y = 1$$
$$3x + 2y = 3$$

are inconsistent. Adding the first two equations together gives $3x + 2y = 2$, which can be subtracted from the third equation to yield 0 = 1. Note that any two of these equations have a common solution. The same phenomenon can occur for any number of equations.

In general, inconsistencies occur if the left-hand sides of the equations in a system are linearly dependent, and the constant terms do not satisfy the dependence relation. A system of equations whose left-hand sides are linearly independent is always consistent.

Putting it another way, according to the Rouché–Capelli theorem, any system of equations (over-determined or otherwise) is inconsistent if the rank of the augmented matrix is greater than the rank of the coefficient matrix. If, on the other hand, the ranks of these two matrices are equal, the system must have at least one solution. The solution is unique if and only if the rank equals the number of variables. Otherwise the general solution has k free parameters where k is the difference between the number of variables and the rank; hence in such a case there are an infinitude of solutions. The rank of a system of equations can never be higher than [the number of variables] + 1, which means that a system with any number of equations can always be reduced to a system that has a number of independent equations that is at most equal to [the number of variables] + 1.

Equivalence

Two linear systems using the same set of variables are equivalent if each of the equations in the second system can be derived algebraically from the equations in the first system, and vice versa. Two systems are equivalent if either both are inconsistent or each equation of each of them is a linear combination of the equations of the other one. It follows that two linear systems are equivalent if and only if they have the same solution set.

Solving a Linear System

There are several algorithms for solving a system of linear equations.

Describing the Solution

When the solution set is finite, it is reduced to a single element. In this case, the unique solution is described by a sequence of equations whose left-hand sides are the names of the unknowns and right-hand sides are the corresponding values, for example $(x = 3, y = -2, z = 6)$. When an order on the unknowns has been fixed, for example the alphabetical order the solution may be described as a vector of values, like $(3, -2, 6)$ for the previous example.

It can be difficult to describe a set with infinite solutions. Typically, some of the variables are designated as free (or independent, or as parameters), meaning that they are allowed to take any value, while the remaining variables are dependent on the values of the free variables.

For example, consider the following system:

$$x + 3y - 2z = 5$$
$$3x + 5y + 6z = 7$$

The solution set to this system can be described by the following equations:

$$x = -7z - 1 \quad \text{and} \quad y = 3z + 2.$$

Here z is the free variable, while x and y are dependent on z. Any point in the solution set can be obtained by first choosing a value for z, and then computing the corresponding values for x and y.

Each free variable gives the solution space one degree of freedom, the number of which is equal to the dimension of the solution set. For example, the solution set for the above equation is a line, since a point in the solution set can be chosen by specifying the value of the parameter z. An infinite solution of higher order may describe a plane, or higher-dimensional set.

Different choices for the free variables may lead to different descriptions of the same solution set. For example, the solution to the above equations can alternatively be described as follows:

$$y = -\frac{3}{7}x + \frac{11}{7} \quad \text{and} \quad z = -\frac{1}{7}x - \frac{1}{7}.$$

Here x is the free variable, and y and z are dependent.

Elimination of Variables

The simplest method for solving a system of linear equations is to repeatedly eliminate variables. This method can be described as follows:

1. In the first equation, solve for one of the variables in terms of the others.

2. Substitute this expression into the remaining equations. This yields a system of equations with one fewer equation and one fewer unknown.

3. Continue until you have reduced the system to a single linear equation.

4. Solve this equation, and then back-substitute until the entire solution is found.

For example, consider the following system:

$$x + 3y - 2z = 5$$
$$3x + 5y + 6z = 7$$
$$2x + 4y + 3z = 8$$

Solving the first equation for x gives $x = 5 + 2z - 3y$, and plugging this into the second and third equation yields

$$-4y + 12z = -8$$
$$-2y + 7z = -2$$

Solving the first of these equations for y yields $y = 2 + 3z$, and plugging this into the second equation yields $z = 2$. We now have:

$$x = 5 + 2z - 3y$$
$$y = 2 + 3z$$
$$z = 2$$

Substituting $z = 2$ into the second equation gives $y = 8$, and substituting $z = 2$ and $y = 8$ into the first equation yields $x = -15$. Therefore, the solution set is the single point $(x, y, z) = (-15, 8, 2)$.

Row Reduction

In row reduction, the linear system is represented as an augmented matrix:

$$\begin{bmatrix} 1 & 3 & -2 & \bigm| & 5 \\ 3 & 5 & 6 & \bigm| & 7 \\ 2 & 4 & 3 & \bigm| & 8 \end{bmatrix}.$$

This matrix is then modified using elementary row operations until it reaches reduced row echelon form. There are three types of elementary row operations:

Type 1: Swap the positions of two rows.

Type 2: Multiply a row by a nonzero scalar.

Type 3: Add to one row a scalar multiple of another.

Because these operations are reversible, the augmented matrix produced always represents a linear system that is equivalent to the original.

There are several specific algorithms to row-reduce an augmented matrix, the simplest of which are Gaussian elimination and Gauss-Jordan elimination. The following computation shows Gauss-Jordan elimination applied to the matrix above:

$$\begin{bmatrix} 1 & 3 & -2 & 5 \\ 3 & 5 & 6 & 7 \\ 2 & 4 & 3 & 8 \end{bmatrix} \sim$$

$$\begin{bmatrix} 1 & 3 & -2 & 5 \\ 0 & -4 & 12 & -8 \\ 2 & 4 & 3 & 8 \end{bmatrix} \sim$$

$$\begin{bmatrix} 1 & 3 & -2 & 5 \\ 0 & -4 & 12 & -8 \\ 0 & -2 & 7 & -2 \end{bmatrix} \sim$$

$$\begin{bmatrix} 1 & 3 & -2 & 5 \\ 0 & 1 & -3 & 2 \\ 0 & -2 & 7 & -2 \end{bmatrix} \sim$$

$$\begin{bmatrix} 1 & 3 & -2 & 5 \\ 0 & 1 & -3 & 2 \\ 0 & 0 & 1 & 2 \end{bmatrix} \sim$$

$$\begin{bmatrix} 1 & 3 & -2 & 5 \\ 0 & 1 & 0 & 8 \\ 0 & 0 & 1 & 2 \end{bmatrix} \sim$$

$$\begin{bmatrix} 1 & 3 & 0 & 9 \\ 0 & 1 & 0 & 8 \\ 0 & 0 & 1 & 2 \end{bmatrix} \sim$$

$$\begin{bmatrix} 1 & 0 & 0 & -15 \\ 0 & 1 & 0 & 8 \\ 0 & 0 & 1 & 2 \end{bmatrix}.$$

The last matrix is in reduced row echelon form, and represents the system $x = -15$, $y = 8$, $z = 2$. A comparison with the example in the previous section on the algebraic elimination of variables shows that these two methods are in fact the same; the difference lies in how the computations are written down.

Cramer's Rule

Cramer's rule is an explicit formula for the solution of a system of linear equations, with each variable given by a quotient of two determinants. For example, the solution to the system

$$x + 3y - 2z = 5$$
$$3x + 5y + 6z = 7$$
$$2x + 4y + 3z = 8$$

is given by

$$
x = \frac{\begin{vmatrix} 5 & 3 & -2 \\ 7 & 5 & 6 \\ 8 & 4 & 3 \end{vmatrix}}{\begin{vmatrix} 1 & 3 & -2 \\ 3 & 5 & 6 \\ 2 & 4 & 3 \end{vmatrix}}, \quad y = \frac{\begin{vmatrix} 1 & 5 & -2 \\ 3 & 7 & 6 \\ 2 & 8 & 3 \end{vmatrix}}{\begin{vmatrix} 1 & 3 & -2 \\ 3 & 5 & 6 \\ 2 & 4 & 3 \end{vmatrix}}, \quad z = \frac{\begin{vmatrix} 1 & 3 & 5 \\ 3 & 5 & 7 \\ 2 & 4 & 8 \end{vmatrix}}{\begin{vmatrix} 1 & 3 & -2 \\ 3 & 5 & 6 \\ 2 & 4 & 3 \end{vmatrix}}.
$$

For each variable, the denominator is the determinant of the matrix of coefficients, while the numerator is the determinant of a matrix in which one column has been replaced by the vector of constant terms.

Though Cramer's rule is important theoretically, it has little practical value for large matrices, since the computation of large determinants is somewhat cumbersome. (Indeed, large determinants are most easily computed using row reduction.) Further, Cramer's rule has very poor numerical properties, making it unsuitable for solving even small systems reliably, unless the operations are performed in rational arithmetic with unbounded precision.

Matrix Solution

If the equation system is expressed in the matrix form $A\mathbf{x} = \mathbf{b}$, , the entire solution set can also be expressed in matrix form. If the matrix A is square (has m rows and $n=m$ columns) and has full rank (all m rows are independent), then the system has a unique solution given by

$$\mathbf{x} = A^{-1}\mathbf{b}$$

where A^{-1} is the inverse of A. More generally, regardless of whether $m=n$ or not and regardless of the rank of A, all solutions (if any exist) are given using the Moore-Penrose pseudoinverse of A, denoted A^g , as follows:

$$\mathbf{x} = A^g\mathbf{b} + (I - A^g A)\mathbf{w}$$

where $\mathbf{w}$ is a vector of free parameters that ranges over all possible $n\times 1$ vectors. A necessary and sufficient condition for any solution(s) to exist is that the potential solution obtained using $\mathbf{w} = \mathbf{0}$ satisfy $A\mathbf{x} = \mathbf{b}$ — that is, that $AA^g\mathbf{b} = \mathbf{b}$. If this condition does not hold, the equation system is inconsistent and has no solution. If the condition holds, the system is consistent and at least one solution exists. For example, in the above-mentioned case in which A is square and of full rank, A^g simply equals A^{-1} and the general solution equation simplifies to $\mathbf{x} = A^{-1}\mathbf{b} + (I - A^{-1}A)\mathbf{w} = A^{-1}\mathbf{b} + (I - I)\mathbf{w} = A^{-1}\mathbf{b}$ as previously stated, where $\mathbf{w}$ has completely dropped out of the solution, leaving only a single solution. In other cases, though, $\mathbf{w}$ remains and hence an infinitude of potential values of the free parameter vector $\mathbf{w}$ give an infinitude of solutions of the equation.

Other Methods

While systems of three or four equations can be readily solved by hand, computers are often used for larger systems. The standard algorithm for solving a system of linear equations is based on Gaussian elimination with some modifications. Firstly, it is essential to avoid division by small numbers, which may lead to inaccurate results. This can be done by reordering the equations if necessary, a process known as *pivoting*. Secondly, the algorithm does not exactly do Gaussian elimination, but it computes the LU decomposition of the matrix A. This is mostly an organizational tool, but it is much quicker if one has to solve several systems with the same matrix A but different vectors b.

If the matrix A has some special structure, this can be exploited to obtain faster or more accurate algorithms. For instance, systems with a symmetric positive definite matrix can be solved twice as fast with the Cholesky decomposition. Levinson recursion is a fast method for Toeplitz matrices. Special methods exist also for matrices with many zero elements (so-called sparse matrices), which appear often in applications.

A completely different approach is often taken for very large systems, which would otherwise take too much time or memory. The idea is to start with an initial approximation to the solution (which does not have to be accurate at all), and to change this approximation in several steps to bring it closer to the true solution. Once the approximation is sufficiently accurate, this is taken to be the solution to the system. This leads to the class of iterative methods.

Homogeneous Systems

A homogeneous system is equivalent to a matrix equation of the form

$$A\mathbf{x} = \mathbf{0}$$

where A is an $m \times n$ matrix, x is a column vector with n entries, and 0 is the zero vector with m entries.

Solution Set

Every homogeneous system has at least one solution, known as the zero solution (or trivial solution), which is obtained by assigning the value of zero to each of the variables. If the system has a non-singular matrix ($\det(A) \neq 0$) then it is also the only solution. If the system has a singular matrix then there is a solution set with an infinite number of solutions. This solution set has the following additional properties:

1. If u and v are two vectors representing solutions to a homogeneous system, then the vector sum u + v is also a solution to the system.

2. If u is a vector representing a solution to a homogeneous system, and r is any scalar, then ru is also a solution to the system.

These are exactly the properties required for the solution set to be a linear subspace of $\mathbb{R}^n$. In particular, the solution set to a homogeneous system is the same as the null space of the corresponding matrix A. Numerical solutions to a homogeneous system can be found with an SVD decomposition.

Relation to Nonhomogeneous Systems

There is a close relationship between the solutions to a linear system and the solutions to the corresponding homogeneous system:

$$A\mathbf{x} = \mathbf{b} \qquad \text{and} \qquad A\mathbf{x} = \mathbf{0}.$$

Specifically, if p is any specific solution to the linear system $A\mathbf{x} = b$, then the entire solution set can be described as

$$\{\mathbf{p} + \mathbf{v} : \mathbf{v} \text{ is any solution to } A\mathbf{x} = \mathbf{0}\}.$$

Geometrically, this says that the solution set for $A\mathbf{x} = b$ is a translation of the solution set for $A\mathbf{x} = 0$. Specifically, the flat for the first system can be obtained by translating the linear subspace for the homogeneous system by the vector p.

This reasoning only applies if the system $A\mathbf{x} = b$ has at least one solution. This occurs if and only if the vector b lies in the image of the linear transformation A.

Cramer's Rule

In linear algebra, Cramer's rule is an explicit formula for the solution of a system of linear equations with as many equations as unknowns, valid whenever the system has a unique solution. It expresses the solution in terms of the determinants of the (square) coefficient matrix and of matrices obtained from it by replacing one column by the vector of right hand sides of the equations. It is named after Gabriel Cramer (1704–1752), who published the rule for an arbitrary number of unknowns in 1750, although Colin Maclaurin also published special cases of the rule in 1748 (and possibly knew of it as early as 1729).

Cramer's rule is computationally very inefficient for systems of more than two or three equations; its asymptotic complexity is O(n·n!) compared to elimination methods that have polynomial time complexity. Cramer's rule is also numerically unstable even for 2×2 systems.

General Case

Consider a system of n linear equations for n unknowns, represented in matrix multiplication form as follows:

$$Ax = b$$

where the $n \times n$ matrix A has a nonzero determinant, and the vector $x = (x_1, \ldots, x_n)^{\mathrm{T}}$ is the column vector of the variables. Then the theorem states that in this case the system has a unique solution, whose individual values for the unknowns are given by:

$$x_i = \frac{\det(A_i)}{\det(A)} \qquad i = 1, \ldots, n$$

where A_i is the matrix formed by replacing the i-th column of A by the column vector b.

A more general version of Cramer's rule considers the matrix equation

$$AX = B$$

where the $n \times n$ matrix A has a nonzero determinant, and X, B are $n \times m$ matrices. Given sequences $1 \le i_1 < i_2 < \ldots < i_k \le n$ and $1 \le j_1 < j_2 < \ldots < j_k \le m$, let $X_{I,J}$ be the $k \times k$ submatrix of X with rows in $I := (i_1, \ldots, i_k)$ and columns in $J := (j_1, \ldots, j_k)$. Let $A_B(I, J)$ be the $n \times n$ matrix formed by replacing the i_s column of A by the j_s column of B, for all $s = 1, \ldots, k$. Then

$$\det X_{I,J} = \frac{\det(A_B(I,J))}{\det(A)}.$$

In the case $k = 1$, this reduces to the normal Cramer's rule.

The rule holds for systems of equations with coefficients and unknowns in any field, not just in the real numbers. It has recently been shown that Cramer's rule can be implemented in O(n^3) time, which is comparable to more common methods of solving systems of linear equations, such as Gaussian elimination (consistently requiring 2.5 times as many arithmetic operations for all matrix sizes, while exhibiting comparable numeric stability in most cases).

Proof

The proof for Cramer's rule uses just two properties of determinants: linearity with respect to any given column (taking for that column a linear combination of column vectors produces as determinant the corresponding linear combination of their determinants), and the fact that the determinant is zero whenever two columns are equal (which is implied by the basic property that the sign of the determinant flips if you switch two columns).

Fix the index j of a column. Linearity means that if we consider only column j as variable (fixing the others arbitrarily), the resulting function $R^n \to R$ (assuming matrix entries are in R) can be given by a matrix, with one row and n columns, that acts on column j. In fact this is precisely what Laplace expansion does, writing det(A) = $C_1 a_{1j} + \ldots + C_n a_{nj}$ for certain coefficients C_1, ..., C_n that depend on the columns of A other than column j (the precise expression for these cofactors is not important here). The value det(A) is then the result of applying the one-line matrix $L_{(j)}$ = ($C_1 \ C_2 \ \ldots \ C_n$) to column j of A. If $L_{(j)}$ is applied to any *other* column k of A, then the result is the determinant of the matrix obtained from A by replacing column j by a copy of column k, so the resulting determinant is 0 (the case of two equal columns).

Now consider a system of n linear equations in n unknowns $x_1, \ldots, x_n$, whose coefficient matrix is A, with det(A) assumed to be nonzero:

$$
\begin{aligned}
a_{11}x_1 + a_{12}x_2 + \cdots + a_{1n}x_n &= b_1 \\
a_{21}x_1 + a_{22}x_2 + \cdots + a_{2n}x_n &= b_2 \\
&\ \ \vdots \\
a_{n1}x_1 + a_{n2}x_2 + \cdots + a_{nn}x_n &= b_n.
\end{aligned}
$$

If one combines these equations by taking C_1 times the first equation, plus C_2 times the second, and so forth until C_n times the last, then the coefficient of x_j will become $C_1 a_{1,j} + \ldots + C_n a_{n,j} = \det(A)$, while the coefficients of all other unknowns become 0; the left hand side becomes simply $\det(A)x_j$. The right hand side is $C_1 b_1 + \ldots + C_n b_n$, which is $L_{(j)}$ applied to the column vector b of the right hand sides b_i. In fact what has been done here is multiply the matrix equation $Ax = b$ on the left by $L_{(j)}$. Dividing by the nonzero number $\det(A)$ one finds the following equation, necessary to satisfy the system:

$$x_j = \frac{L_{(j)} \cdot \mathbf{b}}{\det(A)}.$$

But by construction the numerator is the determinant of the matrix obtained from A by replacing column j by b, so we get the expression of Cramer's rule as a necessary condition for a solution. The same procedure can be repeated for other values of j to find values for the other unknowns.

The only point that remains to prove is that these values for the unknowns, the only possible ones, do indeed together form a solution. But if the matrix A is invertible with inverse A^{-1}, then $x = A^{-1}b$ will be a solution, thus showing its existence. To see that A is invertible when $\det(A)$ is nonzero, consider the $n \times n$ matrix M obtained by stacking the one-line matrices $L_{(j)}$ on top of each other for $j = 1, \ldots, n$ (this gives the adjugate matrix for A). It was shown that $L_{(j)}A = (0 \ldots 0 \det(A) 0 \ldots 0)$ where $\det(A)$ appears at the position j; from this it follows that $MA = \det(A) I_n$. Therefore,

$$\frac{1}{\det(A)} M = A^{-1},$$

completing the proof.

Finding Inverse Matrix

Let A be an $n \times n$ matrix. Then

$$A \; adj \; A = (adj \, A)A = \det(A)I$$

where adj(A) denotes the adjugate matrix of A, det(A) is the determinant, and I is the identity matrix. If det(A) is invertible in R, then the inverse matrix of A is

$$A^{-1} = \frac{1}{\det(A)} \text{adj}(A).$$

If R is a field (such as the field of real numbers), then this gives a formula for the inverse of A, provided det(A) $\neq$ 0. In fact, this formula will work whenever R is a commutative ring, provided that det(A) is a unit. If det(A) is not a unit, then A is not invertible.

Applications

Explicit Formulas for Small Systems

Consider the linear system

$$\begin{cases} a_1x + b_1y &= c_1 \\ a_2x + b_2y &= c_2 \end{cases}$$

which in matrix format is

$$\begin{bmatrix} a_1 & b_1 \\ a_2 & b_2 \end{bmatrix}\begin{bmatrix} x \\ y \end{bmatrix} = \begin{bmatrix} c_1 \\ c_2 \end{bmatrix}.$$

Assume $a_1b_2 - b_1a_2$ nonzero. Then, with help of determinants x and y can be found with Cramer's rule as

$$x = \frac{\begin{vmatrix} c_1 & b_1 \\ c_2 & b_2 \end{vmatrix}}{\begin{vmatrix} a_1 & b_1 \\ a_2 & b_2 \end{vmatrix}} = \frac{c_1b_2 - b_1c_2}{a_1b_2 - b_1a_2}, \quad y = \frac{\begin{vmatrix} a_1 & c_1 \\ a_2 & c_2 \end{vmatrix}}{\begin{vmatrix} a_1 & b_1 \\ a_2 & b_2 \end{vmatrix}} = \frac{a_1c_2 - c_1a_2}{a_1b_2 - b_1a_2}$$

The rules for 3×3 matrices are similar. Given

$$\begin{cases} a_1x + b_1y + c_1z &= d_1 \\ a_2x + b_2y + c_2z &= d_2 \\ a_3x + b_3y + c_3z &= d_3 \end{cases}$$

which in matrix format is

$$\begin{bmatrix} a_1 & b_1 & c_1 \\ a_2 & b_2 & c_2 \\ a_3 & b_3 & c_3 \end{bmatrix}\begin{bmatrix} x \\ y \\ z \end{bmatrix} = \begin{bmatrix} d_1 \\ d_2 \\ d_3 \end{bmatrix}.$$

Then the values of x, y and z can be found as follows:

$$x = \frac{\begin{vmatrix} d_1 & b_1 & c_1 \\ d_2 & b_2 & c_2 \\ d_3 & b_3 & c_3 \end{vmatrix}}{\begin{vmatrix} a_1 & b_1 & c_1 \\ a_2 & b_2 & c_2 \\ a_3 & b_3 & c_3 \end{vmatrix}}, \quad y = \frac{\begin{vmatrix} a_1 & d_1 & c_1 \\ a_2 & d_2 & c_2 \\ a_3 & d_3 & c_3 \end{vmatrix}}{\begin{vmatrix} a_1 & b_1 & c_1 \\ a_2 & b_2 & c_2 \\ a_3 & b_3 & c_3 \end{vmatrix}}, \text{ and } z = \frac{\begin{vmatrix} a_1 & b_1 & d_1 \\ a_2 & b_2 & d_2 \\ a_3 & b_3 & d_3 \end{vmatrix}}{\begin{vmatrix} a_1 & b_1 & c_1 \\ a_2 & b_2 & c_2 \\ a_3 & b_3 & c_3 \end{vmatrix}}.$$

Differential Geometry

Ricci Calculus

Cramer's rule is used in the Ricci calculus in various calculations involving the Christoffel symbols of the first and second kind.

In particular, Cramer's rule can be used to prove that the divergence operator on a Riemannian manifold is invariant with respect to change of coordinates. We give a direct proof, suppressing the role of the Christoffel symbols. Let (M, g) be a Riemannian manifold equipped with local co-ordinates $(x^1, x^2, \ldots, x^n)$. Let $A = A^i \dfrac{\partial}{\partial x^i}$ be a vector field. We use the summation convention throughout.

Theorem.

The divergence *of* A,

$$div A = \frac{1}{\sqrt{\det g}} \frac{\partial}{\partial x^i} \left(A^i \sqrt{\det g} \right),$$

is invariant under change of coordinates.

Proof

Let $(x^1, x^2, \ldots, x^n) \mapsto (\bar{x}^1, \ldots, \bar{x}^n)$ be a coordinate transformation with non-singular Jacobian. Then the classical transformation laws imply that $A = \bar{A}^k \dfrac{\partial}{\partial \bar{x}^k}$ where $\bar{A}^k = \dfrac{\partial \bar{x}^k}{\partial x^j} A^j$. Similarly, if

$g = g_{mk} dx^m \otimes dx^k = \bar{g}_{ij} d\bar{x}^i \otimes d\bar{x}^j$, then $\bar{g}_{ij} = \dfrac{\partial x^m}{\partial \bar{x}^i} \dfrac{\partial x^k}{\partial \bar{x}^j} g_{mk}$.

Writing this transformation law in terms of matrices yields

$$\bar{g} = \left(\frac{\partial x}{\partial \bar{x}} \right)^{\mathrm{T}} g \left(\frac{\partial x}{\partial \bar{x}} \right), \text{ which implies } \det \bar{g} = \left(\det \left(\frac{\partial x}{\partial \bar{x}} \right) \right)^2 \det g.$$

Now one computes

$$\operatorname{div} A = \frac{1}{\sqrt{\det g}} \frac{\partial}{\partial x^i} \left(A^i \sqrt{\det g} \right)$$

$$= \det \left(\frac{\partial x}{\partial \bar{x}} \right) \frac{1}{\sqrt{\det \bar{g}}} \frac{\partial \bar{x}^k}{\partial x^i} \frac{\partial}{\partial \bar{x}^k} \left(\frac{\partial x^i}{\partial \bar{x}^\ell} \bar{A}^\ell \det \left(\frac{\partial x}{\partial \bar{x}} \right)^{-1} \sqrt{\det \bar{g}} \right).$$

In order to show that this equals $\dfrac{1}{\sqrt{\det \bar{g}}} \dfrac{\partial}{\partial \bar{x}^k} \left(\bar{A}^k \sqrt{\det \bar{g}} \right)$, it is necessary and sufficient to show that

$$\frac{\partial \bar{x}^k}{\partial x^i} \frac{\partial}{\partial \bar{x}^k} \left(\frac{\partial x^i}{\partial \bar{x}^\ell} \det \left(\frac{\partial x}{\partial \bar{x}} \right)^{-1} \right) = 0 \qquad \text{for all } \ell,$$

which is equivalent to

$$\frac{\partial}{\partial \overline{x}^\ell} \det\left(\frac{\partial x}{\partial \overline{x}}\right) = \det\left(\frac{\partial x}{\partial \overline{x}}\right) \frac{\partial \overline{x}^k}{\partial x^i} \frac{\partial^2 x^i}{\partial \overline{x}^k \partial \overline{x}^\ell}.$$

Carrying out the differentiation on the left-hand side, we get:

$$\frac{\partial}{\partial \overline{x}^\ell} \det\left(\frac{\partial x}{\partial \overline{x}}\right) = (-1)^{i+j} \frac{\partial^2 x^i}{\partial \overline{x}^\ell \partial \overline{x}^j} \det M(i\,|\,j) = \frac{\partial^2 x^i}{\partial \overline{x}^\ell \partial \overline{x}^j} \det\left(\frac{\partial x}{\partial \overline{x}}\right) \frac{(-1)^{i+j}}{\det\left(\dfrac{\partial x}{\partial \overline{x}}\right)} \det M(i\,|\,j) = (*),$$

where $M(i\,|\,j)$ denotes the matrix obtained from $\left(\dfrac{\partial x}{\partial \overline{x}}\right)$ by deleting the i th row and j th column. But Cramer's Rule says that

$$\frac{(-1)^{i+j}}{\det\left(\dfrac{\partial x}{\partial \overline{x}}\right)} \det M(i\,|\,j)$$

is the (j, i) th entry of the matrix $\left(\dfrac{\partial \overline{x}}{\partial x}\right)$. Thus

$$(*) = \det\left(\frac{\partial x}{\partial \overline{x}}\right) \frac{\partial^2 x^i}{\partial \overline{x}^\ell \partial \overline{x}^j} \frac{\partial \overline{x}^j}{\partial x^i},$$

completing the proof.

Computing Derivatives Implicitly

Consider the two equations $F(x, y, u, v) = 0$ and $G(x, y, u, v) = 0$. When u and v are independent variables, we can define $x = X(u, v)$ and $y = Y(u, v)$.

Finding an equation for $\dfrac{\partial x}{\partial u}$ is a trivial application of Cramer's rule.

Calculation of $\dfrac{\partial x}{\partial u}$

First, calculate the first derivatives of F, G, x, and y:

$$F = \frac{\partial F}{\partial x} dx + \frac{\partial F}{\partial y} dy + \frac{\partial F}{\partial u} du + \frac{\partial F}{\partial v} dv = 0$$

$$dG = \frac{\partial G}{\partial x} dx + \frac{\partial G}{\partial y} dy + \frac{\partial G}{\partial u} du + \frac{\partial G}{\partial v} dv = 0$$

$$dx = \frac{\partial X}{\partial u} du + \frac{\partial X}{\partial v} dv \quad , \quad dy = \frac{\partial Y}{\partial u} du + \frac{\partial Y}{\partial v} dv.$$

Substituting dx, dy into dF and dG, we have:

$$dF = \left(\frac{\partial F}{\partial x}\frac{\partial x}{\partial u} + \frac{\partial F}{\partial y}\frac{\partial y}{\partial u} + \frac{\partial F}{\partial u} \right) du + \left(\frac{\partial F}{\partial x}\frac{\partial x}{\partial v} + \frac{\partial F}{\partial y}\frac{\partial y}{\partial v} + \frac{\partial F}{\partial v} \right) dv = 0$$

$$dG = \left(\frac{\partial G}{\partial x}\frac{\partial x}{\partial u} + \frac{\partial G}{\partial y}\frac{\partial y}{\partial u} + \frac{\partial G}{\partial u} \right) du + \left(\frac{\partial G}{\partial x}\frac{\partial x}{\partial v} + \frac{\partial G}{\partial y}\frac{\partial y}{\partial v} + \frac{\partial G}{\partial v} \right) dv = 0.$$

Since u, v are both independent, the coefficients of du, dv must be zero. So we can write out equations for the coefficients:

$$\frac{\partial F}{\partial x}\frac{\partial x}{\partial u} + \frac{\partial F}{\partial y}\frac{\partial y}{\partial u} = -\frac{\partial F}{\partial u}$$

$$\frac{\partial G}{\partial x}\frac{\partial x}{\partial u} + \frac{\partial G}{\partial y}\frac{\partial y}{\partial u} = -\frac{\partial G}{\partial u}$$

$$\frac{\partial F}{\partial x}\frac{\partial x}{\partial v} + \frac{\partial F}{\partial y}\frac{\partial y}{\partial v} = -\frac{\partial F}{\partial v}$$

$$\frac{\partial G}{\partial x}\frac{\partial x}{\partial v} + \frac{\partial G}{\partial y}\frac{\partial y}{\partial v} = -\frac{\partial G}{\partial v}.$$

Now, by Cramer's rule, we see that:

$$\frac{\partial x}{\partial u} = \frac{\begin{vmatrix} -\dfrac{\partial F}{\partial u} & \dfrac{\partial F}{\partial y} \\[2mm] -\dfrac{\partial G}{\partial u} & \dfrac{\partial G}{\partial y} \end{vmatrix}}{\begin{vmatrix} \dfrac{\partial F}{\partial x} & \dfrac{\partial F}{\partial y} \\[2mm] \dfrac{\partial G}{\partial x} & \dfrac{\partial G}{\partial y} \end{vmatrix}}.$$

This is now a formula in terms of two Jacobians:

$$\frac{\partial x}{\partial u} = -\frac{\left(\dfrac{\partial(F,G)}{\partial(u,y)} \right)}{\left(\dfrac{\partial(F,G)}{\partial(x,y)} \right)}.$$

Similar formulas can be derived for $\dfrac{\partial x}{\partial v}, \dfrac{\partial y}{\partial u}, \dfrac{\partial y}{\partial v}.$

Integer Programming

Cramer's rule can be used to prove that an integer programming problem whose constraint matrix is totally unimodular and whose right-hand side is integer, has integer basic solutions. This makes the integer program substantially easier to solve.

Ordinary Differential Equations

Cramer's rule is used to derive the general solution to an inhomogeneous linear differential equation by the method of variation of parameters.

Geometric Interpretation

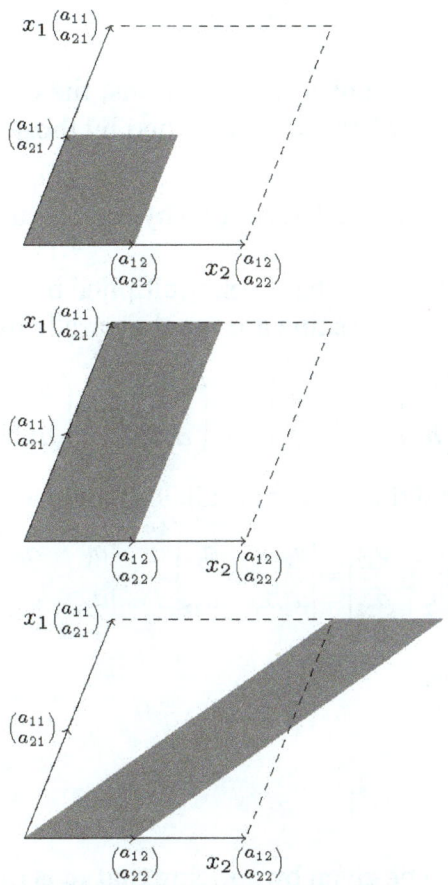

Geometric interpretation of Cramer's rule. The areas of the second and third shaded parallelograms are the same and the second is x_1 times the first. From this equality Cramer's rule follows.

Cramer's rule has a geometric interpretation that can be considered also a proof or simply giving insight about its geometric nature. These geometric arguments work in general and not only in the case of two equations with two unknowns presented here.

Given the system of equations

$$a_{11}x_1 + a_{12}x_2 = b_1$$
$$a_{21}x_1 + a_{22}x_2 = b_2$$

it can be considered as an equation between vectors

$$x_1 \begin{pmatrix} a_{11} \\ a_{21} \end{pmatrix} + x_2 \begin{pmatrix} a_{12} \\ a_{22} \end{pmatrix} = \begin{pmatrix} b_1 \\ b_2 \end{pmatrix}.$$

The area of the parallelogram determined by $\begin{pmatrix} a_{11} \\ a_{21} \end{pmatrix}$ and $\begin{pmatrix} a_{12} \\ a_{22} \end{pmatrix}$ is given by the determinant of the system of equations:

$$\begin{vmatrix} a_{11} & a_{12} \\ a_{21} & a_{22} \end{vmatrix}.$$

In general, when there are more variables and equations, the determinant of n vectors of length n will give the *volume* of the *parallelepiped* determined by those vectors in the n-th dimensional Euclidean space.

Therefore, the area of the parallelogram determined by $x_1\begin{pmatrix} a_{11} \\ a_{21} \end{pmatrix}$ and $\begin{pmatrix} a_{12} \\ a_{22} \end{pmatrix}$ has to be x_1 times the area of the first one since one of the sides has been multiplied by this factor. Now, this last parallelogram, by Cavalieri's principle, has the same area as the parallelogram determined by

$$\begin{pmatrix} b_1 \\ b_2 \end{pmatrix} = x_1\begin{pmatrix} a_{11} \\ a_{21} \end{pmatrix} + x_2\begin{pmatrix} a_{12} \\ a_{22} \end{pmatrix} \text{ and } \begin{pmatrix} a_{12} \\ a_{22} \end{pmatrix}.$$

Equating the areas of this last and the second parallelogram gives the equation

$$\begin{vmatrix} b_1 & a_{12} \\ b_2 & a_{22} \end{vmatrix} = \begin{vmatrix} a_{11}x_1 & a_{12} \\ a_{21}x_1 & a_{22} \end{vmatrix} = x_1\begin{vmatrix} a_{11} & a_{12} \\ a_{21} & a_{22} \end{vmatrix}$$

from which Cramer's rule follows.

Other Proofs

A Short Proof

A short proof of Cramer's rulecan be given by noticing that x_1 is the determinant of the matrix

$$X_1 = \begin{bmatrix} x_1 & 0 & 0 & \dots & 0 \\ x_2 & 1 & 0 & \dots & 0 \\ x_3 & 0 & 1 & \dots & 0 \\ \vdots & \vdots & \vdots & \ddots & \vdots \\ x_n & 0 & 0 & \dots & 1 \end{bmatrix}$$

On the other hand, assuming that our original matrix A is invertible, this matrix X_1 has columns $A^{-1}b, A^{-1}v_2, \dots, A^{-1}v_n$, where v_k is the k-th column of the matrix A. Recall that the matrix A_1 has columns $b, v_2, \dots, v_n$. Hence we have

$$x_1 = \det(X_1) = \det(A^{-1})\det(A_1) = \frac{\det(A_1)}{\det(A)}.$$

The proof for other x_j is similar.

Proof using Clifford Algebra

Consider the system of three scalar equations in three unknown scalars x_1, x_2, x_3

$$a_{11}x_1 + a_{12}x_2 + a_{13}x_3 = c_1$$
$$a_{21}x_1 + a_{22}x_2 + a_{23}x_3 = c_2$$
$$a_{31}x_1 + a_{32}x_2 + a_{33}x_3 = c_3$$

and assign an orthonormal vector basis $\mathbf{e}_1, \mathbf{e}_2, \mathbf{e}_3$ for $\mathcal{G}_3$ as

$$a_{11}\mathbf{e}_1 x_1 + a_{12}\mathbf{e}_1 x_2 + a_{13}\mathbf{e}_1 x_3 = c_1\mathbf{e}_1$$
$$a_{21}\mathbf{e}_2 x_1 + a_{22}\mathbf{e}_2 x_2 + a_{23}\mathbf{e}_2 x_3 = c_2\mathbf{e}_2$$
$$a_{31}\mathbf{e}_3 x_1 + a_{32}\mathbf{e}_3 x_2 + a_{33}\mathbf{e}_3 x_3 = c_3\mathbf{e}_3$$

Let the vectors

$$\mathbf{a}_1 = a_{11}\mathbf{e}_1 + a_{21}\mathbf{e}_2 + a_{31}\mathbf{e}_3$$
$$\mathbf{a}_2 = a_{12}\mathbf{e}_1 + a_{22}\mathbf{e}_2 + a_{32}\mathbf{e}_3$$
$$\mathbf{a}_3 = a_{13}\mathbf{e}_1 + a_{23}\mathbf{e}_2 + a_{33}\mathbf{e}_3$$

Adding the system of equations, it is seen that

$$\mathbf{c} = c_1\mathbf{e}_1 + c_2\mathbf{e}_2 + c_3\mathbf{e}_3$$
$$= x_1\mathbf{a}_1 + x_2\mathbf{a}_2 + x_3\mathbf{a}_3$$

Using the exterior product, each unknown scalar x_k can be solved as

$$\mathbf{c} \wedge \mathbf{a}_2 \wedge \mathbf{a}_3 = x_1 \mathbf{a}_1 \wedge \mathbf{a}_2 \wedge \mathbf{a}_3$$
$$\mathbf{c} \wedge \mathbf{a}_1 \wedge \mathbf{a}_3 = x_2 \mathbf{a}_2 \wedge \mathbf{a}_1 \wedge \mathbf{a}_3$$
$$\mathbf{c} \wedge \mathbf{a}_1 \wedge \mathbf{a}_2 = x_3 \mathbf{a}_3 \wedge \mathbf{a}_1 \wedge \mathbf{a}_2$$

$$x_1 = \frac{\mathbf{c} \wedge \mathbf{a}_2 \wedge \mathbf{a}_3}{\mathbf{a}_1 \wedge \mathbf{a}_2 \wedge \mathbf{a}_3}$$

$$x_2 = \frac{\mathbf{c} \wedge \mathbf{a}_1 \wedge \mathbf{a}_3}{\mathbf{a}_2 \wedge \mathbf{a}_1 \wedge \mathbf{a}_3} = \frac{\mathbf{a}_1 \wedge \mathbf{c} \wedge \mathbf{a}_3}{\mathbf{a}_1 \wedge \mathbf{a}_2 \wedge \mathbf{a}_3}$$

$$x_3 = \frac{\mathbf{c} \wedge \mathbf{a}_1 \wedge \mathbf{a}_2}{\mathbf{a}_3 \wedge \mathbf{a}_1 \wedge \mathbf{a}_2} = \frac{\mathbf{a}_1 \wedge \mathbf{a}_2 \wedge \mathbf{c}}{\mathbf{a}_1 \wedge \mathbf{a}_2 \wedge \mathbf{a}_3}$$

For n equations in n unknowns, the solution for the k-th unknown x_k generalizes to

$$x_k = \frac{\mathbf{a}_1 \wedge \cdots \wedge (\mathbf{c})_k \wedge \cdots \wedge \mathbf{a}_n}{\mathbf{a}_1 \wedge \cdots \wedge \mathbf{a}_k \wedge \cdots \wedge \mathbf{a}_n}$$

$$= (\mathbf{a}_1 \wedge \cdots \wedge (\mathbf{c})_k \wedge \cdots \wedge \mathbf{a}_n)(\mathbf{a}_1 \wedge \cdots \wedge \mathbf{a}_k \wedge \cdots \wedge \mathbf{a}_n)^{-1}$$

$$= \frac{(\mathbf{a}_1 \wedge \cdots \wedge (\mathbf{c})_k \wedge \cdots \wedge \mathbf{a}_n)(\mathbf{a}_1 \wedge \cdots \wedge \mathbf{a}_k \wedge \cdots \wedge \mathbf{a}_n)}{(\mathbf{a}_1 \wedge \cdots \wedge \mathbf{a}_k \wedge \cdots \wedge \mathbf{a}_n)(\mathbf{a}_1 \wedge \cdots \wedge \mathbf{a}_k \wedge \cdots \wedge \mathbf{a}_n)}$$

$$= \frac{(\mathbf{a}_1 \wedge \cdots \wedge (\mathbf{c})_k \wedge \cdots \wedge \mathbf{a}_n) \cdot (\mathbf{a}_1 \wedge \cdots \wedge \mathbf{a}_k \wedge \cdots \wedge \mathbf{a}_n)}{(-1)^{\frac{n(n-1)}{2}} (\mathbf{a}_n \wedge \cdots \wedge \mathbf{a}_k \wedge \cdots \wedge \mathbf{a}_1) \cdot (\mathbf{a}_1 \wedge \cdots \wedge \mathbf{a}_k \wedge \cdots \wedge \mathbf{a}_n)}$$

$$= \frac{(\mathbf{a}_n \wedge \cdots \wedge (\mathbf{c})_k \wedge \cdots \wedge \mathbf{a}_1) \cdot (\mathbf{a}_1 \wedge \cdots \wedge \mathbf{a}_k \wedge \cdots \wedge \mathbf{a}_n)}{(\mathbf{a}_n \wedge \cdots \wedge \mathbf{a}_k \wedge \cdots \wedge \mathbf{a}_1) \cdot (\mathbf{a}_1 \wedge \cdots \wedge \mathbf{a}_k \wedge \cdots \wedge \mathbf{a}_n)}$$

If a_k are linearly independent, then the x_k can be expressed in determinant form identical to Cramer's Rule as

$$x_k = \frac{(\mathbf{a}_n \wedge \cdots \wedge (\mathbf{c})_k \wedge \cdots \wedge \mathbf{a}_1) \cdot (\mathbf{a}_1 \wedge \cdots \wedge \mathbf{a}_k \wedge \cdots \wedge \mathbf{a}_n)}{(\mathbf{a}_n \wedge \cdots \wedge \mathbf{a}_k \wedge \cdots \wedge \mathbf{a}_1) \cdot (\mathbf{a}_1 \wedge \cdots \wedge \mathbf{a}_k \wedge \cdots \wedge \mathbf{a}_n)}$$

$$= \begin{vmatrix} \mathbf{a}_1 \cdot \mathbf{a}_1 & \cdots & \mathbf{a}_1 \cdot (\mathbf{c})_k & \cdots & \mathbf{a}_1 \cdot \mathbf{a}_n \\ \vdots & \ddots & \vdots & \ddots & \vdots \\ \mathbf{a}_k \cdot \mathbf{a}_1 & \cdots & \mathbf{a}_k \cdot (\mathbf{c})_k & \cdots & \mathbf{a}_k \cdot \mathbf{a}_n \\ \vdots & \ddots & \vdots & \ddots & \vdots \\ \mathbf{a}_n \cdot \mathbf{a}_1 & \cdots & \mathbf{a}_n \cdot (\mathbf{c})_k & \cdots & \mathbf{a}_n \cdot \mathbf{a}_n \end{vmatrix} \begin{vmatrix} \mathbf{a}_1 \cdot \mathbf{a}_1 & \cdots & \mathbf{a}_1 \cdot \mathbf{a}_k & \cdots & \mathbf{a}_1 \cdot \mathbf{a}_n \\ \vdots & \ddots & \vdots & \ddots & \vdots \\ \mathbf{a}_k \cdot \mathbf{a}_1 & \cdots & \mathbf{a}_k \cdot \mathbf{a}_k & \cdots & \mathbf{a}_k \cdot \mathbf{a}_n \\ \vdots & \ddots & \vdots & \ddots & \vdots \\ \mathbf{a}_n \cdot \mathbf{a}_1 & \cdots & \mathbf{a}_n \cdot \mathbf{a}_k & \cdots & \mathbf{a}_n \cdot \mathbf{a}_n \end{vmatrix}^{-1}$$

$$= \begin{vmatrix} \mathbf{a}_1 \\ \vdots \\ \mathbf{a}_k \\ \vdots \\ \mathbf{a}_n \end{vmatrix} \begin{vmatrix} \mathbf{a}_1 & \cdots & (\mathbf{c})_k & \cdots & \mathbf{a}_n \end{vmatrix} \begin{vmatrix} \mathbf{a}_1 \\ \vdots \\ \mathbf{a}_k \\ \vdots \\ \mathbf{a}_n \end{vmatrix} \begin{vmatrix} \mathbf{a}_1 & \cdots & \mathbf{a}_k & \cdots & \mathbf{a}_n \end{vmatrix}^{-1}$$

$$= \begin{vmatrix} \mathbf{a}_1 & \cdots & (\mathbf{c})_k & \cdots & \mathbf{a}_n \end{vmatrix} \begin{vmatrix} \mathbf{a}_1 & \cdots & \mathbf{a}_k & \cdots & \mathbf{a}_n \end{vmatrix}^{-1}$$

$$= \begin{vmatrix} a_{11} & \cdots & c_1 & \cdots & a_{1n} \\ \vdots & \ddots & \vdots & \ddots & \vdots \\ a_{k1} & \cdots & c_k & \cdots & a_{kn} \\ \vdots & \ddots & \vdots & \ddots & \vdots \\ a_{n1} & \cdots & c_n & \cdots & a_{nn} \end{vmatrix} \begin{vmatrix} a_{11} & \cdots & a_{1k} & \cdots & a_{1n} \\ \vdots & & & & \end{vmatrix}^{-1}$$

where $(\mathbf{c})_k$ denotes the substitution of vector $\mathbf{a}_k$ with vector $\mathbf{c}$ in the k-th numerator position.

Incompatible and Indeterminate Cases

A system of equations is said to be incompatible or inconsistent when there are no solutions and it is called indeterminate when there is more than one solution. For linear equations, an indeterminate system will have infinitely many solutions (if it is over an infinite field), since the solutions can be expressed in terms of one or more parameters that can take arbitrary values.

Cramer's rule applies to the case where the coefficient determinant is nonzero. In the 2×2 case, if the coefficient determinant is zero, then the system is incompatible if the numerator determinants are nonzero, or indeterminate if the numerator determinants are zero.

For 3×3 or higher systems, the only thing one can say when the coefficient determinant equals zero is that if any of the numerator determinants are nonzero, then the system must be incompatible. However, having all determinants zero does not imply that the system is indeterminate. A simple example where all determinants vanish (equal zero) but the system is still incompatible is the 3×3 system x+y+z=1, x+y+z=2, x+y+z=3.

Gaussian Elimination

In linear algebra, Gaussian elimination (also known as row reduction) is an algorithm for solving systems of linear equations. It is usually understood as a sequence of operations performed on the corresponding matrix of coefficients. This method can also be used to find the rank of a matrix, to calculate the determinant of a matrix, and to calculate the inverse of an invertible square matrix. The method is named after Carl Friedrich Gauss (1777–1855), although it was known to Chinese mathematicians as early as 179 CE.

To perform row reduction on a matrix, one uses a sequence of elementary row operations to modify the matrix until the lower left-hand corner of the matrix is filled with zeros, as much as possible. There are three types of elementary row operations: 1) Swapping two rows, 2) Multiplying a row by a non-zero number, 3) Adding a multiple of one row to another row. Using these operations, a matrix can always be transformed into an upper triangular matrix, and in fact one that is in row echelon form. Once all of the leading coefficients (the left-most non-zero entry in each row) are 1, and every column containing a leading coefficient has zeros elsewhere, the matrix is said to be in reduced row echelon form. This final form is unique; in other words, it is independent of the sequence of row operations used. For example, in the following sequence of row operations (where multiple elementary operations might be done at each step), the third and fourth matrices are the ones in row echelon form, and the final matrix is the unique reduced row echelon form.

$$\begin{bmatrix} 1 & 3 & 1 & 9 \\ 1 & 1 & -1 & 1 \\ 3 & 11 & 5 & 35 \end{bmatrix} \rightarrow \begin{bmatrix} 1 & 3 & 1 & 9 \\ 0 & -2 & -2 & -8 \\ 0 & 2 & 2 & 8 \end{bmatrix} \rightarrow \begin{bmatrix} 1 & 3 & 1 & 9 \\ 0 & -2 & -2 & -8 \\ 0 & 0 & 0 & 0 \end{bmatrix} \rightarrow \begin{bmatrix} 1 & 0 & -2 & -3 \\ 0 & 1 & 1 & 4 \\ 0 & 0 & 0 & 0 \end{bmatrix}$$

Using row operations to convert a matrix into reduced row echelon form is sometimes called Gauss–Jordan elimination. Some authors use the term Gaussian elimination to refer to the process until it has reached its upper triangular, or (non-reduced) row echelon form. For computational reasons, when solving systems of linear equations, it is sometimes preferable to stop row operations before the matrix is completely reduced.

Definitions and Example of Algorithm

The process of row reduction makes use of elementary row operations, and can be divided into two parts. The first part (sometimes called Forward Elimination) reduces a given system to *row echelon form*, from which one can tell whether there are no solutions, a unique solution, or infinitely many solutions. The second part (sometimes called back substitution) continues to use row operations until the solution is found; in other words, it puts the matrix into *reduced* row echelon form.

Another point of view, which turns out to be very useful to analyze the algorithm, is that row reduction produces a matrix decomposition of the original matrix. The elementary row operations may be viewed as the multiplication on the left of the original matrix by elementary matrices. Alternatively, a sequence of elementary operations that reduces a single row may be viewed as multiplication by a Frobenius matrix. Then the first part of the algorithm computes an LU decomposition, while the second part writes the original matrix as the product of a uniquely determined invertible matrix and a uniquely determined reduced row echelon matrix.

Row Operations

There are three types of elementary row operations which may be performed on the rows of a matrix:

> Type 1: Swap the positions of two rows.
>
> Type 2: Multiply a row by a nonzero scalar.
>
> Type 3: Add to one row a scalar multiple of another.

If the matrix is associated to a system of linear equations, then these operations do not change the solution set. Therefore, if one's goal is to solve a system of linear equations, then using these row operations could make the problem easier.

Echelon Form

For each row in a matrix, if the row does not consist of only zeros, then the left-most non-zero entry is called the *leading coefficient* (or *pivot*) of that row. So if two leading coefficients are in the same column, then a row operation of type 3 could be used to make one of those coefficients zero. Then by using the row swapping operation, one can always order the rows so that for every non-zero row, the leading coefficient is to the right of the leading coefficient of the row above. If this is the case, then matrix is said to be in row echelon form. So the lower left part of the matrix contains only zeros, and all of the zero rows are below the non-zero rows. The word "echelon" is used here because one can roughly think of the rows being ranked by their size, with the largest being at the top and the smallest being at the bottom.

For example, the following matrix is in row echelon form, and its leading coefficients are shown in red.

$$\begin{bmatrix} 0 & 2 & 1 & -1 \\ 0 & 0 & 3 & 1 \\ 0 & 0 & 0 & 0 \end{bmatrix}$$

It is in echelon form because the zero row is at the bottom, and the leading coefficient of the second row (in the third column), is to the right of the leading coefficient of the first row (in the second column).

A matrix is said to be in reduced row echelon form if furthermore all of the leading coefficients are equal to 1 (which can be achieved by using the elementary row operation of type 2), and in every column containing a leading coefficient, all of the other entries in that column are zero (which can be achieved by using elementary row operations of type 3).

Example of the Algorithm

Suppose the goal is to find and describe the set of solutions to the following system of linear equations:

$$2x + y - z = 8 \qquad (L_1)$$
$$-3x - y + 2z = -11 \qquad (L_2)$$
$$-2x + y + 2z = -3 \qquad (L_3)$$

The table below is the row reduction process applied simultaneously to the system of equations, and its associated augmented matrix. In practice, one does not usually deal with the systems in terms of equations but instead makes use of the augmented matrix, which is more suitable for computer manipulations. The row reduction procedure may be summarized as follows: eliminate x from all equations below L_1, and then eliminate y from all equations below L_2. This will put the system into triangular form. Then, using back-substitution, each unknown can be solved for.

System of equations	Row operations	Augmented matrix
$2x + y - z = 8$ $-3x - y + 2z = -11$ $-2x + y + 2z = -3$		$\begin{bmatrix} 2 & 1 & -1 & 8 \\ -3 & -1 & 2 & -11 \\ -2 & 1 & 2 & -3 \end{bmatrix}$
$2x + y - z = 8$ $\frac{1}{2}y + \frac{1}{2}z = 1$ $2y + z = 5$	$L_2 + \frac{3}{2}L_1 \to L_2$ $L_3 + L_1 \to L_3$	$\begin{bmatrix} 2 & 1 & -1 & 8 \\ 0 & 1/2 & 1/2 & 1 \\ 0 & 2 & 1 & 5 \end{bmatrix}$
$2x + y - z = 8$ $\frac{1}{2}y + \frac{1}{2}z = 1$ $-z = 1$	$L_3 + -4L_2 \to L_3$	$\begin{bmatrix} 2 & 1 & -1 & 8 \\ 0 & 1/2 & 1/2 & 1 \\ 0 & 0 & -1 & 1 \end{bmatrix}$

The matrix is now in echelon form (also called triangular form)		
$2x + y = 7$ $\dfrac{1}{2}y = \dfrac{3}{2}$ $-z = 1$	$L_2 + \dfrac{1}{2}L_3 \rightarrow L_2$ $L_1 - L_3 \rightarrow L_1$	$\begin{bmatrix} 2 & 1 & 0 & \bigm| & 7 \\ 0 & 1/2 & 0 & \bigm| & 3/2 \\ 0 & 0 & -1 & \bigm| & 1 \end{bmatrix}$
$2x + y = 7$ $y = 3$ $z = -1$	$2L_2 \rightarrow L_2$ $-L_3 \rightarrow L_3$	$\begin{bmatrix} 2 & 1 & 0 & \bigm| & 7 \\ 0 & 1 & 0 & \bigm| & 3 \\ 0 & 0 & 1 & \bigm| & -1 \end{bmatrix}$
$x = 2$ $y = 3$ $z = -1$	$L_1 - L_2 \rightarrow L_1$ $\dfrac{1}{2}L_1 \rightarrow L_1$	$\begin{bmatrix} 1 & 0 & 0 & \bigm| & 2 \\ 0 & 1 & 0 & \bigm| & 3 \\ 0 & 0 & 1 & \bigm| & -1 \end{bmatrix}$

The second column describes which row operations have just been performed. So for the first step, the x is eliminated from L_2 by adding $\dfrac{3}{2}L_1$ to L_2. Next x is eliminated from L_3 by adding L_1 to L_3. These row operations are labelled in the table as

$$L_2 + \frac{3}{2}L_1 \rightarrow L_2$$

$$L_3 + L_1 \rightarrow L_3.$$

Once y is also eliminated from the third row, the result is a system of linear equations in triangular form, and so the first part of the algorithm is complete. From a computational point of view, it is faster to solve the variables in reverse order, a process known as back-substitution. One sees the solution is $z = -1$, $y = 3$, and $x = 2$. So there is a unique solution to the original system of equations.

Instead of stopping once the matrix is in echelon form, one could continue until the matrix is in *reduced* row echelon form, as it is done in the table. The process of row reducing until the matrix is reduced is sometimes referred to as Gauss-Jordan elimination, to distinguish it from stopping after reaching echelon form.

History

The method of Gaussian elimination appears in the Chinese mathematical text Chapter Eight *Rectangular Arrays* of *The Nine Chapters on the Mathematical Art*. Its use is illustrated in eighteen problems, with two to five equations. The first reference to the book by this title is dated to 179 CE, but parts of it were written as early as approximately 150 BCE. It was commented on by Liu Hui in the 3rd century.

The method in Europe stems from the notes of Isaac Newton. In 1670, he wrote that all the algebra books known to him lacked a lesson for solving simultaneous equations, which Newton then supplied. Cambridge University eventually published the notes as *Arithmetica Universalis* in 1707 long after Newton left academic life. The notes were widely imitated, which made (what is now called) Gaussian elimination a standard lesson in algebra textbooks by the end of the 18th century. Carl Friedrich Gauss in 1810 devised a notation for symmetric elimination that was adopted in the 19th century by professional hand computers to solve the normal equations of least-squares problems. The algorithm that is taught in high school was named for Gauss only in the 1950s as a result of confusion over the history of the subject.

Some authors use the term *Gaussian elimination* to refer only to the procedure until the matrix is in echelon form, and use the term Gauss-Jordan elimination to refer to the procedure which ends in reduced echelon form. The name is used because it is a variation of Gaussian elimination as described by Wilhelm Jordan in 1888. However, the method also appears in an article by Clasen published in the same year. Jordan and Clasen probably discovered Gauss–Jordan elimination independently.

Applications

The historically first application of the row reduction method is for solving systems of linear equations. Here are some other important applications of the algorithm.

Computing Determinants

To explain how Gaussian elimination allows the computation of the determinant of a square matrix, we have to recall how the elementary row operations change the determinant:

- Swapping two rows multiplies the determinant by -1

- Multiplying a row by a nonzero scalar multiplies the determinant by the same scalar

- Adding to one row a scalar multiple of another does not change the determinant.

If the Gaussian elimination applied to a square matrix A produces a row echelon matrix B, let d be the product of the scalars by which the determinant has been multiplied, using above rules. Then the determinant of A is the quotient by d of the product of the elements of the diagonal of B: $\det(A) = \prod \text{diag}(B) / d$.

Computationally, for a $n \times n$ matrix, this method needs only $O(n^3)$ arithmetic operations, while solving by elementary methods requires $O(2^n)$ or $O(n!)$ operations. Even on the fastest computers, the elementary methods are impractical for n above 20.

Finding the Inverse of a Matrix

A variant of Gaussian elimination called Gauss–Jordan elimination can be used for finding the inverse of a matrix, if it exists. If A is a n by n square matrix, then one can use row reduction to compute its inverse matrix, if it exists. First, the n by n identity matrix is augmented to the right of A, forming a n by $2n$ block matrix $[A \mid I]$. Now through application of elementary row operations, find the reduced echelon form of this n by $2n$ matrix. The matrix A is invertible if and only if the left block can be reduced to the identity matrix I; in this case the right block of the final matrix is A^{-1}. If the algorithm is unable to reduce the left block to I, then A is not invertible.

For example, consider the following matrix

$$A = \begin{bmatrix} 2 & -1 & 0 \\ -1 & 2 & -1 \\ 0 & -1 & 2 \end{bmatrix}.$$

To find the inverse of this matrix, one takes the following matrix augmented by the identity, and row reduces it as a 3 by 6 matrix:

$$[A \,|\, I] = \begin{bmatrix} 2 & -1 & 0 & 1 & 0 & 0 \\ -1 & 2 & -1 & 0 & 1 & 0 \\ 0 & -1 & 2 & 0 & 0 & 1 \end{bmatrix}.$$

By performing row operations, one can check that the reduced row echelon form of this augmented matrix is:

$$[I \,|\, B] = \begin{bmatrix} 1 & 0 & 0 & \dfrac{3}{4} & \dfrac{1}{2} & \dfrac{1}{4} \\ 0 & 1 & 0 & \dfrac{1}{2} & 1 & \dfrac{1}{2} \\ 0 & 0 & 1 & \dfrac{1}{4} & \dfrac{1}{2} & \dfrac{3}{4} \end{bmatrix}.$$

One can think of each row operation as the left product by an elementary matrix. Denoting by B the product of these elementary matrices, we showed, on the left, that $BA = I$, and therefore, $B = A^{-1}$. On the right, we kept a record of $BI = B$, which we know is the inverse desired. This procedure for finding the inverse works for square matrices of any size.

Computing Ranks and Bases

The Gaussian elimination algorithm can be applied to any $m \times n$ matrix A. In this way, for example, some 6×9 matrices can be transformed to a matrix that has a row echelon form like

$$T = \begin{bmatrix} a & * & * & * & * & * & * & * & * \\ 0 & 0 & b & * & * & * & * & * & * \\ 0 & 0 & 0 & c & * & * & * & * & * \\ 0 & 0 & 0 & 0 & 0 & 0 & d & * & * \\ 0 & 0 & 0 & 0 & 0 & 0 & 0 & 0 & e \\ 0 & 0 & 0 & 0 & 0 & 0 & 0 & 0 & 0 \end{bmatrix}$$

where the *s are arbitrary entries and a, b, c, d, e are nonzero entries. This echelon matrix T contains a wealth of information about A : the rank of A is 5 since there are 5 non-zero rows in T ; the vector space spanned by the columns of A has a basis consisting of the first, third, fourth, seventh and ninth column of A (the columns of a, b, c, d, e in T), and the *s tell you how the other columns of A can be written as linear combinations of the basis columns. This is a consequence of the distributivity of the dot product in the expression of a linear map as a matrix.

All of this applies also to the reduced row echelon form, which is a particular row echelon form.

Computational Efficiency

The number of arithmetic operations required to perform row reduction is one way of measuring the algorithm's computational efficiency. For example, to solve a system of n equations for n unknowns by performing row operations on the matrix until it is in echelon form, and then solving for each unknown in reverse order, requires $n(n-1)/2$ divisions, $(2n^3 + 3n^2 - 5n)/6$ multiplications, and $(2n^3 + 3n^2 - 5n)/6$ subtractions, for a total of approximately $2n^3/3$ operations. Thus it has arithmetic complexity of $O(n^3)$. This arithmetic complexity is a good measure of the time needed for the whole computation when the time for each arithmetic operation is approximately constant. This is the case when the coefficients are represented by floating point numbers or when they belong to a finite field. If the coefficients are integers or rational numbers exactly represented, the intermediate entries can grow exponentially large, so the bit complexity is exponential. However, there is a variant of Gaussian elimination, called Bareiss algorithm that avoids this exponential growth of the intermediate entries, and, with the same arithmetic complexity of $O(n^3)$, has a bit complexity of $O(n^5)$.

This algorithm can be used on a computer for systems with thousands of equations and unknowns. However, the cost becomes prohibitive for systems with millions of equations. These large systems are generally solved using iterative methods. Specific methods exist for systems whose coefficients follow a regular pattern.

To put an n by n matrix into reduced echelon form by row operations, one needs n^3 arithmetic operations; which is approximately 50% more computation steps.

One possible problem is numerical instability, caused by the possibility of dividing by very small numbers. If, for example, the leading coefficient of one of the rows is very close to zero, then to row reduce the matrix one would need to divide by that number so the leading coefficient is 1. This means any error that existed for the number which was close to zero would be amplified. Gaussian elimination is numerically stable for diagonally dominant or positive-definite matrices. For general matrices, Gaussian elimination is usually considered to be stable, when using partial pivoting, even though there are examples of stable matrices for which it is unstable.

Generalizations

The Gaussian elimination can be performed over any field, not just the real numbers.

Gaussian elimination does not generalize in any simple way to higher order tensors (matrices are array representations of order 2 tensors); even computing the rank of a tensor of order greater than 2 is a difficult problem.

Pseudocode

As explained above, Gaussian elimination writes a given $m \times n$ matrix A uniquely as a product of an invertible $m \times m$ matrix S and a row-echelon matrix T. Here, S is the product of the matrices corresponding to the row operations performed.

The formal algorithm to compute T from A follows. We write $A[i, j]$ for the entry in row i, column j in matrix A with 1 being the first index. The transformation is performed *in place*, meaning that the original matrix A is lost and successively replaced by T.

> for k = 1 ... min(m,n):
>
> *Find the k-th pivot:*
>
> i_max:= argmax (i = k ... m, abs(A[i, k]))
>
> if A[i_max, k] = 0
>
> error "Matrix is singular!"
>
> swap rows(k, i_max)
>
> *Do for all rows below pivot:*
>
> for i = k + 1 ... m:
>
> f := A[i, k] / A[k, k]
>
> *Do for all remaining elements in current row:*
>
> for j = k + 1 ... n:
>
> A[i, j]:= A[i, j] - A[k, j] * f
>
> *Fill lower triangular matrix with zeros:*
>
> A[i, k]:= 0

This algorithm differs slightly from the one discussed earlier, because before eliminating a variable, it first exchanges rows to move the entry with the largest absolute value to the pivot position. Such *partial pivoting* improves the numerical stability of the algorithm; some other variants are used.

Upon completion of this procedure the augmented matrix will be in row-echelon form and may be solved by back-substitution.

With modern computers, Gaussian elimination is not always the fastest algorithm to compute the row echelon form of matrix. There are computer libraries, like BLAS, that exploit the specifics of the computer hardware and of the structure of the matrix to choose the best algorithm automatically.

Vector Space

A vector space (also called a linear space) is a collection of objects called vectors, which may be added together and multiplied ("scaled") by numbers, called *scalars* in this context. Scalars are often taken to be real numbers, but there are also vector spaces with scalar multiplication by complex numbers, rational numbers, or generally any field. The operations of vector addition and scalar multiplication must satisfy certain requirements, called *axioms*, listed below.

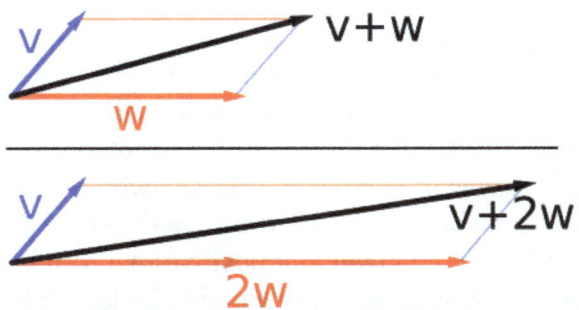

Vector addition and scalar multiplication: a vector v (blue) is added to another
vector w (red, upper illustration). Below, w is stretched by a factor of 2, yielding the sum v + 2w.

Euclidean vectors are an example of a vector space. They represent physical quantities such as forces: any two forces (of the same type) can be added to yield a third, and the multiplication of a force vector by a real multiplier is another force vector. In the same vein, but in a more geometric sense, vectors representing displacements in the plane or in three-dimensional space also form vector spaces. Vectors in vector spaces do not necessarily have to be arrow-like objects as they appear in the mentioned examples: vectors are regarded as abstract mathematical objects with particular properties, which in some cases can be visualized as arrows.

Vector spaces are the subject of linear algebra and are well characterized by their dimension, which, roughly speaking, specifies the number of independent directions in the space. Infinite-dimensional vector spaces arise naturally in mathematical analysis, as function spaces, whose vectors are functions. These vector spaces are generally endowed with additional structure, which may be a topology, allowing the consideration of issues of proximity and continuity. Among these topologies, those that are defined by a norm or inner product are more commonly used, as having a notion of distance between two vectors. This is particularly the case of Banach spaces and Hilbert spaces, which are fundamental in mathematical analysis.

Historically, the first ideas leading to vector spaces can be traced back as far as the 17th century's analytic geometry, matrices, systems of linear equations, and Euclidean vectors. The modern, more abstract treatment, first formulated by Giuseppe Peano in 1888, encompasses more general objects than Euclidean space, but much of the theory can be seen as an extension of classical geometric ideas like lines, planes and their higher-dimensional analogs.

Today, vector spaces are applied throughout mathematics, science and engineering. They are the appropriate linear-algebraic notion to deal with systems of linear equations; offer a framework for Fourier expansion, which is employed in image compression routines; or provide an environment that can be used for solution techniques for partial differential equations. Furthermore, vector spaces furnish an abstract, coordinate-free way of dealing with geometrical and physical objects such as tensors. This in turn allows the examination of local properties of manifolds by linearization techniques. Vector spaces may be generalized in several ways, leading to more advanced notions in geometry and abstract algebra.

Introduction and Definition

The concept of vector space will first be explained by describing two particular examples:

First Example: Arrows in the Plane

The first example of a vector space consists of arrows in a fixed plane, starting at one fixed point. This is used in physics to describe forces or velocities. Given any two such arrows, v and w, the parallelogram spanned by these two arrows contains one diagonal arrow that starts at the origin, too. This new arrow is called the *sum* of the two arrows and is denoted v + w. In the special case of two arrows on the same line, their sum is the arrow on this line whose length is the sum or the difference of the lengths, depending on whether the arrows have the same direction. Another operation that can be done with arrows is scaling: given any positive real number a, the arrow that has the same direction as v, but is dilated or shrunk by multiplying its length by a, is called *multiplication* of v by a. It is denoted av. When a is negative, av is defined as the arrow pointing in the opposite direction, instead.

The following shows a few examples: if $a = 2$, the resulting vector aw has the same direction as w, but is stretched to the double length of w (right image below). Equivalently 2w is the sum w + w. Moreover, (-1)v = $-$v has the opposite direction and the same length as v (blue vector pointing down in the right image).

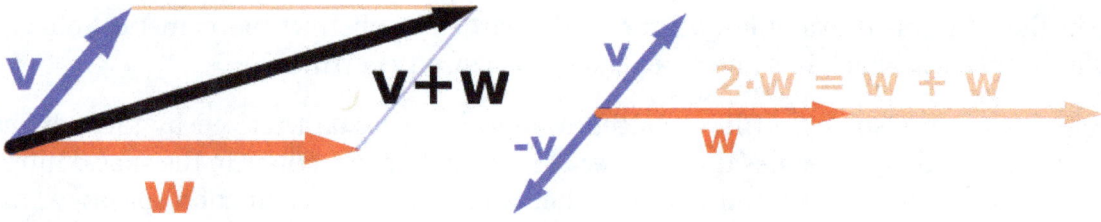

Second Example: Ordered Pairs of Numbers

A second key example of a vector space is provided by pairs of real numbers x and y. (The order of the components x and y is significant, so such a pair is also called an ordered pair.) Such a pair is written as (x, y). The sum of two such pairs and multiplication of a pair with a number is defined as follows:

$$(x_1, y_1) + (x_2, y_2) = (x_1 + x_2, y_1 + y_2)$$

and

$$a(x, y) = (ax, ay).$$

The first example above reduces to this one if the arrows are represented by the pair of Cartesian coordinates of their end points.

Definition

A vector space over a field F is a set V together with two operations that satisfy the eight axioms listed below. Elements of V are commonly called *vectors*. Elements of F are commonly called *sca-*

lars. The first operation, called *vector addition* or simply *addition*, takes any two vectors v and w and assigns to them a third vector which is commonly written as v + w, and called the sum of these two vectors. The second operation, called *scalar multiplication* takes any scalar *a* and any vector v and gives another vector *a*v.

In this chapter, vectors are distinguished from scalars by boldface. In the two examples above, the field is the field of the real numbers and the set of the vectors consists of the planar arrows with fixed starting point and of pairs of real numbers, respectively.

To qualify as a vector space, the set *V* and the operations of addition and multiplication must adhere to a number of requirements called axioms. In the list below, let u, v and w be arbitrary vectors in *V*, and *a* and *b* scalars in *F*.

Axiom	Meaning
Associativity of addition	u + (v + w) = (u + v) + w
Commutativity of addition	u + v = v + u
Identity element of addition	There exists an element o ∈ *V*, called the *zero vector*, such that v + o = v for all v ∈ *V*.
Inverse elements of addition	For every v ∈ *V*, there exists an element −v ∈ *V*, called the *additive inverse* of v, such that v + (−v) = o.
Compatibility of scalar multiplication with field multiplication	*a*(*b*v) = (*ab*)v
Identity element of scalar multiplication	1v = v, where 1 denotes the multiplicative identity in *F*.
Distributivity of scalar multiplication with respect to vector addition	*a*(u + v) = *a*u + *a*v
Distributivity of scalar multiplication with respect to field addition	(*a* + *b*)v = *a*v + *b*v

These axioms generalize properties of the vectors introduced in the above examples. Indeed, the result of addition of two ordered pairs (as in the second example above) does not depend on the order of the summands:

$$(x_v, y_v) + (x_w, y_w) = (x_w, y_w) + (x_v, y_v).$$

Likewise, in the geometric example of vectors as arrows, v + w = w + v since the parallelogram defining the sum of the vectors is independent of the order of the vectors. All other axioms can be checked in a similar manner in both examples. Thus, by disregarding the concrete nature of the particular type of vectors, the definition incorporates these two and many more examples in one notion of vector space.

Subtraction of two vectors and division by a (non-zero) scalar can be defined as

$$v - w = v + (-w),$$

$$v/a = (1/a)v.$$

When the scalar field F is the real numbers R, the vector space is called a *real vector space*. When the scalar field is the complex numbers, it is called a *complex vector space*. These two cases are the ones used most often in engineering. The general definition of a vector space allows scalars to be elements of any fixed field F. The notion is then known as an *F-vector spaces* or a *vector space over F*. A field is, essentially, a set of numbers possessing addition, subtraction, multiplication and division operations. For example, rational numbers also form a field.

In contrast to the intuition stemming from vectors in the plane and higher-dimensional cases, there is, in general vector spaces, no notion of nearness, angles or distances. To deal with such matters, particular types of vector spaces are introduced.

Alternative Formulations and Elementary Consequences

Vector addition and scalar multiplication are operations, satisfying the closure property: $u + v$ and av are in V for all a in F, and u, v in V. Some older sources mention these properties as separate axioms.

In the parlance of abstract algebra, the first four axioms can be subsumed by requiring the set of vectors to be an abelian group under addition. The remaining axioms give this group an F-module structure. In other words, there is a ring homomorphism f from the field F into the endomorphism ring of the group of vectors. Then scalar multiplication av is defined as $(f(a))(v)$.

There are a number of direct consequences of the vector space axioms. Some of them derive from elementary group theory, applied to the additive group of vectors: for example the zero vector 0 of V and the additive inverse $-v$ of any vector v are unique. Other properties follow from the distributive law, for example av equals 0 if and only if a equals 0 or v equals 0.

History

Vector spaces stem from affine geometry via the introduction of coordinates in the plane or three-dimensional space. Around 1636, Descartes and Fermat founded analytic geometry by equating solutions to an equation of two variables with points on a plane curve. In 1804, to achieve geometric solutions without using coordinates, Bolzano introduced certain operations on points, lines and planes, which are predecessors of vectors. His work was then used in the conception of barycentric coordinates by Möbius in 1827. In 1828 C. V. Mourey suggested the existence of an algebra surpassing not only ordinary algebra but also two-dimensional algebra created by him searching a geometrical interpretation of complex numbers.

The definition of vectors was founded on Bellavitis' notion of the bipoint, an oriented segment of which one end is the origin and the other a target, then further elaborated with the presentation of complex numbers by Argand and Hamilton and the introduction of quaternions and biquaternions by the latter. They are elements in R^2, R^4, and R^8; their treatment as linear combinations can be traced back to Laguerre in 1867, who also defined systems of linear equations.

In 1857, Cayley introduced matrix notation, which allows for a harmonization and simplification of linear maps. Around the same time, Grassmann studied the barycentric calculus initiated by Möbius. He envisaged sets of abstract objects endowed with operations. In his work, the concepts of linear independence and dimension, as well as scalar products, are present. In fact, Grassmann's 1844 work exceeds the framework of vector spaces, since his consideration of multiplication led him to what are today called algebras. Peano was the first to give the modern definition of vector spaces and linear maps in 1888.

An important development of vector spaces is due to the construction of function spaces by Lebesgue. This was later formalized by Banach and Hilbert, around 1920. At that time, algebra and the new field of functional analysis began to interact, notably with key concepts such as spaces of p-integrable functions and Hilbert spaces. Vector spaces, including infinite-dimensional ones, then became a firmly established notion, and many mathematical branches started making use of this concept.

Examples

Coordinate Spaces

The simplest example of a vector space over a field F is the field itself, equipped with its standard addition and multiplication. More generally, a vector space can be composed of n-tuples (sequences of length n) of elements of F, such as

$$(a_1, a_2, ..., a_n), \text{ where each } a_i \text{ is an element of } F.$$

A vector space composed of all the n-tuples of a field F is known as a *coordinate space*, usually denoted F^n. The case $n = 1$ is the above-mentioned simplest example, in which the field F is also regarded as a vector space over itself. The case $F = \mathrm{R}$ and $n = 2$ was discussed in the introduction above.

Complex Numbers and other Field Extensions

The set of complex numbers C, i.e., numbers that can be written in the form $x + iy$ for real numbers x and y where i is the imaginary unit, form a vector space over the reals with the usual addition and multiplication: $(x + iy) + (a + ib) = (x + a) + i(y + b)$ and $c \cdot (x + iy) = (c \cdot x) + i(c \cdot y)$ for real numbers x, y, a, b and c. The various axioms of a vector space follow from the fact that the same rules hold for complex number arithmetic.

In fact, the example of complex numbers is essentially the same (i.e., it is *isomorphic*) to the vector space of ordered pairs of real numbers mentioned above: if we think of the complex number $x + iy$ as representing the ordered pair (x, y) in the complex plane then we see that the rules for sum and scalar product correspond exactly to those in the earlier example.

More generally, field extensions provide another class of examples of vector spaces, particularly in algebra and algebraic number theory: a field F containing a smaller field E is an E-vector space, by the given multiplication and addition operations of F. For example, the complex numbers are a vector space over R, and the field extension $\mathbf{Q}(i\sqrt{5})$ is a vector space over Q.

Function Spaces

Functions from any fixed set Ω to a field F also form vector spaces, by performing addition and scalar multiplication pointwise. That is, the sum of two functions f and g is the function $(f + g)$ given by

$$(f + g)(w) = f(w) + g(w),$$

and similarly for multiplication. Such function spaces occur in many geometric situations, when Ω is the real line or an interval, or other subsets of R. Many notions in topology and analysis, such as continuity, integrability or differentiability are well-behaved with respect to linearity: sums and scalar multiples of functions possessing such a property still have that property. Therefore, the set of such functions are vector spaces. They are studied in greater detail using the methods of func-tional analysis. Algebraic constraints also yield vector spaces: the *vector space F[x]* is given by polynomial functions:

$$f(x) = r_0 + r_1 x + \dots + r_{n-1} x^{n-1} + r_n x^n, \text{ where the coefficients } r_0, \dots, r_n \text{ are in } F.$$

Linear Equations

Systems of homogeneous linear equations are closely tied to vector spaces. For example, the solutions of

$$a + 3b + c = 0$$

$$4a + 2b + 2c = 0$$

are given by triples with arbitrary a, $b = a/2$, and $c = -5a/2$. They form a vector space: sums and scalar multiples of such triples still satisfy the same ratios of the three variables; thus they are solutions, too. Matrices can be used to condense multiple linear equations as above into one vector equation, namely

$$Ax = 0,$$

where $A = \begin{bmatrix} 1 & 3 & 1 \\ 4 & 2 & 2 \end{bmatrix}$ is the matrix containing the coefficients of the given equations, x is the vector

(a, b, c), Ax denotes the matrix product, and $0 = (0, 0)$ is the zero vector. In a similar vein, the solutions of homogeneous *linear differential equations* form vector spaces. For example,

$$f''(x) + 2f'(x) + f(x) = 0$$

yields $f(x) = a e^{-x} + bx e^{-x}$, where a and b are arbitrary constants, and e^x is the natural exponential function.

Basis and Dimension

Bases allow to represent vectors by a sequence of scalars called *coordinates* or *components*. A basis is a (finite or infinite) set $B = \{b_i\}_{i \in I}$ of vectors b_i, for convenience often indexed by some index set I, that spans the whole space and is linearly independent. "Spanning the whole space" means that any vector v can be expressed as a finite sum (called a *linear combination*) of the basis elements:

$$\mathbf{v} = a_1 \mathbf{b}_{i_1} + a_2 \mathbf{b}_{i_2} + \dots + a_n \mathbf{b}_{i_n}, \tag{1}$$

where the a_k are scalars, called the coordinates (or the components) of the vector v with respect to the basis B, and b_{ik} ($k = 1, ..., n$) elements of B. Linear independence means that the coordinates a_k are uniquely determined for any vector in the vector space.

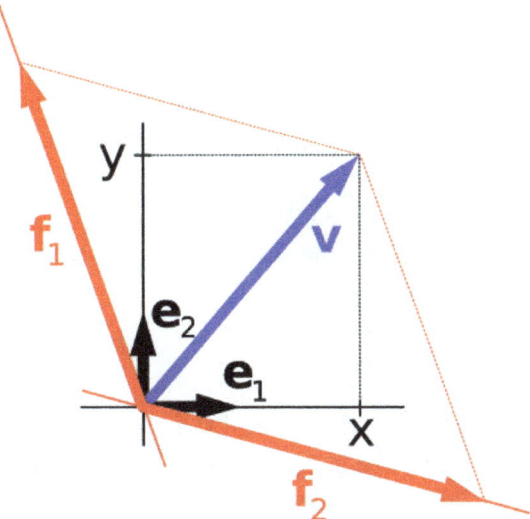

A vector v in R^2 (blue) expressed in terms of different bases: using the standard basis of R^2 v = xe_1 + ye_2 (black), and using a different, non-orthogonal basis: v = f_1 + f_2 (red).

For example, the coordinate vectors e_1 = (1, 0, ..., 0), e_2 = (0, 1, 0, ..., 0), to e_n = (0, 0, ..., 0, 1), form a basis of F^n, called the standard basis, since any vector $(x_1, x_2, ..., x_n)$ can be uniquely expressed as a linear combination of these vectors:

$$(x_1, x_2, ..., x_n) = x_1(1, 0, ..., 0) + x_2(0, 1, 0, ..., 0) + ... + x_n(0, ..., 0, 1) = x_1e_1 + x_2e_2 + ... + x_ne_n.$$

The corresponding coordinates $x_1, x_2, ..., x_n$ are just the Cartesian coordinates of the vector.

Every vector space has a basis. This follows from Zorn's lemma, an equivalent formulation of the Axiom of Choice. Given the other axioms of Zermelo–Fraenkel set theory, the existence of bases is equivalent to the axiom of choice. The ultrafilter lemma, which is weaker than the axiom of choice, implies that all bases of a given vector space have the same number of elements, or cardinality (cf. *Dimension theorem for vector spaces*). It is called the *dimension* of the vector space, denoted dim V. If the space is spanned by finitely many vectors, the above statements can be proven without such fundamental input from set theory.

The dimension of the coordinate space F^n is n, by the basis exhibited above. The dimension of the polynomial ring $F[x]$ introduced above is countably infinite, a basis is given by 1, x, x^2, ... A fortiori, the dimension of more general function spaces, such as the space of functions on some (bounded or unbounded) interval, is infinite. Under suitable regularity assumptions on the coefficients involved, the dimension of the solution space of a homogeneous ordinary differential equation equals the degree of the equation. For example, the solution space for the above equation is generated by e^{-x} and xe^{-x}. These two functions are linearly independent over R, so the dimension of this space is two, as is the degree of the equation.

A field extension over the rationals Q can be thought of as a vector space over Q (by defining vector addition as field addition, defining scalar multiplication as field multiplication by elements of Q, and otherwise ignoring the field multiplication). The dimension (or degree) of the field extension Q(α) over Q depends on α. If α satisfies some polynomial equation

$$q_n \alpha^n + q_{n-1} \alpha^{n-1} + \ldots + q_0 = 0$$

with rational coefficients q_n, ..., q_0 (in other words, if α is algebraic), the dimension is finite. More precisely, it equals the degree of the minimal polynomial having α as a root. For example, the complex numbers C are a two-dimensional real vector space, generated by 1 and the imaginary unit i. The latter satisfies $i^2 + 1 = 0$, an equation of degree two. Thus, C is a two-dimensional R-vector space (and, as any field, one-dimensional as a vector space over itself, C). If α is not algebraic, the dimension of Q(α) over Q is infinite. For instance, for $\alpha = \pi$ there is no such equation, in other words π is transcendental.

Linear Maps and Matrices

The relation of two vector spaces can be expressed by *linear map* or *linear transformation*. They are functions that reflect the vector space structure—i.e., they preserve sums and scalar multiplication:

$$f(x + y) = f(x) + f(y) \text{ and } f(a \cdot x) = a \cdot f(x) \text{ for all x and y in } V, \text{ all } a \text{ in } F.$$

An *isomorphism* is a linear map $f : V \to W$ such that there exists an inverse map $g : W \to V$, which is a map such that the two possible compositions $f \circ g : W \to W$ and $g \circ f : V \to V$ are identity maps. Equivalently, f is both one-to-one (injective) and onto (surjective). If there exists an isomorphism between V and W, the two spaces are said to be *isomorphic*; they are then essentially identical as vector spaces, since all identities holding in V are, via f, transported to similar ones in W, and vice versa via g.

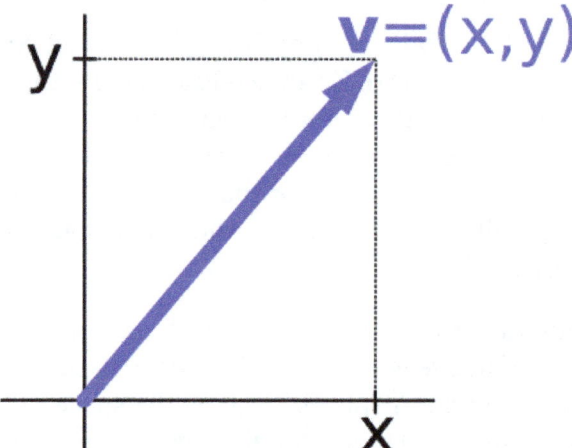

Describing an arrow vector v by its coordinates x and y yields an isomorphism of vector spaces.

For example, the "arrows in the plane" and "ordered pairs of numbers" vector spaces in the introduction are isomorphic: a planar arrow v departing at the origin of some (fixed) coordinate system can be expressed as an ordered pair by considering the x- and y-component of the arrow, as shown in the image at the right. Conversely, given a pair (x, y), the arrow going by x to the right (or to the left, if x is negative), and y up (down, if y is negative) turns back the arrow v.

Linear maps $V \to W$ between two vector spaces form a vector space $\text{Hom}_F(V, W)$, also denoted $L(V, W)$. The space of linear maps from V to F is called the *dual vector space*, denoted V^*. Via the injective natural map $V \to V^{**}$, any vector space can be embedded into its *bidual*; the map is an isomorphism if and only if the space is finite-dimensional.

Once a basis of V is chosen, linear maps $f : V \to W$ are completely determined by specifying the images of the basis vectors, because any element of V is expressed uniquely as a linear combination of them. If $\dim V = \dim W$, a 1-to-1 correspondence between fixed bases of V and W gives rise to a linear map that maps any basis element of V to the corresponding basis element of W. It is an isomorphism, by its very definition. Therefore, two vector spaces are isomorphic if their dimensions agree and vice versa. Another way to express this is that any vector space is *completely classified* (up to isomorphism) by its dimension, a single number. In particular, any n-dimensional F-vector space V is isomorphic to F^n. There is, however, no "canonical" or preferred isomorphism; actually an isomorphism $\varphi : F^n \to V$ is equivalent to the choice of a basis of V, by mapping the standard basis of F^n to V, via φ. The freedom of choosing a convenient basis is particularly useful in the infinite-dimensional context.

Matrices

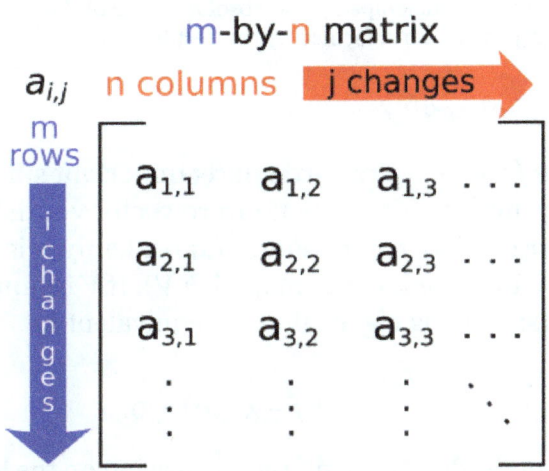

A typical matrix

Matrices are a useful notion to encode linear maps. They are written as a rectangular array of scalars as in the image at the right. Any m-by-n matrix A gives rise to a linear map from F^n to F^m, by the following

$$\mathbf{x} = (x_1, x_2, \cdots, x_n) \mapsto \left(\sum_{j=1}^{n} a_{1j} x_j, \sum_{j=1}^{n} a_{2j} x_j, \cdots, \sum_{j=1}^{n} a_{mj} x_j \right), \text{ where } \sum \text{ denotes summation,}$$

or, using the matrix multiplication of the matrix A with the coordinate vector x:

$$x \mapsto Ax.$$

Moreover, after choosing bases of V and W, *any* linear map $f : V \to W$ is uniquely represented by a matrix via this assignment.

The determinant det (A) of a square matrix A is a scalar that tells whether the associated map is an isomorphism or not: to be so it is sufficient and necessary that the determinant is nonzero. The linear transformation of $\mathbb{R}^n$ corresponding to a real n-by-n matrix is orientation preserving if and only if its determinant is positive.

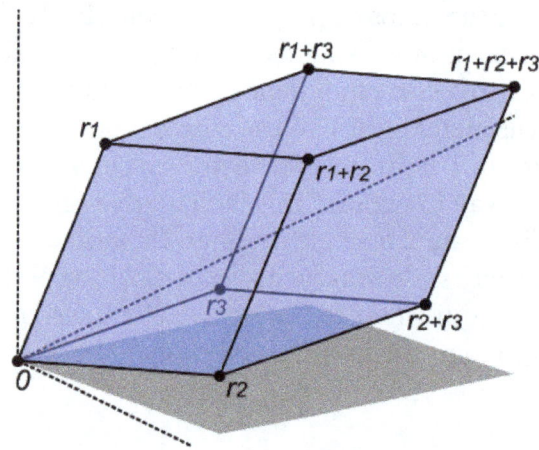

The volume of this parallelepiped is the absolute value of the determinant of the 3-by-3 matrix formed by the vectors r_1, r_2, and r_3.

Eigenvalues and Eigenvectors

Endomorphisms, linear maps $f : V \to V$, are particularly important since in this case vectors v can be compared with their image under f, $f(v)$. Any nonzero vector v satisfying $\lambda v = f(v)$, where λ is a scalar, is called an *eigenvector* of f with *eigenvalue* λ. Equivalently, v is an element of the kernel of the difference $f - \lambda \cdot \mathrm{Id}$ (where Id is the identity map $V \to V$). If V is finite-dimensional, this can be rephrased using determinants: f having eigenvalue λ is equivalent to

$$\det(f - \lambda \cdot \mathrm{Id}) = 0.$$

By spelling out the definition of the determinant, the expression on the left hand side can be seen to be a polynomial function in λ, called the characteristic polynomial of f. If the field F is large enough to contain a zero of this polynomial (which automatically happens for F algebraically closed, such as $F = \mathbb{C}$) any linear map has at least one eigenvector. The vector space V may or may not possess an eigenbasis, a basis consisting of eigenvectors. This phenomenon is governed by the Jordan canonical form of the map. The set of all eigenvectors corresponding to a particular eigenvalue of f forms a vector space known as the *eigenspace* corresponding to the eigenvalue (and f) in question. To achieve the spectral theorem, the corresponding statement in the infinite-dimensional case, the machinery of functional analysis is needed.

Basic Constructions

In addition to the above concrete examples, there are a number of standard linear algebraic constructions that yield vector spaces related to given ones. In addition to the definitions given below, they are also characterized by universal properties, which determine an object X by specifying the linear maps from X to any other vector space.

Subspaces and Quotient Spaces

A nonempty subset W of a vector space V that is closed under addition and scalar multiplication (and therefore contains the 0-vector of V) is called a *linear subspace* of V, or simply a *subspace* of V, when the ambient space is unambiguously a vector space. Subspaces of V are vector spaces (over the same field) in their own right. The intersection of all subspaces containing a given set S of vectors is called its span, and it is the smallest subspace of V containing the set S. Expressed in terms of elements, the span is the subspace consisting of all the linear combinations of elements of S.

A linear subspace of dimension 1 is a vector line. A linear subspace of dimension 2 is a vector plane. A linear subspace that contains all elements but one of a basis of the ambient space is a vector hyperplane. In a vector space of finite dimension n, a vector hyperplane is thus a subspace of dimension $n - 1$.

The counterpart to subspaces are *quotient vector spaces*. Given any subspace $W \subset V$, the quotient space V/W ("V modulo W") is defined as follows: as a set, it consists of $v + W = \{v + w : w \in W\}$, where v is an arbitrary vector in V. The sum of two such elements $v_1 + W$ and $v_2 + W$ is $(v_1 + v_2) + W$, and scalar multiplication is given by $a \cdot (v + W) = (a \cdot v) + W$. The key point in this definition is that $v_1 + W = v_2 + W$ if and only if the difference of v_1 and v_2 lies in W. This way, the quotient space "forgets" information that is contained in the subspace W.

The kernel $\ker(f)$ of a linear map $f : V \to W$ consists of vectors v that are mapped to 0 in W. Both kernel and image $\operatorname{im}(f) = \{f(v) : v \in V\}$ are subspaces of V and W, respectively. The existence of kernels and images is part of the statement that the category of vector spaces (over a fixed field F) is an abelian category, i.e. a corpus of mathematical objects and structure-preserving maps between them (a category) that behaves much like the category of abelian groups. Because of this, many statements such as the first isomorphism theorem (also called rank–nullity theorem in matrix-related terms)

$$V / \ker(f) \equiv \operatorname{im}(f).$$

and the second and third isomorphism theorem can be formulated and proven in a way very similar to the corresponding statements for groups.

An important example is the kernel of a linear map $x \mapsto Ax$ for some fixed matrix A, as above. The kernel of this map is the subspace of vectors x such that $Ax = 0$, which is precisely the set of solutions to the system of homogeneous linear equations belonging to A. This concept also extends to linear differential equations

$$a_0 f + a_1 \frac{df}{dx} + a_2 \frac{d^2 f}{dx^2} + \cdots + a_n \frac{d^n f}{dx^n} = 0, \text{ where the coefficients } a_i \text{ are functions in } x, \text{ too.}$$

In the corresponding map

$$f \mapsto D(f) = \sum_{i=0}^{n} a_i \frac{d^i f}{dx^i},$$

the derivatives of the function f appear linearly (as opposed to $f''(x)^2$, for example). Since differentiation is a linear procedure (i.e., $(f + g)' = f' + g'$ and $(c \cdot f)' = c \cdot f'$ for a constant c) this assignment is linear, called a linear differential operator. In particular, the solutions to the differential equation $D(f) = 0$ form a vector space (over R or C).

Direct Product and Direct Sum

The *direct product* of vector spaces and the *direct sum* of vector spaces are two ways of combining an indexed family of vector spaces into a new vector space.

The *direct product* $\prod_{i \in I} V_i$ of a family of vector spaces V_i consists of the set of all tuples $(v_i)_{i \in I}$, which specify for each index i in some index set I an element v_i of V_i. Addition and scalar multiplication is performed componentwise. A variant of this construction is the *direct sum* $\oplus_{i \in I} V_i$ (also called coproduct and denoted $\coprod_{i \in I} V_i$), where only tuples with finitely many nonzero vectors are allowed. If the index set I is finite, the two constructions agree, but in general they are different.

Tensor Product

The *tensor product* $V \otimes_F W$, or simply $V \otimes W$, of two vector spaces V and W is one of the central notions of multilinear algebra which deals with extending notions such as linear maps to several variables. A map $g : V \times W \to X$ is called bilinear if g is linear in both variables v and w. That is to say, for fixed w the map $v \mapsto g(v, w)$ is linear in the sense above and likewise for fixed v.

The tensor product is a particular vector space that is a *universal* recipient of bilinear maps g, as follows. It is defined as the vector space consisting of finite (formal) sums of symbols called tensors

$$v_1 \otimes w_1 + v_2 \otimes w_2 + \dots + v_n \otimes w_n,$$

subject to the rules

$$a \cdot (v \otimes w) = (a \cdot v) \otimes w = v \otimes (a \cdot w), \text{ where } a \text{ is a scalar,}$$

$$(v_1 + v_2) \otimes w = v_1 \otimes w + v_2 \otimes w, \text{ and}$$

$$v \otimes (w_1 + w_2) = v \otimes w_1 + v \otimes w_2.$$

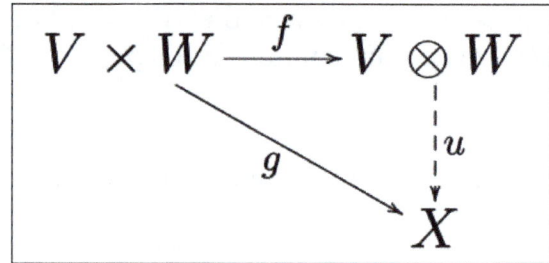

Commutative diagram depicting the universal property of the tensor product.

These rules ensure that the map f from the $V \times W$ to $V \otimes W$ that maps a tuple (v, w) to $v \otimes w$ is bilinear. The universality states that given *any* vector space X and *any* bilinear map $g : V \times W \to X$, there exists a unique map u, shown in the diagram with a dotted arrow, whose composition with f equals g: $u(v \otimes w) = g(v, w)$. This is called the universal property of the tensor product, an instance of the method—much used in advanced abstract algebra—to indirectly define objects by specifying maps from or to this object.

Vector Spaces with Additional Structure

From the point of view of linear algebra, vector spaces are completely understood insofar as any vector space is characterized, up to isomorphism, by its dimension. However, vector spaces *per se* do not offer a framework to deal with the question—crucial to analysis—whether a sequence of functions converges to another function. Likewise, linear algebra is not adapted to deal with infinite series, since the addition operation allows only finitely many terms to be added. *Therefore, the needs of functional analysis require considering additional structures.*

A vector space may be given a partial order $\leq$, under which some vectors can be compared. For example, n-dimensional real space R^n can be ordered by comparing its vectors componentwise. Ordered vector spaces, for example Riesz spaces, are fundamental to Lebesgue integration, which relies on the ability to express a function as a difference of two positive functions

$$f = f^+ - f^-,$$

where f^+ denotes the positive part of f and f^- the negative part.

Normed Vector Spaces and Inner Product Spaces

"Measuring" vectors is done by specifying a norm, a datum which measures lengths of vectors, or by an inner product, which measures angles between vectors. Norms and inner products are denoted $|\mathbf{v}|$ and $\langle \mathbf{v}, \mathbf{w} \rangle$, respectively. The datum of an inner product entails that lengths of vectors can be defined too, by defining the associated norm $|\mathbf{v}| := \sqrt{\langle \mathbf{v}, \mathbf{v} \rangle}$. Vector spaces endowed with such data are known as *normed vector spaces* and *inner product spaces*, respectively.

Coordinate space F^n can be equipped with the standard dot product:

$$\langle \mathbf{x}, \mathbf{y} \rangle = \mathbf{x} \cdot \mathbf{y} = x_1 y_1 + \cdots + x_n y_n.$$

In R^2, this reflects the common notion of the angle between two vectors x and y, by the law of cosines:

$$\mathbf{x} \cdot \mathbf{y} = \cos\big(\angle(\mathbf{x}, \mathbf{y})\big) \cdot |\mathbf{x}| \cdot |\mathbf{y}|.$$

Because of this, two vectors satisfying $\langle \mathbf{x}, \mathbf{y} \rangle = 0$ are called orthogonal. An important variant of the standard dot product is used in Minkowski space: R^4 endowed with the Lorentz product

$$\langle \mathbf{x} \,|\, \mathbf{y} \rangle = x_1 y_1 + x_2 y_2 + x_3 y_3 - x_4 y_4.$$

In contrast to the standard dot product, it is not positive definite: $\langle \mathbf{x} \,|\, \mathbf{x} \rangle$ also takes negative values, for example for $\mathbf{x} = (0, 0, 0, 1)$. Singling out the fourth coordinate—corresponding to time, as opposed to three space-dimensions—makes it useful for the mathematical treatment of special relativity.

Topological Vector Spaces

Convergence questions are treated by considering vector spaces V carrying a compatible topology, a structure that allows one to talk about elements being close to each other. Compatible here means that addition and scalar multiplication have to be continuous maps. Roughly, if x and y in V, and a in F vary by a bounded amount, then so do x + y and ax. To make sense of specifying the amount a scalar changes, the field F also has to carry a topology in this context; a common choice are the reals or the complex numbers.

In such *topological vector spaces* one can consider series of vectors. The infinite sum

$$\sum_{i=0}^{\infty} f_i$$

denotes the limit of the corresponding finite partial sums of the sequence $(f_i)_{i\in\mathbb{N}}$ of elements of V. For example, the f_i could be (real or complex) functions belonging to some function space V, in which case the series is a function series. The mode of convergence of the series depends on the topology imposed on the function space. In such cases, pointwise convergence and uniform convergence are two prominent examples.

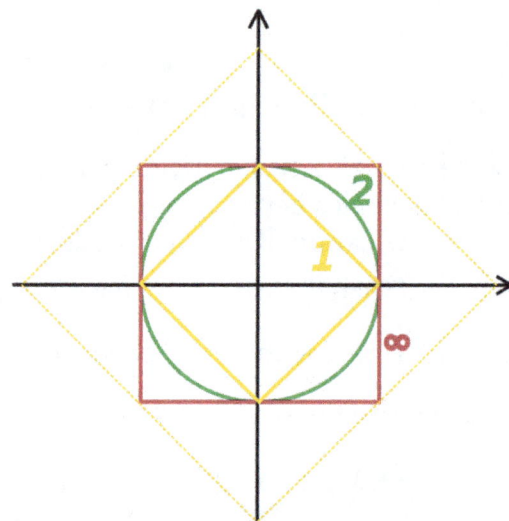

Unit "spheres" in $\mathbb{R}^2$ consist of plane vectors of norm 1. Depicted are the unit spheres in different p-norms, for $p = 1, 2,$ and ∞. The bigger diamond depicts points of 1-norm equal to .

A way to ensure the existence of limits of certain infinite series is to restrict attention to spaces where any Cauchy sequence has a limit; such a vector space is called complete. Roughly, a vector space is complete provided that it contains all necessary limits. For example, the vector space of polynomials on the unit interval [0,1], equipped with the topology of uniform convergence is not complete because any continuous function on [0,1] can be uniformly approximated by a sequence of polynomials, by the Weierstrass approximation theorem. In contrast, the space of *all* continuous functions on [0,1] with the same topology is complete. A norm gives rise to a topology by defining that a sequence of vectors v_n converges to v if and only if

$$\lim_{n \to \infty} |\mathbf{v}_n - \mathbf{v}| = 0.$$

Banach and Hilbert spaces are complete topological vector spaces whose topologies are given, respectively, by a norm and an inner product. Their study—a key piece of functional analysis—focusses on infinite-dimensional vector spaces, since all norms on finite-dimensional topological vector spaces give rise to the same notion of convergence. The image at the right shows the equivalence of the 1-norm and ∞-norm on $\mathbb{R}^2$: as the unit "balls" enclose each other, a sequence converges to zero in one norm if and only if it so does in the other norm. In the infinite-dimensional case, however, there will generally be inequivalent topologies, which makes the study of topological vector spaces richer than that of vector spaces without additional data.

From a conceptual point of view, all notions related to topological vector spaces should match the topology. For example, instead of considering all linear maps (also called functionals) $V \to W$, maps between topological vector spaces are required to be continuous. In particular, the *(topological) dual space* V^* consists of continuous functionals $V \to \mathbb{R}$ (or to $\mathbb{C}$). The fundamental Hahn–Banach theorem is concerned with separating subspaces of appropriate topological vector spaces by continuous functionals.

Banach Spaces

Banach spaces, introduced by Stefan Banach, are complete normed vector spaces. A first example is the vector space ℓ^p consisting of infinite vectors with real entries $x = (x_1, x_2, ...)$ whose p-norm ($1 \le p \le \infty$) given by

$$|\mathbf{x}|_p := \left(\sum_i |x_i|^p \right)^{1/p} \text{ for } p < \infty \text{ and } |\mathbf{x}|_\infty := \sup_i |x_i|$$

is finite. The topologies on the infinite-dimensional space ℓ^p are inequivalent for different p. E.g. the sequence of vectors $x_n = (2^{-n}, 2^{-n}, ..., 2^{-n}, 0, 0, ...)$, i.e. the first 2^n components are 2^{-n}, the following ones are 0, converges to the zero vector for $p = \infty$, but does not for $p = 1$:

$$|x_n|_\infty = \sup(2^{-n}, 0) = 2^{-n} \to 0, \text{, but } |x_n|_1 = \sum_{i=1}^{2^n} 2^{-n} = 2^n \cdot 2^{-n} = 1.$$

More generally than sequences of real numbers, functions $f: \Omega \to \mathbb{R}$ are endowed with a norm that replaces the above sum by the Lebesgue integral

$$|f|_p := \left(\int_\Omega |f(x)|^p \, dx \right)^{1/p}.$$

The space of integrable functions on a given domain Ω (for example an interval) satisfying $|f|_p < \infty$, and equipped with this norm are called Lebesgue spaces, denoted $L^p(\Omega)$. These spaces are complete. (If one uses the Riemann integral instead, the space is *not* complete, which may be seen as a justification for Lebesgue's integration theory.) Concretely this means that for any sequence of Lebesgue-integrable functions $f_1, f_2, ...$ with $|f_n|_p < \infty$, satisfying the condition

$$\lim_{k, n \to \infty} \int_\Omega |f_k(x) - f_n(x)|^p \, dx = 0$$

there exists a function $f(x)$ belonging to the vector space $L^p(\Omega)$ such that

$$\lim_{k\to\infty} \int_{\Omega} |f(x) - f_k(x)|^p \, dx = 0.$$

Imposing boundedness conditions not only on the function, but also on its derivatives leads to Sobolev spaces.

Hilbert Spaces

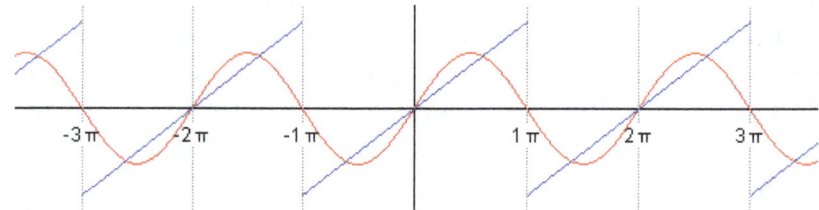

The succeeding snapshots show summation of 1 to 5 terms in approximating a periodic function (blue) by finite sum of sine functions (red).

Complete inner product spaces are known as *Hilbert spaces*, in honor of David Hilbert. The Hilbert space $L^2(\Omega)$, with inner product given by

$$\langle f, g \rangle = \int_{\Omega} f(x)\overline{g(x)}dx,$$

where $\overline{g(x)}$ denotes the complex conjugate of $g(x)$, is a key case.

By definition, in a Hilbert space any Cauchy sequence converges to a limit. Conversely, finding a sequence of functions f_n with desirable properties that approximates a given limit function, is equally crucial. Early analysis, in the guise of the Taylor approximation, established an approximation of differentiable functions f by polynomials. By the *Stone–Weierstrass theorem*, every continuous function on $[a, b]$ can be approximated as closely as desired by a polynomial. A similar approximation technique by trigonometric functions is commonly called Fourier expansion, and is much applied in engineering. *More generally, and more conceptually, the theorem yields a simple description of what "basic functions", or, in abstract Hilbert spaces, what basic vectors suffice to generate a Hilbert space H, in the sense that the closure of their span (i.e., finite linear combinations and limits of those) is the whole space.* Such a set of functions is called a *basis of H*, its cardinality is known as the Hilbert space dimension. Not only does the theorem exhibit suitable basis functions as sufficient for approximation purposes, but together with the Gram-Schmidt process, it enables one to construct a basis of orthogonal vectors. Such orthogonal bases are the Hilbert space generalization of the coordinate axes in finite-dimensional Euclidean space.

The solutions to various differential equations can be interpreted in terms of Hilbert spaces. For example, a great many fields in physics and engineering lead to such equations and frequently solutions with particular physical properties are used as basis functions, often orthogonal. As an example from physics, the time-dependent Schrödinger equation in quantum mechanics describes the change of physical properties in time by means of a partial differential equation, whose solutions are called wavefunctions. Definite values for physical properties such as energy, or momentum, correspond to eigenvalues of a certain (linear) differential operator and the associated wave-functions are called eigenstates. The *spectral theorem decomposes a linear compact operator acting on functions in terms of these eigenfunctions and their eigenvalues.*

Algebras Over Fields

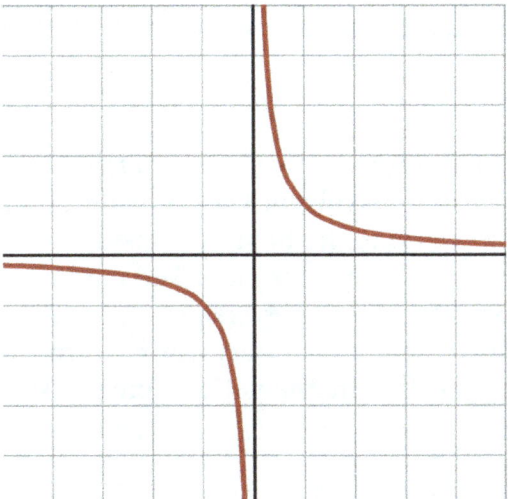

A hyperbola, given by the equation $x \cdot y = 1$. The coordinate ring of functions on this hyperbola is given by $R[x, y] / (x \cdot y - 1)$, an infinite-dimensional vector space over R.

General vector spaces do not possess a multiplication between vectors. A vector space equipped with an additional bilinear operator defining the multiplication of two vectors is an *algebra over a field*. Many algebras stem from functions on some geometrical object: since functions with values in a given field can be multiplied pointwise, these entities form algebras. The Stone–Weierstrass theorem mentioned above, for example, relies on Banach algebras which are both Banach spaces and algebras.

Commutative algebra makes great use of rings of polynomials in one or several variables, introduced above. Their multiplication is both commutative and associative. These rings and their quotients form the basis of algebraic geometry, because they are rings of functions of algebraic geometric objects.

Another crucial example are *Lie algebras*, which are neither commutative nor associative, but the failure to be so is limited by the constraints ($[x, y]$ denotes the product of x and y):

- $[x, y] = -[y, x]$ (anticommutativity), and

- $[x, [y, z]] + [y, [z, x]] + [z, [x, y]] = 0$ (Jacobi identity).

Examples include the vector space of n-by-n matrices, with $[x, y] = xy - yx$, the commutator of two matrices, and R^3, endowed with the cross product.

The tensor algebra $T(V)$ is a formal way of adding products to any vector space V to obtain an algebra. As a vector space, it is spanned by symbols, called simple tensors

$$v_1 \otimes v_2 \otimes \ldots \otimes v_n, \text{ where the degree } n \text{ varies.}$$

The multiplication is given by concatenating such symbols, imposing the distributive law under addition, and requiring that scalar multiplication commute with the tensor product $\otimes$, much the same way as with the tensor product of two vector spaces introduced above. In general, there are no relations between $v_1 \otimes v_2$ and $v_2 \otimes v_1$. Forcing two such elements to be equal leads to the symmetric algebra, whereas forcing $v_1 \otimes v_2 = -v_2 \otimes v_1$ yields the exterior algebra.

When a field, F is explicitly stated, a common term used is F-algebra.

Applications

Vector spaces have manifold applications as they occur in many circumstances, namely wherever functions with values in some field are involved. They provide a framework to deal with analytical and geometrical problems, or are used in the Fourier transform. This list is not exhaustive: many more applications exist, for example in optimization. The minimax theorem of game theory stating the existence of a unique payoff when all players play optimally can be formulated and proven using vector spaces methods. Representation theory fruitfully transfers the good understanding of linear algebra and vector spaces to other mathematical domains such as group theory.

Distributions

A *distribution* (or *generalized function*) is a linear map assigning a number to each "test" function, typically a smooth function with compact support, in a continuous way: in the above terminology the space of distributions is the (continuous) dual of the test function space. The latter space is endowed with a topology that takes into account not only f itself, but also all its higher derivatives. A standard example is the result of integrating a test function f over some domain Ω:

$$I(f) = \int_\Omega f(x)dx.$$

When $\Omega = \{p\}$, the set consisting of a single point, this reduces to the Dirac distribution, denoted by δ, which associates to a test function f its value at the p: $\delta(f) = f(p)$. Distributions are a powerful instrument to solve differential equations. Since all standard analytic notions such as derivatives are linear, they extend naturally to the space of distributions. Therefore, the equation in question can be transferred to a distribution space, which is bigger than the underlying function space, so that more flexible methods are available for solving the equation. For example, Green's functions and fundamental solutions are usually distributions rather than proper functions, and can then be used to find solutions of the equation with prescribed boundary conditions. The found solution can then in some cases be proven to be actually a true function, and a solution to the original equation (e.g., using the Lax–Milgram theorem, a consequence of the Riesz representation theorem).

Fourier Analysis

Resolving a periodic function into a sum of trigonometric functions forms a *Fourier series*, a technique much used in physics and engineering. The underlying vector space is usually the Hilbert space $L^2(0, 2\pi)$, for which the functions $\sin mx$ and $\cos mx$ (m an integer) form an orthogonal basis. The Fourier expansion of an L^2 function f is

$$\frac{a_0}{2} + \sum_{m=1}^{\infty} \left[a_m \cos(mx) + b_m \sin(mx) \right].$$

The coefficients a_m and b_m are called Fourier coefficients of f, and are calculated by the formulas

$$a_m = \frac{1}{\pi} \int_0^{2\pi} f(t) \cos(mt) dt, \quad b_m = \frac{1}{\pi} \int_0^{2\pi} f(t) \sin(mt) dt.$$

In physical terms the function is represented as a superposition of sine waves and the coefficients give information about the function's frequency spectrum. A complex-number form of Fourier series is also commonly used. The concrete formulae above are consequences of a more general mathematical duality called Pontryagin duality. Applied to the group R, it yields the classical Fourier transform; an application in physics are reciprocal lattices, where the underlying group is a finite-dimensional real vector space endowed with the additional datum of a lattice encoding positions of atoms in crystals.

Fourier series are used to solve boundary value problems in partial differential equations. In 1822, Fourier first used this technique to solve the heat equation. A discrete version of the Fourier series can be used in sampling applications where the function value is known only at a finite number of equally spaced points. In this case the Fourier series is finite and its value is equal to the sampled values at all points. The set of coefficients is known as the discrete Fourier transform (DFT) of the given sample sequence. The DFT is one of the key tools of digital signal processing, a field whose applications include radar, speech encoding, image compression. The JPEG image format is an application of the closely related discrete cosine transform.

The fast Fourier transform is an algorithm for rapidly computing the discrete Fourier transform. It is used not only for calculating the Fourier coefficients but, using the convolution theorem, also for computing the convolution of two finite sequences. They in turn are applied in digital filters and as a rapid multiplication algorithm for polynomials and large integers (Schönhage-Strassen algorithm).

Differential Geometry

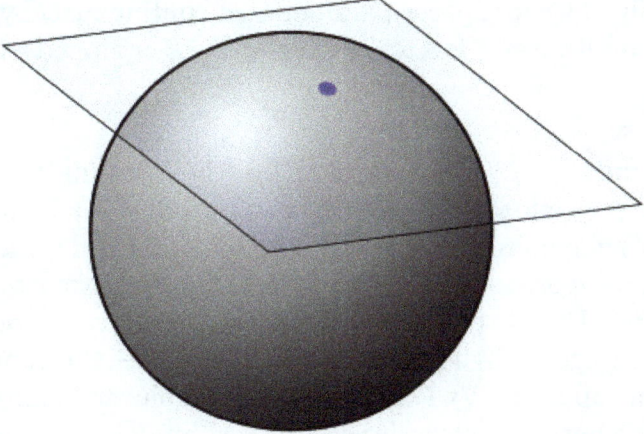

The tangent space to the 2-sphere at some point is the infinite plane touching the sphere in this point.

The tangent plane to a surface at a point is naturally a vector space whose origin is identified with the point of contact. The tangent plane is the best linear approximation, or linearization, of a surface at a point. Even in a three-dimensional Euclidean space, there is typically no natural way to prescribe a basis of the tangent plane, and so it is conceived of as an abstract vector space rather than a real coordinate space. The *tangent space* is the generalization to higher-dimensional differentiable manifolds.

Riemannian manifolds are manifolds whose tangent spaces are endowed with a suitable inner product. Derived therefrom, the Riemann curvature tensor encodes all curvatures of a manifold in one object, which finds applications in general relativity, for example, where the Einstein curvature tensor describes the matter and energy content of space-time. The tangent space of a Lie group can be given naturally the structure of a Lie algebra and can be used to classify compact Lie groups.

Generalizations

Vector Bundles

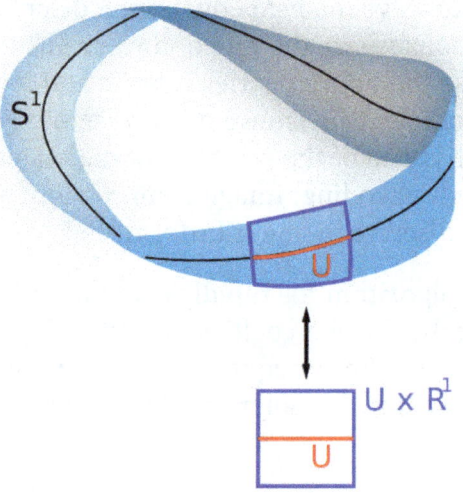

A Möbius strip. Locally, it looks like $U \times R$.

A *vector bundle* is a family of vector spaces parametrized continuously by a topological space X. More precisely, a vector bundle over X is a topological space E equipped with a continuous map

$$\pi : E \to X$$

such that for every x in X, the fiber $\pi^{-1}(x)$ is a vector space. The case dim $V = 1$ is called a line bundle. For any vector space V, the projection $X \times V \to X$ makes the product $X \times V$ into a "trivial" vector bundle. Vector bundles over X are required to be locally a product of X and some (fixed) vector space V: for every x in X, there is a neighborhood U of x such that the restriction of π to $\pi^{-1}(U)$ is isomorphic to the trivial bundle $U \times V \to U$. Despite their locally trivial character, vector bundles may (depending on the shape of the underlying space X) be "twisted" in the large (i.e., the bundle need not be (globally isomorphic to) the trivial bundle $X \times V$). For example, the Möbius strip can be seen as a line bundle over the circle S^1 (by identifying open intervals with the real line). It is, however, different from the cylinder $S^1 \times R$, because the latter is orientable whereas the former is not.

Properties of certain vector bundles provide information about the underlying topological space. For example, the tangent bundle consists of the collection of tangent spaces parametrized by the points of a differentiable manifold. The tangent bundle of the circle S^1 is globally isomorphic to $S^1 \times R$, since there is a global nonzero vector field on S^1. In contrast, by the hairy ball theorem, there is no (tangent) vector field on the 2-sphere S^2 which is everywhere nonzero. K-theory studies the isomorphism classes of all vector bundles over some topological space. In addition to deepening topological and geometrical insight, it has purely algebraic consequences, such as the classification of finite-dimensional real division algebras: R, C, the quaternions H and the octonions O.

The cotangent bundle of a differentiable manifold consists, at every point of the manifold, of the dual of the tangent space, the cotangent space. Sections of that bundle are known as differential one-forms.

Modules

Modules are to rings what vector spaces are to fields: the same axioms, applied to a ring R instead of a field F, yield modules. The theory of modules, compared to that of vector spaces, is complicated by the presence of ring elements that do not have multiplicative inverses. For example, modules need not have bases, as the Z-module (i.e., abelian group) Z/2Z shows; those modules that do (including all vector spaces) are known as free modules. Nevertheless, a vector space can be compactly defined as a module over a ring which is a field with the elements being called vectors. Some authors use the term *vector space* to mean modules over a division ring. The algebro-geometric interpretation of commutative rings via their spectrum allows the development of concepts such as locally free modules, the algebraic counterpart to vector bundles.

Affine and Projective Spaces

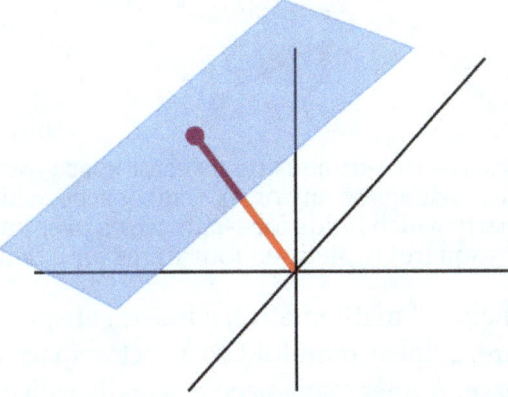

An affine plane (light blue) in R³. It is a two-dimensional subspace shifted by a vector x (red).

Roughly, *affine spaces* are vector spaces whose origins are not specified. More precisely, an affine space is a set with a free transitive vector space action. In particular, a vector space is an affine space over itself, by the map

$$V \times V \to V, (v, a) \mapsto a + v.$$

If W is a vector space, then an affine subspace is a subset of W obtained by translating a linear subspace V by a fixed vector $x \in W$; this space is denoted by $x + V$ (it is a coset of V in W) and consists of all vectors of the form x + v for v ∈ V. An important example is the space of solutions of a system of inhomogeneous linear equations

$$Ax = b$$

generalizing the homogeneous case b = 0 above. The space of solutions is the affine subspace x + V where x is a particular solution of the equation, and V is the space of solutions of the homogeneous equation (the nullspace of A).

The set of one-dimensional subspaces of a fixed finite-dimensional vector space V is known as *projective space*; it may be used to formalize the idea of parallel lines intersecting at infinity. Grassmannians and flag manifolds generalize this by parametrizing linear subspaces of fixed dimension k and flags of subspaces, respectively.

Linear Subspace

If the reduced row echelon form of A is

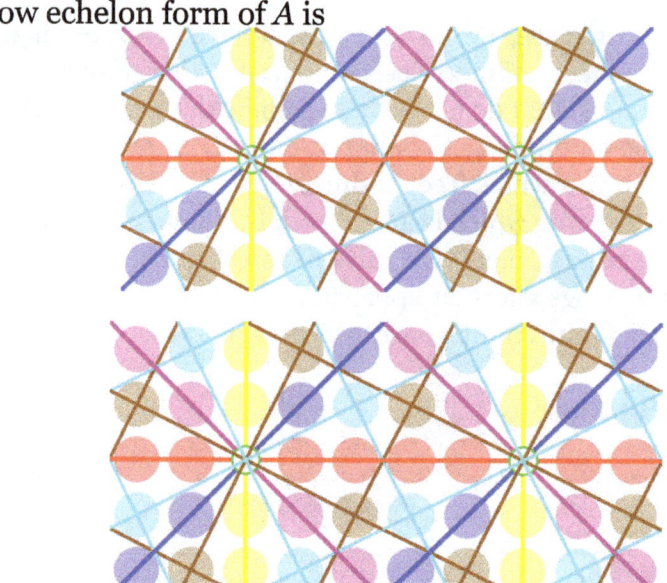

One-dimensional subspaces in the two-dimensional vector space over the finite field F_5. The origin (0, 0), marked with green circles, belongs to any of six 1-subspaces, while each of 24 remaining points belongs to exactly one; a property which holds for 1-subspaces over any field and in all dimensions. All F_5^2 (i.e. a 5 × 5 square) is pictured four times for a better visualization

In linear algebra and related fields of mathematics, a linear subspace, also known as a vector subspace, or, in the older literature, a linear manifold, is a vector space that is a subset of some other (higher-dimension) vector space. A linear subspace is usually called simply a *subspace* when the context serves to distinguish it from other kinds of subspaces.

Definition and Useful Characterization of Subspace

Let K be a field (such as the real numbers), and let V be a vector space over K. As usual, we call elements of V *vectors* and call elements of K *scalars*. Ignoring the full extent of mathematical generalization, scalars can be understood simply as numbers. Suppose that W is a subset of V. If W is a vector space itself (which means that it is closed under operations of addition and scalar multiplication), with the same vector space operations as V has, then W is a subspace of V.

To use this definition, we don't have to prove that all the properties of a vector space hold for W. Instead, we can prove a theorem that gives us an easier way to show that a subset of a vector space is a subspace.

Theorem: Let V be a vector space over the field K, and let W be a subset of V. Then W is a subspace if and only if W satisfies the following three conditions:

1. The zero vector, 0, is in W.

2. If u and v are elements of W, then the sum u + v is an element of W;

3. If u is an element of W and c is a scalar from K, then the product cu is an element of W.

Proof: Firstly, property 1 ensures W is nonempty. Looking at the definition of a vector space, we see that properties 2 and 3 above assure closure of W under addition and scalar multiplication, so the vector space operations are well defined. Since elements of W are necessarily elements of V, axioms 1, 2 and 5–8 of a vector space are satisfied. By the closure of W under scalar multiplication (specifically by 0 and –1), the vector space's definitional axiom identity element of addition and axiom inverse element of addition are satisfied.

Conversely, if W is a subspace of V, then W is itself a vector space under the operations induced by V, so properties 2 and 3 are satisfied. By property 3, –w is in W whenever w is, and it follows that W is closed under subtraction as well. Since W is nonempty, there is an element x in W, and $\mathbf{x} - \mathbf{x} = \mathbf{0}$ is in W, so property 1 is satisfied. One can also argue that since W is nonempty, there is an element x in W, and 0 is in the field K so $0\mathbf{x} = \mathbf{0}$ and therefore property 1 is satisfied.

Examples

Example I: Let the field K be the set R of real numbers, and let the vector space V be the real co-ordinate space R³. Take W to be the set of all vectors in V whose last component is 0. Then W is a subspace of V.

Proof:

1. Given u and v in W, then they can be expressed as u = $(u_1, u_2, 0)$ and v = $(v_1, v_2, 0)$. Then u + v = $(u_1+v_1, u_2+v_2, 0+0) = (u_1+v_1, u_2+v_2, 0)$. Thus, u + v is an element of W, too.

2. Given u in W and a scalar c in R, if u = $(u_1, u_2, 0)$ again, then cu = $(cu_1, cu_2, c0) = (cu_1, cu_2, 0)$. Thus, cu is an element of W too.

Example II: Let the field be R again, but now let the vector space be the Cartesian plane R². Take W to be the set of points (x, y) of R² such that $x = y$. Then W is a subspace of R².

Proof:

1. Let p = (p_1, p_2) and q = (q_1, q_2) be elements of W, that is, points in the plane such that $p_1 = p_2$ and $q_1 = q_2$. Then p + q = (p_1+q_1, p_2+q_2); since $p_1 = p_2$ and $q_1 = q_2$, then $p_1 + q_1 = p_2 + q_2$, so p + q is an element of W.

2. Let p = (p_1, p_2) be an element of W, that is, a point in the plane such that $p_1 = p_2$, and let

c be a scalar in R. Then cp = (cp_1, cp_2); since p_1 = p_2, then cp_1 = cp_2, so cp is an element of *W*.

In general, any subset of the real coordinate space R^n that is defined by a system of homogeneous linear equations will yield a subspace. (The equation in example I was $z = 0$, and the equation in example II was $x = y$.) Geometrically, these subspaces are points, lines, planes, and so on, that pass through the point 0.

Examples Related to Calculus

Example III: Again take the field to be R, but now let the vector space *V* be the set R^R of all functions from R to R. Let C(R) be the subset consisting of continuous functions. Then C(R) is a subspace of R^R.

Proof:

1. We know from calculus that $0 \in C(R) \subset R^R$.

2. We know from calculus that the sum of continuous functions is continuous.

3. Again, we know from calculus that the product of a continuous function and a number is continuous.

Example IV: Keep the same field and vector space as before, but now consider the set Diff(R) of all differentiable functions. The same sort of argument as before shows that this is a subspace too.

Examples that extend these themes are common in functional analysis.

Properties of Subspaces

A way to characterize subspaces is that they are closed under linear combinations. That is, a non-empty set *W* is a subspace if and only if every linear combination of (finitely many) elements of *W* also belongs to *W*. Conditions 2 and 3 for a subspace are simply the most basic kinds of linear combinations.

In a topological vector space *X*, a subspace *W* need not be closed in general, but a finite-dimensional subspace is always closed. The same is true for subspaces of finite codimension, i.e. determined by a finite number of continuous linear functionals.

Descriptions

Descriptions of subspaces include the solution set to a homogeneous system of linear equations, the subset of Euclidean space described by a system of homogeneous linear parametric equations, the span of a collection of vectors, and the null space, column space, and row space of a matrix. Geometrically (especially, over the field of real numbers and its subfields), a subspace is a flat in an *n*-space that passes through the origin.

A natural description of an 1-subspace is the scalar multiplication of one non-zero vector v to all possible scalar values. 1-subspaces specified by two vectors are equal if and only if one vector can be obtained from another with scalar multiplication:

$$\exists c \in K : \mathbf{v}' = c\mathbf{v} \ (\text{or } \mathbf{v} = \frac{1}{c}\mathbf{v}')$$

This idea is generalized for higher dimensions with linear span, but criteria for equality of k-spaces specified by sets of k vectors are not so simple.

A dual description is provided with linear functionals (usually implemented as linear equations). One non-zero linear functional F specifies its kernel subspace F = 0 of codimension 1. Subspaces of codimension 1 specified by two linear functionals are equal if and only if one functional can be obtained from another with scalar multiplication (in the dual space):

$$\exists c \in K : \mathbf{F}' = c\mathbf{F} \ (\text{or } \mathbf{F} = \frac{1}{c}\mathbf{F}')$$

It is generalized for higher codimensions with a system of equations. The following two subsections will present this latter description in details, and the remaining four subsections further describe the idea of linear span.

Systems of Linear Equations

The solution set to any homogeneous system of linear equations with n variables is a subspace in the coordinate space K^n:

$$\left\{ \begin{bmatrix} x_1 \\ x_2 \\ \vdots \\ x_n \end{bmatrix} \in K^n : \begin{array}{ccccc} a_{11}x_1 + & a_{12}x_2 & + \cdots + & a_{1n}x_n & = 0 \\ a_{21}x_1 + & a_{22}x_2 & + \cdots + & a_{2n}x_n & = 0 \\ \vdots & \vdots & & \vdots & \vdots \\ a_{m1}x_1 + & a_{m2}x_2 & + \cdots + & a_{mn}x_n & = 0 \end{array} \right\}.$$

For example (over real or rational numbers), the set of all vectors (x, y, z) satisfying the equations

$$x + 3y + 2z = 0 \quad \text{and} \quad 2x - 4y + 5z = 0$$

is a one-dimensional subspace. More generally, that is to say that given a set of n independent functions, the dimension of the subspace in K^k will be the dimension of the null set of A, the composite matrix of the n functions.

Null Space of a Matrix

In a finite-dimensional space, a homogeneous system of linear equations can be written as a single matrix equation:

$$A\mathbf{x} = \mathbf{0}.$$

The set of solutions to this equation is known as the null space of the matrix. For example, the subspace described above is the null space of the matrix

$$A = \begin{bmatrix} 1 & 3 & 2 \\ 3 & -4 & 5 \end{bmatrix}.$$

Every subspace of K^n can be described as the null space of some matrix.

Linear Parametric Equations

The subset of K^n described by a system of homogeneous linear parametric equations is a subspace:

$$\left\{ \begin{bmatrix} x_1 \\ x_2 \\ \vdots \\ x_n \end{bmatrix} \in K^n : \begin{matrix} x_1 = & a_{11}t_1 & + a_{12}t_2 & + \cdots + & a_{1m}t_m \\ x_2 = & a_{21}t_1 & + a_{22}t_2 & + \cdots + & a_{2m}t_m \\ \vdots & \vdots & \vdots & & \vdots \\ x_n = & a_{n1}t_1 & + a_{n2}t_2 & + \cdots + & a_{nm}t_m \end{matrix} \text{ for some } t_1, \ldots, t_m \in K \right\}.$$

For example, the set of all vectors (x, y, z) parameterized by the equations

$$x = 2t_1 + 3t_2, \quad y = 5t_1 - 4t_2, \quad \text{and} \quad z = -t_1 + 2t_2$$

is a two-dimensional subspace of K^3, if K is a number field (such as real or rational numbers).

Span of Vectors

In linear algebra, the system of parametric equations can be written as a single vector equation:

$$\begin{bmatrix} x \\ y \\ z \end{bmatrix} = t_1 \begin{bmatrix} 2 \\ 5 \\ -1 \end{bmatrix} + t_2 \begin{bmatrix} 3 \\ -4 \\ 2 \end{bmatrix}.$$

The expression on the right is called a linear combination of the vectors $(2, 5, -1)$ and $(3, -4, 2)$. These two vectors are said to span the resulting subspace.

In general, a linear combination of vectors $v_1, v_2, \ldots, v_k$ is any vector of the form

$$t_1 v_1 + \cdots + t_k v_k.$$

The set of all possible linear combinations is called the span:

$$\text{Span}\{v_1, \ldots, v_k\} = \{t_1 v_1 + \cdots + t_k v_k : t_1, \ldots, t_k \in K\}.$$

If the vectors $v_1, \ldots, v_k$ have n components, then their span is a subspace of K^n. Geometrically, the span is the flat through the origin in n-dimensional space determined by the points $v_1, \ldots, v_k$.

Example

　　　The xz-plane in $\mathbb{R}^3$ can be parameterized by the equations

$$x = t_1, \quad y = 0, \quad z = t_2.$$

As a subspace, the xz-plane is spanned by the vectors (1, 0, 0) and (0, 0, 1). Every vector in the xz-plane can be written as a linear combination of these two:

$$(t_1, 0, t_2) = t_1(1,0,0) + t_2(0,0,1).$$

Geometrically, this corresponds to the fact that every point on the xz-plane can be reached from the origin by first moving some distance in the direction of (1, 0, 0) and then moving some distance in the direction of (0, 0, 1).

Column Space and Row Space

A system of linear parametric equations in a finite-dimensional space can also be written as a single matrix equation:

$$\mathbf{x} = A\mathbf{t} \quad \text{where} \quad A = \begin{bmatrix} 2 & 3 \\ 5 & -4 \\ -1 & 2 \end{bmatrix}$$

In this case, the subspace consists of all possible values of the vector x. In linear algebra, this subspace is known as the column space (or image) of the matrix A. It is precisely the subspace of K^n spanned by the column vectors of A.

The row space of a matrix is the subspace spanned by its row vectors. The row space is interesting because it is the orthogonal complement of the null space.

Independence, Basis, and Dimension

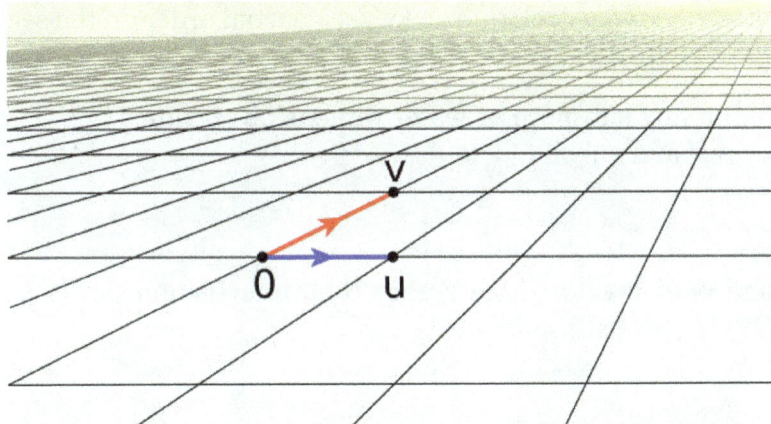

The vectors u and v are a basis for this two-dimensional subspace of R³.

In general, a subspace of K^n determined by k parameters (or spanned by k vectors) has dimension k. However, there are exceptions to this rule. For example, the subspace of K^3 spanned by the three vectors (1, 0, 0), (0, 0, 1), and (2, 0, 3) is just the xz-plane, with each point on the plane described by infinitely many different values of t_1, t_2, t_3.

In general, vectors $v_1, \ldots, v_k$ are called linearly independent if

$$t_1 \mathbf{v}_1 + \cdots + t_k \mathbf{v}_k \neq u_1 \mathbf{v}_1 + \cdots + u_k \mathbf{v}_k$$

for $(t_1, t_2, \ldots, t_k) \neq (u_1, u_2, \ldots, u_k)$. If $v_1, \ldots, v_k$ are linearly independent, then the coordinates $t_1, \ldots, t_k$ for a vector in the span are uniquely determined.

A basis for a subspace S is a set of linearly independent vectors whose span is S. The number of elements in a basis is always equal to the geometric dimension of the subspace. Any spanning set for a subspace can be changed into a basis by removing redundant vectors.

Example

Let S be the subspace of $\mathbf{R}^4$ defined by the equations

$$x_1 = 2x_2 \quad \text{and} \quad x_3 = 5x_4.$$

Then the vectors $(2, 1, 0, 0)$ and $(0, 0, 5, 1)$ are a basis for S. In particular, every vector that satisfies the above equations can be written uniquely as a linear combination of the two basis vectors:

$$(2t_1, t_1, 5t_2, t_2) = t_1(2,1,0,0) + t_2(0,0,5,1).$$

The subspace S is two-dimensional. Geometrically, it is the plane in $\mathbf{R}^4$ passing through the points $(0, 0, 0, 0)$, $(2, 1, 0, 0)$, and $(0, 0, 5, 1)$.

Operations and Relations on Subspaces

Inclusion

The set-theoretical inclusion binary relation specifies a partial order on the set of all subspaces (of any dimension).

A subspace cannot lie in any subspace of lesser dimension. If dim $U = k$, a finite number, and $U \subset W$, then dim $W = k$ if and only if $U = W$.

Intersection

Given subspaces U and W of a vector space V, then their intersection $U \cap W := \{v \in V : v \text{ is an element of both } U \text{ and } W\}$ is also a subspace of V.

Proof:

1. Let v and w be elements of $U \cap W$. Then v and w belong to both U and W. Because U is a subspace, then v + w belongs to U. Similarly, since W is a subspace, then v + w belongs to W. Thus, v + w belongs to $U \cap W$.

2. Let v belong to $U \cap W$, and let c be a scalar. Then v belongs to both U and W. Since U and W are subspaces, cv belongs to both U and W.

3. Since U and W are vector spaces, then o belongs to both sets. Thus, o belongs to $U \cap W$.

For every vector space V, the set $\{o\}$ and V itself are subspaces of V.

Sum

If U and W are subspaces, their sum is the subspace

$$U + W = \{\mathbf{u} + \mathbf{w} : \mathbf{u} \in U, \mathbf{w} \in W\}.$$

For example, the sum of two lines is the plane that contains them both. The dimension of the sum satisfies the inequality

$$\max(\dim U, \dim W) \leq \dim(U + W) \leq \dim(U) + \dim(W).$$

Here the minimum only occurs if one subspace is contained in the other, while the maximum is the most general case. The dimension of the intersection and the sum are related:

$$\dim(U + W) = \dim(U) + \dim(W) - \dim(U \cap W).$$

Lattice of Subspaces

The operations intersection and sum make the set of all subspaces a bounded modular lattice, where the $\{o\}$ subspace, the least element, is an identity element of the sum operation, and the identical subspace V, the greatest element, is an identity element of the intersection operation.

Other

If V is an inner product space, then the orthogonal complement $\perp$ of any subspace of V is again a subspace. This operation, understood as negation ($\neg$), makes the lattice of subspaces a (possibly infinite) orthocomplemented lattice (it is not a distributive lattice).

In a pseudo-Euclidean space there are orthogonal complements too, but such operation does not form a Boolean algebra (nor a Heyting algebra) because of null subspaces, for which $N \cap N^{\perp} = N \neq \{o\}$. The same case presents the $\perp$ operation in symplectic vector spaces.

Algorithms

Most algorithms for dealing with subspaces involve row reduction. This is the process of applying elementary row operations to a matrix until it reaches either row echelon form or reduced row echelon form. Row reduction has the following important properties:

1. The reduced matrix has the same null space as the original.

2. Row reduction does not change the span of the row vectors, i.e. the reduced matrix has the same row space as the original.

3. Row reduction does not affect the linear dependence of the column vectors.

Basis for a Row Space

Input An $m \times n$ matrix A.

Output A basis for the row space of A.

1. Use elementary row operations to put A into row echelon form.

2. The nonzero rows of the echelon form are a basis for the row space of A.

If we instead put the matrix A into reduced row echelon form, then the resulting basis for the row space is uniquely determined. This provides an algorithm for checking whether two row spaces are equal and, by extension, whether two subspaces of K^n are equal.

Subspace Membership

Input A basis $\{b_1, b_2, ..., b_k\}$ for a subspace S of K^n, and a vector v with n components.

Output Determines whether v is an element of S

1. Create a $(k + 1) \times n$ matrix A whose rows are the vectors $b_1, ... , b_k$ and v.

2. Use elementary row operations to put A into row echelon form.

3. If the echelon form has a row of zeroes, then the vectors $\{b_1, ..., b_k, v\}$ are linearly dependent, and therefore $v \in S$.

Basis for a Column Space

Input An $m \times n$ matrix A

Output A basis for the column space of A

1. Use elementary row operations to put A into row echelon form.

2. Determine which columns of the echelon form have pivots. The corresponding columns of the original matrix are a basis for the column space.

This produces a basis for the column space that is a subset of the original column vectors. It works because the columns with pivots are a basis for the column space of the echelon form, and row reduction does not change the linear dependence relationships between the columns.

Coordinates for a Vector

Input A basis $\{b_1, b_2, ..., b_k\}$ for a subspace S of K^n, and a vector $v \in S$

Output Numbers $t_1, t_2, ..., t_k$ such that $v = t_1 b_1 + \cdots + t_k b_k$

1. Create an augmented matrix A whose columns are $b_1,...,b_k$, with the last column being v.

2. Use elementary row operations to put A into reduced row echelon form.

3. Express the final column of the reduced echelon form as a linear combination of the first k columns. The coefficients used are the desired numbers $t_1, t_2, ..., t_k$. (These should be precisely the first k entries in the final column of the reduced echelon form.)

If the final column of the reduced row echelon form contains a pivot, then the input vector v does not lie in S.

Basis for a Null Space

Input An $m \times n$ matrix A.

Output A basis for the null space of A

1. Use elementary row operations to put A in reduced row echelon form.

2. Using the reduced row echelon form, determine which of the variables $x_1, x_2, ..., x_n$ are free. Write equations for the dependent variables in terms of the free variables.

3. For each free variable x_i, choose a vector in the null space for which $x_i = 1$ and the remaining free variables are zero. The resulting collection of vectors is a basis for the null space of A.

Basis for the Sum and Intersection of two Subspaces

Given two subspaces U and W of V, a basis of the sum $U + W$ and the intersection $U \cap W$ can be calculated using the Zassenhaus algorithm

Equations for a Subspace

Input A basis $\{b_1, b_2, ..., b_k\}$ for a subspace S of K^n

Output An $(n - k) \times n$ matrix whose null space is S.

1. Create a matrix A whose rows are $b_1, b_2, ..., b_k$.

2. Use elementary row operations to put A into reduced row echelon form.

3. Let $c_1, c_2, ..., c_n$ be the columns of the reduced row echelon form. For each column without a pivot, write an equation expressing the column as a linear combination of the columns with pivots.

4. This results in a homogeneous system of $n - k$ linear equations involving the variables $c_1, ..., c_n$. The $(n - k) \times n$ matrix corresponding to this system is the desired matrix with nullspace S.

Example

If the reduced row echelon form of A is

$$\begin{bmatrix} 1 & 0 & -3 & 0 & 2 & 0 \\ 0 & 1 & 5 & 0 & -1 & 4 \\ 0 & 0 & 0 & 1 & 7 & -9 \\ 0 & 0 & 0 & 0 & 0 & 0 \end{bmatrix}$$

then the column vectors $c_1, ..., c_6$ satisfy the equations

$$\mathbf{c}_3 = -3\mathbf{c}_1 + 5\mathbf{c}_2$$
$$\mathbf{c}_5 = 2\mathbf{c}_1 - \mathbf{c}_2 + 7\mathbf{c}_4$$
$$\mathbf{c}_6 = 4\mathbf{c}_2 - 9\mathbf{c}_4$$

It follows that the row vectors of A satisfy the equations

$$x_3 = -3x_1 + 5x_2$$
$$x_5 = 2x_1 - x_2 + 7x_4$$
$$x_6 = 4x_2 - 9x_4.$$

In particular, the row vectors of A are a basis for the null space of the corresponding matrix.

Linear Span

In linear algebra, the linear span (also called the linear hull or just span) of a set of vectors in a vector space is the intersection of all subspaces containing that set. The linear span of a set of vectors is therefore a vector space. Spans can be generalized to matroids and modules.

Definition

Given a vector space V over a field K, the span of a set S of vectors (not necessarily finite) is defined to be the intersection W of all subspaces of V that contain S. W is referred to as the subspace *spanned by S*, or by the vectors in S. Conversely, S is called a *spanning set* of W, and we say that *S spans W*.

Alternatively, the span of S may be defined as the set of all finite linear combinations of elements of S, which follows from the above definition.

$$\text{span}(S) = \left\{ \sum_{i=1}^{k} \lambda_i v_i \,\middle|\, k \in \mathbb{N}, v_i \in S, \lambda_i \in \mathbf{K} \right\}.$$

In particular, if S is a finite subset of V, then the span of S is the set of all linear combinations of the elements of S. In the case of infinite S, infinite linear combinations (i.e. where a combination may involve an infinite sum, assuming such sums are defined somehow, e.g. if V is a Banach space) are excluded by the definition; a generalization that allows these is not equivalent.

Examples

The real vector space $\mathbb{R}^3$ has $\{(1,0,0), (0,1,0), (0,0,1)\}$ as a spanning set. This particular spanning set is also a basis. If $(1,0,0)$ were replaced by $(-1,0,0)$, it would also form the canonical basis of $\mathbb{R}^3$.

Another spanning set for the same space is given by $\{(1,2,3), (0,1,2), (-1,1/2,3), (1,1,1)\}$, but this set is not a basis, because it is linearly dependent.

The set $\{(1,0,0), (0,1,0), (1,1,0)\}$ is not a spanning set of $\mathbb{R}^3$; instead its span is the space of all vectors in $\mathbb{R}^3$ whose last component is zero. That space (the space of all vectors in $\mathbb{R}^3$ whose last component is zero) is also spanned by the set $\{(1,0,0), (0,1,0)\}$, as $(1,1,0)$ is a linear combination of $(1,0,0)$ and $(0,1,0)$. It does, however, span $\mathbb{R}^2$.

The set of functions x^n where n is a non-negative integer spans the space of polynomials.

Theorems

Theorem 1: The subspace spanned by a non-empty subset S of a vector space V is the set of all linear combinations of vectors in S.

This theorem is so well known that at times it is referred to as the definition of span of a set.

Theorem 2: Every spannin set S of a vector space V must contain at least as many elements as any linearly independent set of vectors from V.

Theorem 3: Let V be a finite-dimensional vector space. Any set of vectors that spans V can be reduced to a basis for V by discarding vectors if necessary (i.e. if there are linearly dependent vectors in the set). If the axiom of choice holds, this is true without the assumption that V has finite dimension.

This also indicates that a basis is a minimal spanning set when V is finite-dimensional.

Generalizations

Generalizing the definition of the span of points in space, a subset X of the ground set of a matroid is called a *spanning set* if the rank of X equals the rank of the entire ground set.

The vector space definition can also be generalized to modules. Given an R-module A and any collection of elements $a_1,...,a_n$ of A, then the sum of cyclic modules,

$$Ra_1 + \cdots + Ra_n = \left\{ \sum_{k=1}^{n} r_k a_k \;\middle|\; r_k \in R \right\}$$

consisting of all R-linear combinations of the given elements a_i, is called the submodule of A spanned by $a_1,...,a_n$. As with the case of vector spaces, the submodule of A spanned by any subset of A is the intersection of all the submodules containing that subset.

Closed Linear Span (Functional Analysis)

In functional analysis, a closed linear span of a set of vectors is the minimal closed set which contains the linear span of that set.

Suppose that X is a normed vector space and let E be any non-empty subset of X. The closed linear span of E, denoted by $\overline{Sp}(E)$ or $\overline{Span}(E)$, is the intersection of all the closed linear subspaces of X which contain E.

One mathematical formulation of this is

$$\overline{Sp}(E) = \{u \in X \mid \forall \epsilon > 0 \exists x \in Sp(E) : \| x - u \| < \epsilon\}.$$

The closed linear span of the set of functions x^n on the interval [0, 1], where n is a non-negative integer, depends on the norm used. If the L^2 norm is used, then the closed linear span is the Hilbert space of square-integrable functions on the interval. But if the maximum norm is used, the closed linear span will be the space of continuous functions on the interval. In either case, the closed linear span contains functions that are not polynomials, and so are not in the linear span itself. However, the cardinality of the set of functions in the closed linear span is the cardinality of the continuum, which is the same cardinality as for the set of polynomials.

Notes

The linear span of a set is dense in the closed linear span. Moreover, as stated in the lemma below, the closed linear span is indeed the closure of the linear span.

Closed linear spans are important when dealing with closed linear subspaces (which are themselves highly important, consider Riesz's lemma).

A Useful Lemma

Let X be a normed space and let E be any non-empty subset of X. Then

(a)　$\overline{Sp}(E)$ is a closed linear subspace of X which contains E,

(b)　$\overline{Sp}(E) = \overline{Sp(E)}$, viz. $\overline{Sp}(E)$ is the closure of $Sp(E)$,

(c)　$E^{\perp} = (Sp(E))^{\perp} = (\overline{Sp(E)})^{\perp}$.

(So the usual way to find the closed linear span is to find the linear span first, and then the closure of that linear span.)

Linear Map

In mathematics, a linear map (also called a linear mapping, linear transformation or, in some contexts, linear function) is a mapping $V \to W$ between two modules (including vector spaces) that preserves (in the sense defined below) the operations of addition and scalar multiplication.

An important special case is when $V = W$, in which case the map is called a linear operator, or an endomorphism of V. Sometimes the term *linear function* has the same meaning as *linear map*, while in analytic geometry it does not.

A linear map always maps linear subspaces onto linear subspaces (possibly of a lower dimension); for instance it maps a plane through the origin to a plane, straight line or point. Linear maps can often be represented as matrices, and simple examples include rotation and reflection linear transformations.

In the language of abstract algebra, a linear map is a module homomorphism. In the language of category theory it is a morphism in the category of modules over a given ring.

Definition and First Consequences

Let V and W be vector spaces over the same field K. A function $f : V \to W$ is said to be a *linear map* if for any two vectors x and y in V and any scalar α in K, the following two conditions are satisfied:

$$f(\mathbf{x}+\mathbf{y}) = f(\mathbf{x})+f(\mathbf{y}) \qquad \text{additivity}$$

$$f(\alpha \mathbf{x}) = \alpha f(\mathbf{x}) \qquad \text{homogeneity of degree 1}$$

This is equivalent to requiring the same for any linear combination of vectors, i.e. that for any vectors $x_1, ..., x_m \in V$ and scalars $a_1, ..., a_m \in K$, the following equality holds:

$$f(a_1\mathbf{x}_1 + \cdots + a_m\mathbf{x}_m) = a_1 f(\mathbf{x}_1) + \cdots + a_m f(\mathbf{x}_m).$$

Denoting the zero elements of the vector spaces V and W by 0_V and 0_W respectively, it follows that $f(0_V) = 0_W$ because letting $\alpha = 0$ in the equation for homogeneity of degree 1,

$$f(\mathbf{0}_V) = f(0 \cdot \mathbf{0}_V) = 0 \cdot f(\mathbf{0}_V) = \mathbf{0}_W.$$

Occasionally, V and W can be considered to be vector spaces over different fields. It is then necessary to specify which of these ground fields is being used in the definition of "linear". If V and W are considered as spaces over the field K as above, we talk about K-linear maps. For example, the conjugation of complex numbers is an R-linear map $C \to C$, but it is not C-linear.

A linear map from V to K (with K viewed as a vector space over itself) is called a linear functional.

These statements generalize to any left-module $_R M$ over a ring R without modification, and to any right-module upon reversing of the scalar multiplication.

Examples

- The zero map between two left-modules (or two right-modules) over the same ring is always linear.

- The identity map on any module is a linear operator.

- Any homothecy centered in the origin of a vector space, $v \mapsto cv$ where c is a scalar, is a linear operator. This does not hold in general for modules, where such a map might only be semilinear.

- For real numbers, the map $x \mapsto x^2$ is not linear.

- For real numbers, the map $x \mapsto x + 1$ is not linear (but is an affine transformation; $y = x + 1$ is a linear equation, as the term is used in analytic geometry.)

- If A is a real $m \times n$ matrix, then A defines a linear map from R^n to R^m by sending the column vector $x \in R^n$ to the column vector $Ax \in R^m$. Conversely, any linear map between finite-dimensional vector spaces can be represented in this manner.

- Differentiation defines a linear map from the space of all differentiable functions to the space of all functions. It also defines a linear operator on the space of all smooth functions.

- The (definite) integral over some interval I is a linear map from the space of all real-valued integrable functions on I to R

- The (indefinite) integral (or antiderivative) with a fixed starting point defines a linear map from the space of all real-valued integrable functions on R to the space of all real-valued, differentiable, functions on R. Without a fixed starting point, an exercise in group theory will show that the antiderivative maps to the quotient group of the differentiables over the equivalence relation, "differ by a constant", which yields an identity class of the constant

  valued functions $\left(\int : I(\Re) \rightarrow D(\Re)/\Re \right)$.

- If V and W are finite-dimensional vector spaces over a field F, then functions that send linear maps $f : V \rightarrow W$ to $\dim_F(W) \times \dim_F(V)$ matrices in the way described in the sequel are themselves linear maps (indeed linear isomorphisms).

- The expected value of a random variable (which is in fact a function, and as such a member of a vector space) is linear, as for random variables X and Y we have $E[X + Y] = E[X] + E[Y]$ and $E[aX] = aE[X]$, but the variance of a random variable is not linear.

Matrices

If V and W are finite-dimensional vector spaces and a basis is defined for each vector space, then every linear map from V to W can be represented by a matrix. This is useful because it allows concrete calculations. Matrices yield examples of linear maps: if A is a real $m \times n$ matrix, then $f(x) = Ax$ describes a linear map $R^n \rightarrow R^m$.

Let $\{v_1, ..., v_n\}$ be a basis for V. Then every vector v in V is uniquely determined by the coefficients $c_1, ..., c_n$ in the field R:

$$c_1 \mathbf{v}_1 + \cdots + c_n \mathbf{v}_n.$$

If $f : V \to W$ is a linear map,

$$f(c_1 \mathbf{v}_1 + \cdots + c_n \mathbf{v}_n) = c_1 f(\mathbf{v}_1) + \cdots + c_n f(\mathbf{v}_n),$$

which implies that the function f is entirely determined by the vectors $f(v_1), ..., f(v_n)$. Now let $\{w_1, ..., w_m\}$ be a basis for W. Then we can represent each vector $f(v_j)$ as

$$f(\mathbf{v}_j) = a_{1j} \mathbf{w}_1 + \cdots + a_{mj} \mathbf{w}_m.$$

Thus, the function f is entirely determined by the values of a_{ij}. If we put these values into an $m \times n$ matrix M, then we can conveniently use it to compute the vector output of f for any vector in V. To get M, every column j of M is a vector

$$(a_{1j}, ..., a_{mj})^{\mathrm{T}}$$

corresponding to $f(v_j)$ as defined above. To define it more clearly, for some column j that corresponds to the mapping $f(v_j)$,

$$M = \begin{pmatrix} & & a_{1j} & & \\ & & \cdot & & \\ * & & \cdot & & * \\ & & \cdot & & \\ & & a_{mj} & & \end{pmatrix}$$

where M is the matrix of f. The symbol $*$ denotes that there are other columns which together with column j make up a total of n columns of M. In other words, every column $j = 1, ..., n$ has a corresponding vector $f(v_j)$ whose coordinates $a_{1j}, ..., a_{mj}$ are the elements of column j. A single linear map may be represented by many matrices. This is because the values of the elements of a matrix depend on the bases chosen.

The matrices of a linear transformation can be represented visually:

1. Matrix for T relative to B : A

2. Matrix for T relative to B' : A'

3. Transition matrix from B' to B : P

4. Transition matrix from B to B' : P^{-1}

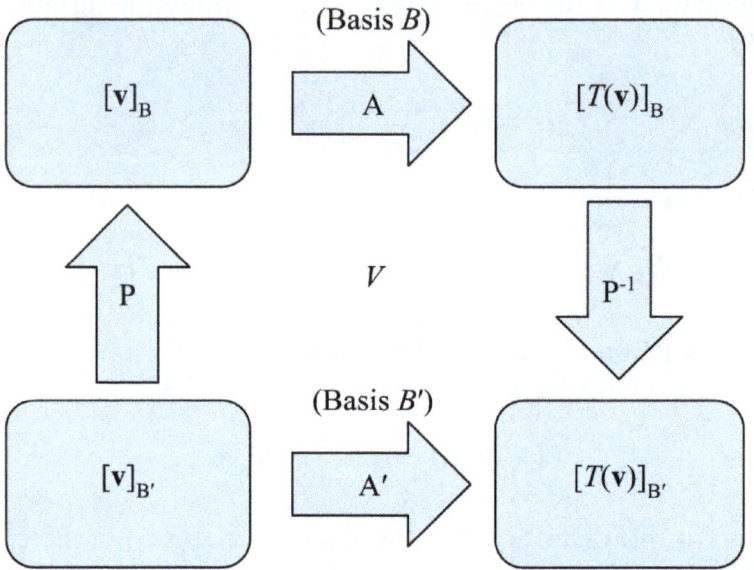

Such that starting in the bottom left corner $[\vec{v}]_{B'}$ and looking for the bottom right corner $[T(\vec{v})]_{B'}$, one would left-multiply—that is, $A'[\vec{v}]_{B'} = [T(\vec{v})]_{B'}$. The equivalent method would be the "longer" method going clockwise from the same point such that $[\vec{v}]_{B'}$ is left-multiplied with $P^{-1}AP$, or $P^{-1}AP[\vec{v}]_{B'} = [T(\vec{v})]_{B'}$.

Examples of Linear Transformation Matrices

In two-dimensional space R² linear maps are described by 2 × 2 real matrices. These are some examples:

- rotation by 90 degrees counterclockwise:

$$\mathbf{A} = \begin{pmatrix} 0 & -1 \\ 1 & 0 \end{pmatrix}$$

- rotation by angle θ counterclockwise:

$$\mathbf{A} = \begin{pmatrix} \cos\theta & -\sin\theta \\ \sin\theta & \cos\theta \end{pmatrix}$$

- reflection against the x axis:

$$\mathbf{A} = \begin{pmatrix} 1 & 0 \\ 0 & -1 \end{pmatrix}$$

- reflection against the y axis:

$$A = \begin{pmatrix} -1 & 0 \\ 0 & 1 \end{pmatrix}$$

- scaling by 2 in all directions:

$$A = \begin{pmatrix} 2 & 0 \\ 0 & 2 \end{pmatrix}$$

- horizontal shear mapping:

$$A = \begin{pmatrix} 1 & m \\ 0 & 1 \end{pmatrix}$$

- squeeze mapping:

$$A = \begin{pmatrix} k & 0 \\ 0 & 1/k \end{pmatrix}$$

- projection onto the y axis:

$$A = \begin{pmatrix} 0 & 0 \\ 0 & 1 \end{pmatrix}.$$

Forming new Linear Maps from Given Ones

The composition of linear maps is linear: if $f : V \to W$ and $g : W \to Z$ are linear, then so is their composition $g \circ f : V \to Z$. It follows from this that the class of all vector spaces over a given field K, together with K-linear maps as morphisms, forms a category.

The inverse of a linear map, when defined, is again a linear map.

If $f_1 : V \to W$ and $f_2 : V \to W$ are linear, then so is their pointwise sum $f_1 + f_2$ (which is defined by $(f_1 + f_2)(x) = f_1(x) + f_2(x)$).

If $f : V \to W$ is linear and a is an element of the ground field K, then the map af, defined by $(af)(x) = a(f(x))$, is also linear.

Thus the set $L(V, W)$ of linear maps from V to W itself forms a vector space over K, sometimes denoted Hom(V, W). Furthermore, in the case that $V = W$, this vector space (denoted End(V)) is an associative algebra under composition of maps, since the composition of two linear maps is again a linear map, and the composition of maps is always associative. This case is discussed in more detail below.

Given again the finite-dimensional case, if bases have been chosen, then the composition of linear maps corresponds to the matrix multiplication, the addition of linear maps corresponds to the matrix addition, and the multiplication of linear maps with scalars corresponds to the multiplication of matrices with scalars.

Endomorphisms and Automorphisms

A linear transformation $f: V \to V$ is an endomorphism of V; the set of all such endomorphisms End(V) together with addition, composition and scalar multiplication as defined above forms an associative algebra with identity element over the field K (and in particular a ring). The multiplicative identity element of this algebra is the identity map id: $V \to V$.

An endomorphism of V that is also an isomorphism is called an automorphism of V. The composition of two automorphisms is again an automorphism, and the set of all automorphisms of V forms a group, the automorphism group of V which is denoted by Aut(V) or GL(V). Since the automorphisms are precisely those endomorphisms which possess inverses under composition, Aut(V) is the group of units in the ring End(V).

If V has finite dimension n, then End(V) is isomorphic to the associative algebra of all $n \times n$ matrices with entries in K. The automorphism group of V is isomorphic to the general linear group GL(n, K) of all $n \times n$ invertible matrices with entries in K.

Kernel, Image and the Rank–nullity Theorem

If $f: V \to W$ is linear, we define the kernel and the image or range of f by

$$\ker(f) = \{x \in V : f(x) = 0\}$$
$$\operatorname{im}(f) = \{w \in W : w = f(x), x \in V\}$$

$\ker(f)$ is a subspace of V and $\operatorname{im}(f)$ is a subspace of W. The following dimension formula is known as the rank–nullity theorem:

$$\dim(\ker(f)) + \dim(\operatorname{im}(f)) = \dim(V)$$

The number $\dim(\operatorname{im}(f))$ is also called the *rank of* f and written as rank(f), or sometimes, $\rho(f)$; the number $\dim(\ker(f))$ is called the *nullity of* f and written as null(f) or $v(f)$. If V and W are finite-dimensional, bases have been chosen and f is represented by the matrix A, then the rank and nullity of f are equal to the rank and nullity of the matrix A, respectively.

Cokernel

A subtler invariant of a linear transformation $f : V \to W$ is the *cokernel*, which is defined as

$$\operatorname{coker} f := W / f(V) = W / \operatorname{im}(f).$$

This is the *dual* notion to the kernel: just as the kernel is a *sub*space of the *domain,* the co-kernel is a *quotient* space of the *target.* Formally, one has the exact sequence

$$0 \to \ker f \to V \to W \to \operatorname{coker} f \to 0.$$

These can be interpreted thus: given a linear equation $f(v) = w$ to solve,

- the kernel is the space of *solutions* to the *homogeneous* equation $f(v) = 0$, and its dimension is the number of *degrees of freedom* in a solution, if it exists;

- the co-kernel is the space of *constraints* that must be satisfied if the equation is to have a solution, and its dimension is the number of constraints that must be satisfied for the equation to have a solution.

The dimension of the co-kernel and the dimension of the image (the rank) add up to the dimension of the target space. For finite dimensions, this means that the dimension of the quotient space $W/f(V)$ is the dimension of the target space minus the dimension of the image.

As a simple example, consider the map $f: R^2 \to R^2$, given by $f(x, y) = (0, y)$. Then for an equation $f(x, y) = (a, b)$ to have a solution, we must have $a = 0$ (one constraint), and in that case the solution space is (x, b) or equivalently stated, $(0, b) + (x, 0)$, (one degree of freedom). The kernel may be expressed as the subspace $(x, 0) < V$: the value of x is the freedom in a solution – while the cokernel may be expressed via the map $W \to R$, $(a, b) \mapsto (a)$: given a vector (a, b), the value of a is the *obstruction* to there being a solution.

An example illustrating the infinite-dimensional case is afforded by the map $f: R^\infty \to R^\infty$, $\{a_n\} \mapsto \{b_n\}$ with $b_1 = 0$ and $b_{n+1} = a_n$ for $n > 0$. Its image consists of all sequences with first element 0, and thus its cokernel consists of the classes of sequences with identical first element. Thus, whereas its kernel has dimension 0 (it maps only the zero sequence to the zero sequence), its co-kernel has dimension 1. Since the domain and the target space are the same, the rank and the dimension of the kernel add up to the same sum as the rank and the dimension of the co-kernel ($\aleph_0 + 0 = \aleph_0 + 1$), but in the infinite-dimensional case it cannot be inferred that the kernel and the co-kernel of an endomorphism have the same dimension ($0 \neq 1$). The reverse situation obtains for the map $h: R^\infty \to R^\infty$, $\{a_n\} \mapsto \{c_n\}$ with $c_n = a_{n+1}$. Its image is the entire target space, and hence its co-kernel has dimension 0, but since it maps all sequences in which only the first element is non-zero to the zero sequence, its kernel has dimension 1.

Index

For a linear operator with finite-dimensional kernel and co-kernel, one may define *index* as:

$$\operatorname{ind} f := \dim \ker f - \dim \operatorname{coker} f,$$

namely the degrees of freedom minus the number of constraints.

For a transformation between finite-dimensional vector spaces, this is just the difference $\dim(V) - \dim(W)$, by rank–nullity. This gives an indication of how many solutions or how many constraints one has: if mapping from a larger space to a smaller one, the map may be onto, and thus will have degrees of freedom even without constraints. Conversely, if mapping from a smaller space to a larger one, the map cannot be onto, and thus one will have constraints even without degrees of freedom.

The index of an operator is precisely the Euler characteristic of the 2-term complex $0 \to V \to W \to 0$. In operator theory, the index of Fredholm operators is an object of study, with a major result being the Atiyah–Singer index theorem.

Algebraic Classifications of Linear Transformations

No classification of linear maps could hope to be exhaustive. The following incomplete list enumerates some important classifications that do not require any additional structure on the vector space.

Let V and W denote vector spaces over a field, F. Let $T: V \rightarrow W$ be a linear map.

- T is said to be *injective* or a *monomorphism* if any of the following equivalent conditions are true:

  o T is one-to-one as a map of sets.

  o $\ker T = \{0_V\}$

  o T is monic or left-cancellable, which is to say, for any vector space U and any pair of linear maps $R: U \rightarrow V$ and $S: U \rightarrow V$, the equation $TR = TS$ implies $R = S$.

  o T is left-invertible, which is to say there exists a linear map $S: W \rightarrow V$ such that ST is the identity map on V.

- T is said to be *surjective* or an *epimorphism* if any of the following equivalent conditions are true:

  o T is onto as a map of sets.

  o $\operatorname{coker} T = \{0_W\}$

  o T is epic or right-cancellable, which is to say, for any vector space U and any pair of linear maps $R: W \rightarrow U$ and $S: W \rightarrow U$, the equation $RT = ST$ implies $R = S$.

  o T is right-invertible, which is to say there exists a linear map $S: W \rightarrow V$ such that TS is the identity map on W.

- T is said to be an *isomorphism* if it is both left- and right-invertible. This is equivalent to T being both one-to-one and onto (a bijection of sets) or also to T being both epic and monic, and so being a bimorphism.

- If $T: V \rightarrow V$ is an endomorphism, then:

  o If, for some positive integer n, the n-th iterate of T, T^n, is identically zero, then T is said to be nilpotent.

  o If $T^2 = T$, then T is said to be idempotent

  o If $T = kI$, where k is some scalar, then T is said to be a scaling transformation or scalar multiplication map.

Change of Basis

Given a linear map which is an endomorphism whose matrix is A, in the basis B of the space it transforms vector coordinates [u] as [v] = A[u]. As vectors change with the inverse of B (vectors are contravariant) its inverse transformation is [v] = B[v'].

Substituting this in the first expression

$$B[v'] = AB[u']$$

hence

$$[v'] = B^{-1}AB[u'] = A'[u'].$$

Therefore, the matrix in the new basis is $A' = B^{-1}AB$, being B the matrix of the given basis.

Therefore, linear maps are said to be 1-co 1-contra -variant objects, or type (1, 1) tensors.

Continuity

A *linear transformation* between topological vector spaces, for example normed spaces, may be continuous. If its domain and codomain are the same, it will then be a continuous linear operator. A linear operator on a normed linear space is continuous if and only if it is bounded, for example, when the domain is finite-dimensional. An infinite-dimensional domain may have discontinuous linear operators.

An example of an unbounded, hence discontinuous, linear transformation is differentiation on the space of smooth functions equipped with the supremum norm (a function with small values can have a derivative with large values, while the derivative of 0 is 0). For a specific example, $\sin(nx)/n$ converges to 0, but its derivative $\cos(nx)$ does not, so differentiation is not continuous at 0 (and by a variation of this argument, it is not continuous anywhere).

Applications

A specific application of linear maps is for geometric transformations, such as those performed in computer graphics, where the translation, rotation and scaling of 2D or 3D objects is performed by the use of a transformation matrix. Linear mappings also are used as a mechanism for describing change: for example in calculus correspond to derivatives; or in relativity, used as a device to keep track of the local transformations of reference frames.

Another application of these transformations is in compiler optimizations of nested-loop code, and in parallelizing compiler techniques.

Matrix (Mathematics)

In mathematics, a matrix (plural matrices) is a rectangular *array* of numbers, symbols, or expressions, arranged in *rows* and *columns*. For example, the dimensions of the matrix below are 2×3 (read "two by three"), because there are two rows and three columns.

$$\begin{bmatrix} 1 & 9 & -13 \\ 20 & 5 & -6 \end{bmatrix}.$$

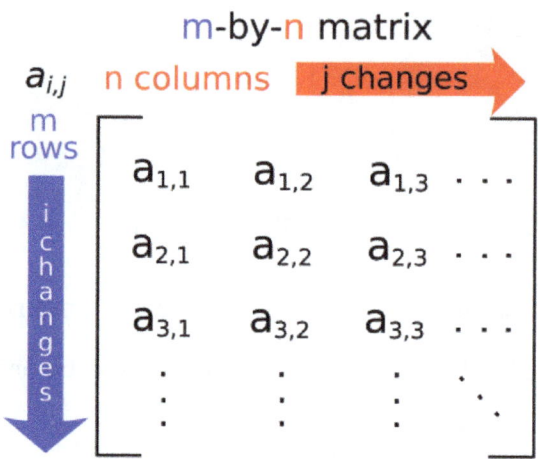

Each element of a matrix is often denoted by a variable with two subscripts. For example, $a_{2,1}$ represents the element at the second row and first column of a matrix A.

The individual items in a matrix are called its *elements* or *entries*. Provided that they are the same size (have the same number of rows and the same number of columns), two matrices can be added or subtracted element by element. The rule for matrix multiplication, however, is that two matrices can be multiplied only when the number of columns in the first equals the number of rows in the second. Any matrix can be multiplied element-wise by a scalar from its associated field. A major application of matrices is to represent linear transformations, that is, generalizations of linear functions such as $f(x) = 4x$. For example, the rotation of vectors in three dimensional space is a linear transformation which can be represented by a rotation matrix R: if v is a column vector (a matrix with only one column) describing the position of a point in space, the product Rv is a column vector describing the position of that point after a rotation. The product of two transformation matrices is a matrix that represents the composition of two linear transformations. Another application of matrices is in the solution of systems of linear equations. If the matrix is square, it is possible to deduce some of its properties by computing its determinant. For example, a square matrix has an inverse if and only if its determinant is not zero. Insight into the geometry of a linear transformation is obtainable (along with other information) from the matrix's eigenvalues and eigenvectors.

Applications of matrices are found in most scientific fields. In every branch of physics, including classical mechanics, optics, electromagnetism, quantum mechanics, and quantum electrodynamics, they are used to study physical phenomena, such as the motion of rigid bodies. In computer graphics, they are used to project a 3D model onto a 2 dimensional screen. In probability theory and statistics, stochastic matrices are used to describe sets of probabilities; for instance, they are used within the PageRank algorithm that ranks the pages in a Google search. Matrix calculus generalizes classical analytical notions such as derivatives and exponentials to higher dimensions.

A major branch of numerical analysis is devoted to the development of efficient algorithms for matrix computations, a subject that is centuries old and is today an expanding area of research. Matrix decomposition methods simplify computations, both theoretically and practically. Algorithms that are tailored to particular matrix structures, such as sparse matrices and near-diagonal matrices, expedite computations in finite element method and other computations. Infinite matrices occur in planetary theory and in atomic theory. A simple example of an infinite matrix is the matrix representing the derivative operator, which acts on the Taylor series of a function.

Definition

A *matrix* is a rectangular array of numbers or other mathematical objects for which operations such as addition and multiplication are defined. Most commonly, a matrix over a field F is a rectangular array of scalars each of which is a member of F. Most of this chapter focuses on *real* and *complex matrices*, that is, matrices whose elements are real numbers or complex numbers, respectively. More general types of entries are discussed below. For instance, this is a real matrix:

$$\mathbf{A} = \begin{bmatrix} -1.3 & 0.6 \\ 20.4 & 5.5 \\ 9.7 & -6.2 \end{bmatrix}.$$

The numbers, symbols or expressions in the matrix are called its *entries* or its *elements*. The horizontal and vertical lines of entries in a matrix are called *rows* and *columns*, respectively.

Size

The size of a matrix is defined by the number of rows and columns that it contains. A matrix with m rows and n columns is called an $m \times n$ matrix or m-by-n matrix, while m and n are called its *dimensions*. For example, the matrix A above is a 3 × 2 matrix.

Matrices which have a single row are called *row vectors*, and those which have a single column are called *column vectors*. A matrix which has the same number of rows and columns is called a *square matrix*. A matrix with an infinite number of rows or columns (or both) is called an *infinite matrix*. In some contexts, such as computer algebra programs, it is useful to consider a matrix with no rows or no columns, called an *empty matrix*.

Name	Size	Example	Description
Row vector	$1 \times n$	$\begin{bmatrix} 3 & 7 & 2 \end{bmatrix}$	A matrix with one row, sometimes used to represent a vector
Column vector	$n \times 1$	$\begin{bmatrix} 4 \\ 1 \\ 8 \end{bmatrix}$	A matrix with one column, sometimes used to represent a vector
Square matrix	$n \times n$	$\begin{bmatrix} 9 & 13 & 5 \\ 1 & 11 & 7 \\ 2 & 6 & 3 \end{bmatrix}$	A matrix with the same number of rows and columns, sometimes used to represent a linear transformation from a vector space to itself, such as reflection, rotation, or shearing.

Notation

Matrices are commonly written in box brackets or parentheses:

$$\mathbf{A} = \begin{bmatrix} a_{11} & a_{12} & \cdots & a_{1n} \\ a_{21} & a_{22} & \cdots & a_{2n} \\ \vdots & \vdots & \ddots & \vdots \\ a_{m1} & a_{m2} & \cdots & a_{mn} \end{bmatrix} = \begin{pmatrix} a_{11} & a_{12} & \cdots & a_{1n} \\ a_{21} & a_{22} & \cdots & a_{2n} \\ \vdots & \vdots & \ddots & \vdots \\ a_{m1} & a_{m2} & \cdots & a_{mn} \end{pmatrix} = \left(a_{ij} \right) \in \mathbb{R}^{m \times n}.$$

The specifics of symbolic matrix notation vary widely, with some prevailing trends. Matrices are usually symbolized using upper-case letters (such as A in the examples above), while the corresponding lower-case letters, with two subscript indices (for example, a_{11}, or $a_{1,1}$), represent the entries. In addition to using upper-case letters to symbolize matrices, many authors use a special typographical style, commonly boldface upright (non-italic), to further distinguish matrices from other mathematical objects. An alternative notation involves the use of a double-underline with the variable name, with or without boldface style, (for example, $\underline{\underline{A}}$).

The entry in the i-th row and j-th column of a matrix A is sometimes referred to as the i,j, (i,j), or $(i,j)^{\text{th}}$ entry of the matrix, and most commonly denoted as $a_{i,j}$, or a_{ij}. Alternative notations for that entry are $A[i,j]$ or $A_{i,j}$. For example, the (1,3) entry of the following matrix A is 5 (also denoted a_{13}, $a_{1,3}$, $A[1,3]$ or $A_{1,3}$):

$$A = \begin{bmatrix} 4 & -7 & 5 & 0 \\ -2 & 0 & 11 & 8 \\ 19 & 1 & -3 & 12 \end{bmatrix}$$

Sometimes, the entries of a matrix can be defined by a formula such as $a_{ij} = f(i, j)$. For example, each of the entries of the following matrix A is determined by $a_{ij} = i - j$.

$$A = \begin{bmatrix} 0 & -1 & -2 & -3 \\ 1 & 0 & -1 & -2 \\ 2 & 1 & 0 & -1 \end{bmatrix}$$

In this case, the matrix itself is sometimes defined by that formula, within square brackets or double parentheses. For example, the matrix above is defined as A = [i-j], or A = ((i-j)). If matrix size is $m \times n$, the above-mentioned formula $f(i, j)$ is valid for any $i = 1, ..., m$ and any $j = 1, ..., n$. This can be either specified separately, or using $m \times n$ as a subscript. For instance, the matrix A above is 3 × 4 and can be defined as A = $[i - j]$ ($i = 1, 2, 3; j = 1, ..., 4$), or A = $[i - j]_{3 \times 4}$.

Some programming languages utilize doubly subscripted arrays (or arrays of arrays) to represent an m-×-n matrix. Some programming languages start the numbering of array indexes at zero, in which case the entries of an m-by-n matrix are indexed by $0 \le i \le m - 1$ and $0 \le j \le n - 1$. This section follows the more common convention in mathematical writing where enumeration starts from 1.

An asterisk is occasionally used to refer to whole rows or columns in a matrix. For example, $a_{i,*}$ refers to the ith row of A, and $a_{*,j}$ refers to the jth column of A. The set of all m-by-n matrices is denoted $\mathbb{M}(m, n)$.

Basic Operations

There are a number of basic operations that can be applied to modify matrices, called *matrix addition*, *scalar multiplication*, *transposition*, *matrix multiplication*, *row operations*, and *submatrix*.

Addition, Scalar Multiplication and Transposition

Operation	Definition	Example
Addition	The *sum* A+B of two *m*-by-*n* matrices A and B is calculated entrywise: $(A + B)_{i,j} = A_{i,j} + B_{i,j}$, where $1 \leq i \leq m$ and $1 \leq j \leq n$.	$\begin{bmatrix} 1 & 3 & 1 \\ 1 & 0 & 0 \end{bmatrix} + \begin{bmatrix} 0 & 0 & 5 \\ 7 & 5 & 0 \end{bmatrix} = \begin{bmatrix} 1+0 & 3+0 & 1+5 \\ 1+7 & 0+5 & 0+0 \end{bmatrix} = \begin{bmatrix} 1 & 3 & 6 \\ 8 & 5 & 0 \end{bmatrix}$
Scalar multiplication	The product *cA* of a number *c* (also called a scalar in the parlance of abstract algebra) and a matrix A is computed by multiplying every entry of A by *c*: $(cA)_{i,j} = c \cdot A_{i,j}$. This operation is called *scalar multiplication*, but its result is not named "scalar product" to avoid confusion, since "scalar product" is sometimes used as a synonym for "inner product".	$2 \cdot \begin{bmatrix} 1 & 8 & -3 \\ 4 & -2 & 5 \end{bmatrix} = \begin{bmatrix} 2 \cdot 1 & 2 \cdot 8 & 2 \cdot -3 \\ 2 \cdot 4 & 2 \cdot -2 & 2 \cdot 5 \end{bmatrix} = \begin{bmatrix} 2 & 16 & -6 \\ 8 & -4 & 10 \end{bmatrix}$
Transposition	The *transpose* of an *m*-by-*n* matrix A is the *n*-by-*m* matrix A^T (also denoted A^{tr} or ^tA) formed by turning rows into columns and vice versa: $(A^T)_{i,j} = A_{j,i}$.	$\begin{bmatrix} 1 & 2 & 3 \\ 0 & -6 & 7 \end{bmatrix}^T = \begin{bmatrix} 1 & 0 \\ 2 & -6 \\ 3 & 7 \end{bmatrix}$

Familiar properties of numbers extend to these operations of matrices: for example, addition is commutative, that is, the matrix sum does not depend on the order of the summands: A + B = B + A.

The transpose is compatible with addition and scalar multiplication, as expressed by $(cA)^T = c(A^T)$ and $(A + B)^T = A^T + B^T$. Finally, $(A^T)^T = A$.

Matrix Multiplication

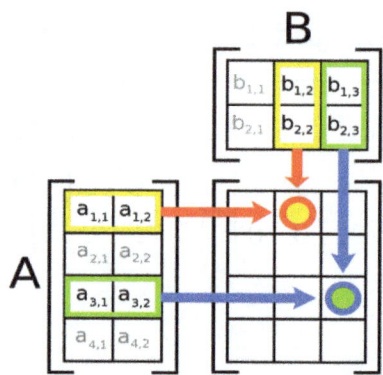

Schematic depiction of the matrix product AB of two matrices A and B.

Multiplication of two matrices is defined if and only if the number of columns of the left matrix is the same as the number of rows of the right matrix. If A is an m-by-n matrix and B is an n-by-p matrix, then their *matrix product* AB is the m-by-p matrix whose entries are given by dot product of the corresponding row of A and the corresponding column of B:

$$[\mathbf{AB}]_{i,j} = A_{i,1}B_{1,j} + A_{i,2}B_{2,j} + \cdots + A_{i,n}B_{n,j} = \sum_{r=1}^{n} A_{i,r}B_{r,j},$$

where $1 \le i \le m$ and $1 \le j \le p$. For example, the underlined entry 2340 in the product is calculated as $(2 \times 1000) + (3 \times 100) + (4 \times 10) = 2340$:

$$\begin{bmatrix} \underline{2} & \underline{3} & \underline{4} \\ 1 & 0 & 0 \end{bmatrix} \begin{bmatrix} 0 & \underline{1000} \\ 1 & \underline{100} \\ 0 & \underline{10} \end{bmatrix} = \begin{bmatrix} 3 & \underline{2340} \\ 0 & 1000 \end{bmatrix}.$$

Matrix multiplication satisfies the rules (AB)C = A(BC) (associativity), and (A+B)C = AC+BC as well as C(A+B) = CA+CB (left and right distributivity), whenever the size of the matrices is such that the various products are defined. The product AB may be defined without BA being defined, namely if A and B are m-by-n and n-by-k matrices, respectively, and $m \ne k$. Even if both products are defined, they need not be equal, that is, generally

$$AB \ne BA,$$

that is, *matrix multiplication is not commutative,* in marked contrast to (rational, real, or complex) numbers whose product is independent of the order of the factors. An example of two matrices not commuting with each other is:

$$\begin{bmatrix} 1 & 2 \\ 3 & 4 \end{bmatrix} \begin{bmatrix} 0 & 1 \\ 0 & 0 \end{bmatrix} = \begin{bmatrix} 0 & 1 \\ 0 & 3 \end{bmatrix},$$

whereas

$$\begin{bmatrix} 0 & 1 \\ 0 & 0 \end{bmatrix} \begin{bmatrix} 1 & 2 \\ 3 & 4 \end{bmatrix} = \begin{bmatrix} 3 & 4 \\ 0 & 0 \end{bmatrix}.$$

Besides the ordinary matrix multiplication just described, there exist other less frequently used operations on matrices that can be considered forms of multiplication, such as the Hadamard product and the Kronecker product. They arise in solving matrix equations such as the Sylvester equation.

Row Operations

There are three types of row operations:

1. row addition, that is adding a row to another.

2. row multiplication, that is multiplying all entries of a row by a non-zero constant;

3. row switching, that is interchanging two rows of a matrix;

These operations are used in a number of ways, including solving linear equations and finding matrix inverses.

Submatrix

A submatrix of a matrix is obtained by deleting any collection of rows and/or columns. For example, from the following 3-by-4 matrix, we can construct a 2-by-3 submatrix by removing row 3 and column 2:

$$\mathbf{A} = \begin{bmatrix} 1 & 2 & 3 & 4 \\ 5 & 6 & 7 & 8 \\ 9 & 10 & 11 & 12 \end{bmatrix} \rightarrow \begin{bmatrix} 1 & 3 & 4 \\ 5 & 7 & 8 \end{bmatrix}.$$

The minors and cofactors of a matrix are found by computing the determinant of certain submatrices.

A principal submatrix is a square submatrix obtained by removing certain rows and columns. The definition varies from author to author. According to some authors, a principal submatrix is a submatrix in which the set of row indices that remain is the same as the set of column indices that remain. Other authors define a principal submatrix to be one in which the first k rows and columns, for some number k, are the ones that remain; this type of submatrix has also been called a leading principal submatrix.

Linear Equations

Matrices can be used to compactly write and work with multiple linear equations, that is, systems of linear equations. For example, if A is an m-by-n matrix, x designates a column vector (that is, $n \times 1$-matrix) of n variables $x_1, x_2, ..., x_n$, and b is an $m \times 1$-column vector, then the matrix equation

$$Ax = b$$

is equivalent to the system of linear equations

$$A_{1,1}x_1 + A_{1,2}x_2 + \ldots + A_{1,n}x_n = b_1$$

$$\ldots$$

$$A_{m,1}x_1 + A_{m,2}x_2 + \ldots + A_{m,n}x_n = b_m .$$

Linear Transformations

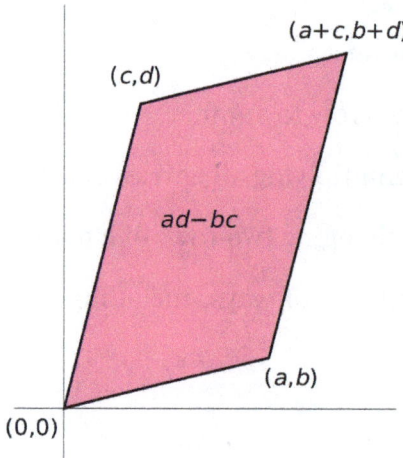

The vectors represented by a 2-by-2 matrix correspond to the sides of a unit square transformed into a parallelogram.

Matrices and matrix multiplication reveal their essential features when related to *linear transformations*, also known as *linear maps*. A real *m-by-n matrix A gives rise to a linear transformation $R^n \to R^m$ mapping each vector x in R^n to the (matrix) product Ax, which is a vector in R^m. Conversely, each linear transformation f: $R^n \to R^m$ arises from a unique m-by-n matrix A: explicitly, the (i, j)-entry of A is the ith coordinate of $f(e_j)$, where e_j = (0,...,0,1,0,...,0) is the unit vector with 1 in the jth position and 0 elsewhere.* The matrix A is said to represent the linear map f, and A is called the *transformation matrix* of f.

For example, the 2×2 matrix

$$\mathbf{A} = \begin{bmatrix} a & c \\ b & d \end{bmatrix}$$

can be viewed as the transform of the unit square into a parallelogram with vertices at (0, 0), (a, b), (a + c, b + d), and (c, d). The parallelogram pictured at the right is obtained by multiplying A with each of the column vectors $\begin{bmatrix} 0 \\ 0 \end{bmatrix}, \begin{bmatrix} 1 \\ 0 \end{bmatrix}, \begin{bmatrix} 1 \\ 1 \end{bmatrix}$ and $\begin{bmatrix} 0 \\ 1 \end{bmatrix}$ in turn. These vectors define the vertices of the unit square.

The following table shows a number of 2-by-2 matrices with the associated linear maps of R^2. The blue original is mapped to the green grid and shapes. The origin (0,0) is marked with a black point.

Horizontal shear with m=1.25.	Reflection through the vertical axis	Squeeze mapping with r=3/2	Scaling by a factor of 3/2	*Rotation* by $\pi/6^R = 30°$
$\begin{bmatrix} 1 & 1.25 \\ 0 & 1 \end{bmatrix}$	$\begin{bmatrix} -1 & 0 \\ 0 & 1 \end{bmatrix}$	$\begin{bmatrix} 3/2 & 0 \\ 0 & 2/3 \end{bmatrix}$	$\begin{bmatrix} 3/2 & 0 \\ 0 & 3/2 \end{bmatrix}$	$\begin{bmatrix} \cos(\pi/6^R) & -\sin(\pi/6^R) \\ \sin(\pi/6^R) & \cos(\pi/6^R) \end{bmatrix}$

Under the 1-to-1 correspondence between matrices and linear maps, matrix multiplication corresponds to composition of maps: if a *k*-by-*m* matrix B represents another linear map $g : R^m \to R^k$, then the composition $g \circ f$ is represented by BA since

$$(g \circ f)(x) = g(f(x)) = g(Ax) = B(Ax) = (BA)x.$$

The last equality follows from the above-mentioned associativity of matrix multiplication.

The rank of a matrix A is the maximum number of linearly independent row vectors of the matrix, which is the same as the maximum number of linearly independent column vectors. Equivalently it is the dimension of the image of the linear map represented by A. The rank-nullity theorem states that the dimension of the kernel of a matrix plus the rank equals the number of columns of the matrix.

Square Matrix

A square matrix is a matrix with the same number of rows and columns. An *n*-by-*n* matrix is known as a square matrix of order *n*. Any two square matrices of the same order can be added and multiplied. The entries a_{ii} form the main diagonal of a square matrix. They lie on the imaginary line which runs from the top left corner to the bottom right corner of the matrix.

Main Types

Name	Example with *n* = 3
Diagonal matrix	$\begin{bmatrix} a_{11} & 0 & 0 \\ 0 & a_{22} & 0 \\ 0 & 0 & a_{33} \end{bmatrix}$
Lower triangular matrix	$\begin{bmatrix} a_{11} & 0 & 0 \\ a_{21} & a_{22} & 0 \\ a_{31} & a_{32} & a_{33} \end{bmatrix}$
Upper triangular matrix	$\begin{bmatrix} a_{11} & a_{12} & a_{13} \\ 0 & a_{22} & a_{23} \\ 0 & 0 & a_{33} \end{bmatrix}$

Diagonal and Triangular Matrix

If all entries of A below the main diagonal are zero, A is called an *upper triangular matrix*. Similarly if all entries of *A* above the main diagonal are zero, A is called a *lower triangular matrix*. If all off-diagonal elements are zero, A is called a diagonal matrix.

Identity Matrix

The *identity matrix* I_n of size n is the n-by-n matrix in which all the elements on the main diagonal are equal to 1 and all other elements are equal to 0, for example,

$$I_1 = [1], I_2 = \begin{bmatrix} 1 & 0 \\ 0 & 1 \end{bmatrix}, \cdots, I_n = \begin{bmatrix} 1 & 0 & \cdots & 0 \\ 0 & 1 & \cdots & 0 \\ \vdots & \vdots & \ddots & \vdots \\ 0 & 0 & \cdots & 1 \end{bmatrix}$$

It is a square matrix of order n, and also a special kind of diagonal matrix. It is called an identity matrix because multiplication with it leaves a matrix unchanged:

$$AI_n = I_m A = A \text{ for any } m\text{-by-}n \text{ matrix A.}$$

A nonzero scalar multiple of an identity matrix is called a *scalar* matrix. If the matrix entries come from a field, the scalar matrices form a group, under matrix multiplication, that is isomorphic to the multiplicative group of nonzero elements of the field.

Symmetric or Skew-symmetric Matrix

A square matrix A that is equal to its transpose, that is, $A = A^T$, is a symmetric matrix. If instead, A is equal to the negative of its transpose, that is, $A = -A^T$, then A is a skew-symmetric matrix. In complex matrices, symmetry is often replaced by the concept of Hermitian matrices, which satisfy $A^* = A$, where the star or asterisk denotes the conjugate transpose of the matrix, that is, the transpose of the complex conjugate of A.

By the spectral theorem, real symmetric matrices and complex Hermitian matrices have an eigenbasis; that is, every vector is expressible as a linear combination of eigenvectors. In both cases, all eigenvalues are real. This theorem can be generalized to infinite-dimensional situations related to matrices with infinitely many rows and columns.

Invertible Matrix and its Inverse

A square matrix A is called *invertible* or *non-singular* if there exists a matrix B such that

$$AB = BA = I_n.$$

If B exists, it is unique and is called the *inverse matrix* of A, denoted A^{-1}.

Definite Matrix

Positive definite matrix	Indefinite matrix
$\begin{bmatrix} 1/4 & 0 \\ 0 & 1 \end{bmatrix}$	$\begin{bmatrix} 1/4 & 0 \\ 0 & -1/4 \end{bmatrix}$
$Q(x,y) = 1/4\,x^2 + y^2$	$Q(x,y) = 1/4\,x^2 - 1/4\,y^2$
Points such that $Q(x,y)=1$ (Ellipse).	Points such that $Q(x,y)=1$ (Hyperbola).

A symmetric $n \times n$-matrix A is called *positive-definite* (respectively *negative-definite*; *indefinite*), if for all nonzero vectors $x \in R^n$ the associated quadratic form given by

$$Q(x) = x^T A x$$

takes only positive values (respectively only negative values; both some negative and some positive values). If the quadratic form takes only non-negative (respectively only non-positive) values, the symmetric matrix is called positive-semidefinite (respectively negative-semidefinite); hence the matrix is indefinite precisely when it is neither positive-semidefinite nor negative-semidefinite.

A symmetric matrix is positive-definite if and only if all its eigenvalues are positive, that is, the matrix is positive-semidefinite and it is invertible. The table at the right shows two possibilities for 2-by-2 matrices.

Allowing as input two different vectors instead yields the bilinear form associated to A:

$$B_A (x, y) = x^T A y.$$

Orthogonal Matrix

An *orthogonal matrix* is a square matrix with real entries whose columns and rows are orthogonal unit vectors (that is, orthonormal vectors). Equivalently, a matrix A is orthogonal if its transpose is equal to its inverse:

$$A^T = A^{-1},$$

which entails

$$A^{\mathrm{T}}A = AA^{\mathrm{T}} = I_n,$$

where I is the identity matrix of size n.

An orthogonal matrix A is necessarily invertible (with inverse $A^{-1} = A^{\mathrm{T}}$), unitary ($A^{-1} = A^*$), and normal ($A^*A = AA^*$). The determinant of any orthogonal matrix is either $+1$ or -1. A *special orthogonal matrix* is an orthogonal matrix with determinant $+1$. As a linear transformation, every orthogonal matrix with determinant $+1$ is a pure rotation, while every orthogonal matrix with determinant -1 is either a pure reflection, or a composition of reflection and rotation.

The complex analogue of an orthogonal matrix is a unitary matrix.

Main Operations

Trace

The trace, tr(A) of a square matrix A is the sum of its diagonal entries. While matrix multiplication is not commutative as mentioned above, the trace of the product of two matrices is independent of the order of the factors:

$$\mathrm{tr}(AB) = \mathrm{tr}(BA).$$

This is immediate from the definition of matrix multiplication:

$$\mathrm{tr}(\mathsf{AB}) = \sum_{i=1}^{m}\sum_{j=1}^{n} A_{ij}B_{ji} = \mathrm{tr}(\mathsf{BA}).$$

Also, the trace of a matrix is equal to that of its transpose, that is,

$$\mathrm{tr}(A) = \mathrm{tr}(A^{\mathrm{T}}).$$

Determinant

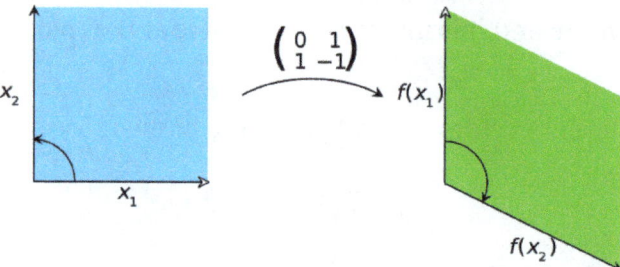

A linear transformation on $\mathbb{R}^2$ given by the indicated matrix. The determinant of this matrix is -1, as the area of the green parallelogram at the right is 1, but the map reverses the orientation, since it turns the counterclockwise orientation of the vectors to a clockwise one.

The *determinant* det(A) or |A| of a square matrix A is a number encoding certain properties of the matrix. A matrix is invertible if and only if its determinant is nonzero. Its absolute value equals the area (in $\mathbb{R}^2$) or volume (in $\mathbb{R}^3$) of the image of the unit square (or cube), while its sign corresponds to the orientation of the corresponding linear map: the determinant is positive if and only if the orientation is preserved.

The determinant of 2-by-2 matrices is given by

$$\det\begin{bmatrix} a & b \\ c & d \end{bmatrix} = ad - bc.$$

The determinant of 3-by-3 matrices involves 6 terms (rule of Sarrus). The more lengthy Leibniz formula generalises these two formulae to all dimensions.

The determinant of a product of square matrices equals the product of their determinants:

$$\det(AB) = \det(A) \cdot \det(B).$$

Adding a multiple of any row to another row, or a multiple of any column to another column, does not change the determinant. Interchanging two rows or two columns affects the determinant by multiplying it by -1. Using these operations, any matrix can be transformed to a lower (or upper) triangular matrix, and for such matrices the determinant equals the product of the entries on the main diagonal; this provides a method to calculate the determinant of any matrix. Finally, the Laplace expansion expresses the determinant in terms of minors, that is, determinants of smaller matrices. This expansion can be used for a recursive definition of determinants (taking as starting case the determinant of a 1-by-1 matrix, which is its unique entry, or even the determinant of a 0-by-0 matrix, which is 1), that can be seen to be equivalent to the Leibniz formula. Determinants can be used to solve linear systems using Cramer's rule, where the division of the determinants of two related square matrices equates to the value of each of the system's variables.

Eigenvalues and Eigenvectors

A number λ and a non-zero vector v satisfying

$$Av = \lambda v$$

are called an *eigenvalue* and an *eigenvector* of A, respectively. The number λ is an eigenvalue of an $n \times n$-matrix A if and only if $A - \lambda I_n$ is not invertible, which is equivalent to

$$\det(A - \lambda I) = 0.$$

The polynomial p_A in an indeterminate X given by evaluation the determinant $\det(XI_n - A)$ is called the characteristic polynomial of A. It is a monic polynomial of degree n. Therefore the polynomial equation $p_A(\lambda) = 0$ has at most n different solutions, that is, eigenvalues of the matrix. They may be complex even if the entries of A are real. According to the Cayley–Hamilton theorem, $p_A(A) = 0$, that is, the result of substituting the matrix itself into its own characteristic polynomial yields the zero matrix.

Computational Aspects

Matrix calculations can be often performed with different techniques. Many problems can be solved by both direct algorithms or iterative approaches. For example, the eigenvectors of a square matrix can be obtained by finding a sequence of vectors x_n converging to an eigenvector when n tends to infinity.

To be able to choose the more appropriate algorithm for each specific problem, it is important to determine both the effectiveness and precision of all the available algorithms. The domain studying these matters is called numerical linear algebra. As with other numerical situations, two main aspects are the complexity of algorithms and their numerical stability.

Determining the complexity of an algorithm means finding upper bounds or estimates of how many elementary operations such as additions and multiplications of scalars are necessary to perform some algorithm, for example, multiplication of matrices. For example, calculating the matrix product of two n-by-n matrix using the definition given above needs n^3 multiplications, since for any of the n^2 entries of the product, n multiplications are necessary. The Strassen algorithm outperforms this "naive" algorithm; it needs only $n^{2.807}$ multiplications. A refined approach also incorporates specific features of the computing devices.

In many practical situations additional information about the matrices involved is known. An important case are sparse matrices, that is, matrices most of whose entries are zero. There are specifically adapted algorithms for, say, solving linear systems Ax = b for sparse matrices A, such as the conjugate gradient method.

An algorithm is, roughly speaking, numerically stable, if little deviations in the input values do not lead to big deviations in the result. For example, calculating the inverse of a matrix via Laplace's formula (Adj (A) denotes the adjugate matrix of A)

$$A^{-1} = Adj(A) / det(A)$$

may lead to significant rounding errors if the determinant of the matrix is very small. The norm of a matrix can be used to capture the conditioning of linear algebraic problems, such as computing a matrix's inverse.

Although most computer languages are not designed with commands or libraries for matrices, as early as the 1970s, some engineering desktop computers such as the HP 9830 had ROM cartridges to add BASIC commands for matrices. Some computer languages such as APL were designed to manipulate matrices, and various mathematical programs can be used to aid computing with matrices.

Decomposition

There are several methods to render matrices into a more easily accessible form. They are generally referred to as *matrix decomposition* or *matrix factorization* techniques. The interest of all these techniques is that they preserve certain properties of the matrices in question, such as determinant, rank or inverse, so that these quantities can be calculated after applying the transformation, or that certain matrix operations are algorithmically easier to carry out for some types of matrices.

The LU decomposition factors matrices as a product of lower (L) and an upper triangular matrices (U). Once this decomposition is calculated, linear systems can be solved more efficiently, by a simple technique called forward and back substitution. Likewise, inverses of triangular matrices are algorithmically easier to calculate. The *Gaussian elimination* is a similar algorithm; it transforms any matrix to row echelon form. Both methods proceed by multiplying the matrix by suitable elementary matrices, which correspond to permuting rows or columns and adding multiples of one row to another row. Singular value decomposition expresses any matrix A as a product UDV*, where U and V are unitary matrices and D is a diagonal matrix.

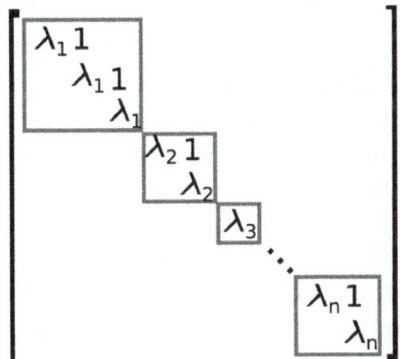

An example of a matrix in Jordan normal form. The grey blocks are called Jordan blocks.

The eigendecomposition or *diagonalization* expresses A as a product VDV^{-1}, where D is a diagonal matrix and V is a suitable invertible matrix. If A can be written in this form, it is called diagonalizable. More generally, and applicable to all matrices, the Jordan decomposition transforms a matrix into Jordan normal form, that is to say matrices whose only nonzero entries are the eigenvalues λ_1 to λ_n of A, placed on the main diagonal and possibly entries equal to one directly above the main diagonal, as shown at the right. Given the eigendecomposition, the nth power of A (that is, n-fold iterated matrix multiplication) can be calculated via

$$A^n = (VDV^{-1})^n = VDV^{-1}VDV^{-1}...VDV^{-1} = VD^nV^{-1}$$

and the power of a diagonal matrix can be calculated by taking the corresponding powers of the diagonal entries, which is much easier than doing the exponentiation for A instead. This can be used to compute the matrix exponential e^A, a need frequently arising in solving linear differential equations, matrix logarithms and square roots of matrices. To avoid numerically ill-conditioned situations, further algorithms such as the Schur decomposition can be employed.

Abstract Algebraic Aspects and Generalizations

Matrices can be generalized in different ways. Abstract algebra uses matrices with entries in more general fields or even rings, while linear algebra codifies properties of matrices in the notion of linear maps. It is possible to consider matrices with infinitely many columns and rows. Another extension are tensors, which can be seen as higher-dimensional arrays of numbers, as opposed to vectors, which can often be realised as sequences of numbers, while matrices are rectangular or two-dimensional arrays of numbers. Matrices, subject to certain requirements tend to form groups known as matrix groups. Similarly under certain conditions matrices form rings known as matrix rings. Though the product of matrices is not in general commutative yet certain matrices form fields known as matrix fields .

Matrices with More General Entries

This chapter focuses on matrices whose entries are real or complex numbers. *However, matrices can be considered with much more general types of entries than real or complex numbers. As a first step of generalization, any field, that is, a set where addition, subtraction, multiplication and* division operations are defined and well-behaved, may be used instead of R or C, for example rational numbers or finite fields. For example, coding theory makes use of matrices over finite fields. Wherever eigenvalues are considered, as these are roots of a polynomial they may exist only in a larger field than that of the entries of the matrix; for instance they may be complex in case of a matrix with real entries. The possibility to reinterpret the entries of a matrix as elements of a larger field (for example, to view a real matrix as a complex matrix whose entries happen to be all real) then allows considering each square matrix to possess a full set of eigenvalues. Alternatively one can consider only matrices with entries in an algebraically closed field, such as C, from the outset.

More generally, abstract algebra makes great use of matrices with entries in a ring R. Rings are a more general notion than fields in that a division operation need not exist. The very same addition and multiplication operations of matrices extend to this setting, too. The set $M(n, R)$ of all square n-by-n matrices over R is a ring called matrix ring, isomorphic to the endomorphism ring of the left R-module R^n. If the ring R is commutative, that is, its multiplication is commutative, then $M(n, R)$ is a unitary noncommutative (unless $n = 1$) associative algebra over R. The determinant of square matrices over a commutative ring R can still be defined using the Leibniz formula; such a matrix is invertible if and only if its determinant is invertible in R, generalising the situation over a field F, where every nonzero element is invertible. Matrices over superrings are called supermatrices.

Matrices do not always have all their entries in the same ring – or even in any ring at all. One special but common case is block matrices, which may be considered as matrices whose entries themselves are matrices. The entries need not be quadratic matrices, and thus need not be members of any ordinary ring; but their sizes must fulfil certain compatibility conditions.

Relationship to Linear Maps

Linear maps $R^n \rightarrow R^m$ are equivalent to m-by-n matrices, as described above. More generally, any linear map $f: V \rightarrow W$ between finite-dimensional vector spaces can be described by a matrix $A = (a_{ij})$, after choosing bases $v_1, ..., v_n$ of V, and $w_1, ..., w_m$ of W (so n is the dimension of V and m is the dimension of W), which is such that

$$f(\mathbf{v}_j) = \sum_{i=1}^{m} a_{i,j}\mathbf{w}_i \qquad \text{for } j = 1, ..., n.$$

In other words, column j of A expresses the image of v_j in terms of the basis vectors w_i of W; thus this relation uniquely determines the entries of the matrix A. Note that the matrix depends on the choice of the bases: different choices of bases give rise to different, but equivalent matrices. Many of the above concrete notions can be reinterpreted in this light, for example, the transpose matrix A^T describes the transpose of the linear map given by A, with respect to the dual bases.

These properties can be restated in a more natural way: the category of all matrices with entries in a field k with multiplication as composition is equivalent to the category of finite dimensional vector spaces and linear maps over this field.

More generally, the set of $m \times n$ matrices can be used to represent the R-linear maps between the free modules R^m and R^n for an arbitrary ring R with unity. When $n = m$ composition of these maps is possible, and this gives rise to the matrix ring of $n \times n$ matrices representing the endomorphism ring of R^n.

Matrix Groups

A group is a mathematical structure consisting of a set of objects together with a binary operation, that is, an operation combining any two objects to a third, subject to certain requirements. A group in which the objects are matrices and the group operation is matrix multiplication is called a *matrix group*. Since in a group every element has to be invertible, the most general matrix groups are the groups of all invertible matrices of a given size, called the general linear groups.

Any property of matrices that is preserved under matrix products and inverses can be used to define further matrix groups. For example, matrices with a given size and with a determinant of 1 form a subgroup of (that is, a smaller group contained in) their general linear group, called a special linear group. Orthogonal matrices, determined by the condition

$$M^TM = I,$$

form the orthogonal group. Every orthogonal matrix has determinant 1 or −1. Orthogonal matrices with determinant 1 form a subgroup called *special orthogonal group*.

Every finite group is isomorphic to a matrix group, as one can see by considering the regular representation of the symmetric group. General groups can be studied using matrix groups, which are comparatively well-understood, by means of representation theory.

Infinite Matrices

It is also possible to consider matrices with infinitely many rows and/or columns even if, being infinite objects, one cannot write down such matrices explicitly. All that matters is that for every element in the set indexing rows, and every element in the set indexing columns, there is a well-defined entry (these index sets need not even be subsets of the natural numbers). The basic operations of addition, subtraction, scalar multiplication and transposition can still be defined without problem; however matrix multiplication may involve infinite summations to define the resulting entries, and these are not defined in general.

If R is any ring with unity, then the ring of endomorphisms of $M = \bigoplus_{i \in I} R$ as a right R module is

isomorphic to the ring of column finite matrices $\mathbb{CFM}_I(R)$ whose entries are indexed by $I \times I$, , and whose columns each contain only finitely many nonzero entries. The endomorphisms of M considered as a left R module result in an analogous object, the row finite matrices $\mathbb{RFM}_I(R)$ whose rows each only have finitely many nonzero entries.

If infinite matrices are used to describe linear maps, then only those matrices can be used all of whose columns have but a finite number of nonzero entries, for the following reason. For a matrix A to describe a linear map $f: V \rightarrow W$, bases for both spaces must have been chosen; recall that by definition this means that every vector in the space can be written uniquely as a (finite) linear combination of basis vectors, so that written as a (column) vector v of coefficients, only finitely many entries v_i are nonzero. Now the columns of A describe the images by f of individual basis vectors of V in the basis of W, which is only meaningful if these columns have only finitely many nonzero entries. There is no restriction on the rows of A however: in the product $A \cdot v$ there are only finitely many nonzero coefficients of v involved, so every one of its entries, even if it is given as an infinite sum of products, involves only finitely many nonzero terms and is therefore well defined. Moreover, this amounts to forming a linear combination of the columns of A that effectively involves only finitely many of them, whence the result has only finitely many nonzero entries, because each of those columns do. One also sees that products of two matrices of the given type is well defined (provided as usual that the column-index and row-index sets match), is again of the same type, and corresponds to the composition of linear maps.

If R is a normed ring, then the condition of row or column finiteness can be relaxed. With the norm in place, absolutely convergent series can be used instead of finite sums. For example, the matrices whose column sums are absolutely convergent sequences form a ring. Analogously of course, the matrices whose row sums are absolutely convergent series also form a ring.

In that vein, infinite matrices can also be used to describe operators on Hilbert spaces, where convergence and continuity questions arise, which again results in certain constraints that have to be imposed. However, the explicit point of view of matrices tends to obfuscate the matter, and the abstract and more powerful tools of functional analysis can be used instead.

Empty Matrices

An *empty matrix* is a matrix in which the number of rows or columns (or both) is zero. Empty matrices help dealing with maps involving the zero vector space. For example, if A is a 3-by-0 matrix and B is a 0-by-3 matrix, then AB is the 3-by-3 zero matrix corresponding to the null map from a 3-dimensional space V to itself, while BA is a 0-by-0 matrix. There is no common notation for empty matrices, but most computer algebra systems allow creating and computing with them. The determinant of the 0-by-0 matrix is 1 as follows from regarding the empty product occurring in the Leibniz formula for the determinant as 1. This value is also consistent with the fact that the identity map from any finite dimensional space to itself has determinant 1, a fact that is often used as a part of the characterization of determinants.

Applications

There are numerous applications of matrices, both in mathematics and other sciences. Some of them merely take advantage of the compact representation of a set of numbers in a matrix. For example, in game theory and economics, the payoff matrix encodes the payoff for two players, depending on which out of a given (finite) set of alternatives the players choose. Text mining and automated thesaurus compilation makes use of document-term matrices such as tf-idf to track frequencies of certain words in several documents.

Complex numbers can be represented by particular real 2-by-2 matrices via

$$a + ib \leftrightarrow \begin{bmatrix} a & -b \\ b & a \end{bmatrix},$$

under which addition and multiplication of complex numbers and matrices correspond to each other. For example, 2-by-2 rotation matrices represent the multiplication with some complex number of absolute value 1, as above. A similar interpretation is possible for quaternions and Clifford algebras in general.

Early encryption techniques such as the Hill cipher also used matrices. However, due to the linear nature of matrices, these codes are comparatively easy to break. Computer graphics uses matrices both to represent objects and to calculate transformations of objects using affine rotation matrices to accomplish tasks such as projecting a three-dimensional object onto a two-dimensional screen, corresponding to a theoretical camera observation. Matrices over a polynomial ring are important in the study of control theory.

Chemistry makes use of matrices in various ways, particularly since the use of quantum theory to discuss molecular bonding and spectroscopy. Examples are the overlap matrix and the Fock matrix used in solving the Roothaan equations to obtain the molecular orbitals of the Hartree–Fock method.

Graph Theory

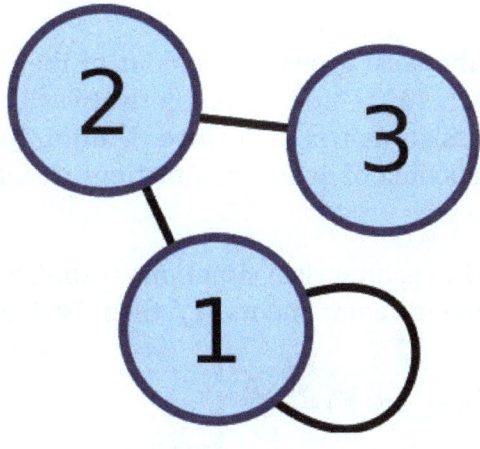

An undirected graph with adjacency matrix $\begin{bmatrix} 1 & 1 & 0 \\ 1 & 0 & 1 \\ 0 & 1 & 0 \end{bmatrix}$.

The adjacency matrix of a finite graph is a basic notion of graph theory. It records which vertices of the graph are connected by an edge. Matrices containing just two different values (1 and 0 meaning for example "yes" and "no", respectively) are called logical matrices. The distance (or cost) matrix contains information about distances of the edges. These concepts can be applied to websites connected by hyperlinks or cities connected by roads etc., in which case (unless the connection network is extremely dense) the matrices tend to be sparse, that is, contain few nonzero entries. Therefore, specifically tailored matrix algorithms can be used in network theory.

Analysis and Geometry

The Hessian matrix of a differentiable function $f: \mathrm{R}^n \to \mathrm{R}$ consists of the second derivatives of f with respect to the several coordinate directions, that is,

$$H(f) = \left[\frac{\partial^2 f}{\partial x_i \partial x_j} \right].$$

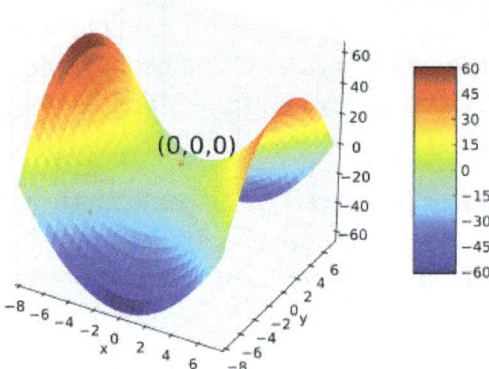

At the saddle point ($x = 0$, $y = 0$) (red) of the function $f(x,-y) = x^2 - y^2$, the Hessian matrix $\begin{bmatrix} 2 & 0 \\ 0 & -2 \end{bmatrix}$ is indefinite.

It encodes information about the local growth behaviour of the function: given a critical point $x = (x_1, ..., x_n)$, that is, a point where the first partial derivatives $\partial f / \partial x_i$ of f vanish, the function has a local minimum if the Hessian matrix is positive definite. Quadratic programming can be used to find global minima or maxima of quadratic functions closely related to the ones attached to matrices.

Another matrix frequently used in geometrical situations is the *Jacobi matrix* of a differentiable map $f: \mathbb{R}^n \to \mathbb{R}^m$. If $f_1, ..., f_m$ denote the components of f, then the Jacobi matrix is defined as

$$J(f) = \left[\frac{\partial f_i}{\partial x_j} \right]_{1 \leq i \leq m, 1 \leq j \leq n} .$$

If $n > m$, and if the rank of the Jacobi matrix attains its maximal value m, f is locally invertible at that point, by the implicit function theorem.

Partial differential equations can be classified by considering the matrix of coefficients of the highest-order differential operators of the equation. For elliptic partial differential equations this matrix is positive definite, which has decisive influence on the set of possible solutions of the equation in question.

The finite element method is an important numerical method to solve partial differential equations, widely applied in simulating complex physical systems. It attempts to approximate the solution to some equation by piecewise linear functions, where the pieces are chosen with respect to a sufficiently fine grid, which in turn can be recast as a matrix equation.

Probability Theory and Statistics

Stochastic matrices are square matrices whose rows are probability vectors, that is, whose entries are non-negative and sum up to one. Stochastic matrices are used to define Markov chains with finitely many states. A row of the stochastic matrix gives the probability distribution for the next position of some particle currently in the state that corresponds to the row. Properties of the Markov chain like absorbing states, that is, states that any particle attains eventually, can be read off the eigenvectors of the transition matrices.

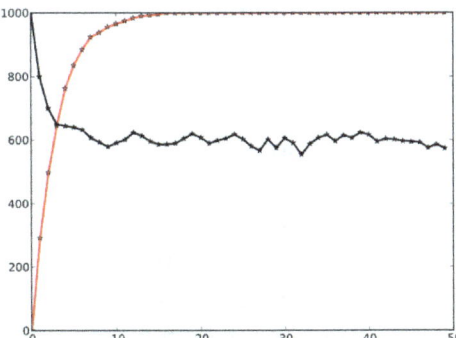

Two different Markov chains. The chart depicts the number of particles (of a total of 1000) in
state "2". Both limiting values can be determined from the transition matrices, which are given by

$$\begin{bmatrix} .7 & 0 \\ .3 & 1 \end{bmatrix} \text{(red) and} \begin{bmatrix} .7 & .2 \\ .3 & .8 \end{bmatrix} \text{(black)}.$$

Statistics also makes use of matrices in many different forms. Descriptive statistics is concerned
with describing data sets, which can often be represented as data matrices, which may then be sub-
jected to dimensionality reduction techniques. The covariance matrix encodes the mutual variance
of several random variables. Another technique using matrices are linear least squares, a method
that approximates a finite set of pairs $(x_1, y_1), (x_2, y_2), ..., (x_N, y_N)$, by a linear function

$$y_i \approx ax_i + b, \, i = 1, ..., N$$

which can be formulated in terms of matrices, related to the singular value decomposition of matrices.

Random matrices are matrices whose entries are random numbers, subject to suitable probability
distributions, such as matrix normal distribution. Beyond probability theory, they are applied in
domains ranging from number theory to physics.

Symmetries and Transformations in Physics

Linear transformations and the associated symmetries play a key role in modern physics. For example, elementary particles in quantum field theory are classified as representations of the Lorentz
group of special relativity and, more specifically, by their behavior under the spin group. Concrete
representations involving the Pauli matrices and more general gamma matrices are an integral
part of the physical description of fermions, which behave as spinors. For the three lightest quarks,
there is a group-theoretical representation involving the special unitary group SU(3); for their
calculations, physicists use a convenient matrix representation known as the Gell-Mann matrices,
which are also used for the SU(3) gauge group that forms the basis of the modern description of
strong nuclear interactions, quantum chromodynamics. The Cabibbo–Kobayashi–Maskawa matrix, in turn, expresses the fact that the basic quark states that are important for weak interactions
are not the same as, but linearly related to the basic quark states that define particles with specific
and distinct masses.

Linear Combinations of Quantum States

The first model of quantum mechanics (Heisenberg, 1925) represented the theory's operators by infinite-dimensional matrices acting on quantum states. This is also referred to as matrix mechanics. One particular example is the density matrix that characterizes the "mixed" state of a quantum system as a linear combination of elementary, "pure" eigenstates.

Another matrix serves as a key tool for describing the scattering experiments that form the cornerstone of experimental particle physics: Collision reactions such as occur in particle accelerators, where non-interacting particles head towards each other and collide in a small interaction zone, with a new set of non-interacting particles as the result, can be described as the scalar product of outgoing particle states and a linear combination of ingoing particle states. The linear combination is given by a matrix known as the S-matrix, which encodes all information about the possible interactions between particles.

Normal Modes

A general application of matrices in physics is to the description of linearly coupled harmonic systems. The equations of motion of such systems can be described in matrix form, with a mass matrix multiplying a generalized velocity to give the kinetic term, and a force matrix multiplying a displacement vector to characterize the interactions. The best way to obtain solutions is to determine the system's eigenvectors, its normal modes, by diagonalizing the matrix equation. Techniques like this are crucial when it comes to the internal dynamics of molecules: the internal vibrations of systems consisting of mutually bound component atoms. They are also needed for describing mechanical vibrations, and oscillations in electrical circuits.

Geometrical Optics

Geometrical optics provides further matrix applications. In this approximative theory, the wave nature of light is neglected. The result is a model in which light rays are indeed geometrical rays. If the deflection of light rays by optical elements is small, the action of a lens or reflective element on a given light ray can be expressed as multiplication of a two-component vector with a two-by-two matrix called ray transfer matrix: the vector's components are the light ray's slope and its distance from the optical axis, while the matrix encodes the properties of the optical element. Actually, there are two kinds of matrices, viz. a *refraction matrix* describing the refraction at a lens surface, and a *translation matrix*, describing the translation of the plane of reference to the next refracting surface, where another refraction matrix applies. The optical system, consisting of a combination of lenses and/or reflective elements, is simply described by the matrix resulting from the product of the components' matrices.

Electronics

Traditional mesh analysis and nodal analysis in electronics lead to a system of linear equations that can be described with a matrix.

The behaviour of many electronic components can be described using matrices. Let A be a 2-dimensional vector with the component's input voltage v_1 and input current i_1 as its elements, and let B be a 2-dimensional vector with the component's output voltage v_2 and output current i_2 as its elements. Then the behaviour of the electronic component can be described by $B = H \cdot A$, where H is a 2 x 2 matrix containing one impedance element (h_{12}), one admittance element (h_{21}) and two dimensionless elements (h_{11} and h_{22}). Calculating a circuit now reduces to multiplying matrices.

History

Matrices have a long history of application in solving linear equations but they were known as arrays until the 1800s. The Chinese text *The Nine Chapters on the Mathematical Art* written in 10th–2nd century BCE is the first example of the use of array methods to solve simultaneous equations, including the concept of determinants. In 1545 Italian mathematician Girolamo Cardano brought the method to Europe when he published *Ars Magna*. The Japanese mathematician Seki used the same array methods to solve simultaneous equations in 1683. The Dutch Mathematician Jan de Witt represented transformations using arrays in his 1659 book *Elements of Curves* (1659). Between 1700 and 1710 Gottfried Wilhelm Leibniz publicized the use of arrays for recording information or solutions and experimented with over 50 different systems of arrays. Cramer presented his rule in 1750.

The term "matrix" (Latin for "womb", derived from *mater*—mother) was coined by James Joseph Sylvester in 1850, who understood a matrix as an object giving rise to a number of determinants today called minors, that is to say, determinants of smaller matrices that derive from the original one by removing columns and rows. In an 1851 paper, Sylvester explains:

> I have in previous papers defined a "Matrix" as a rectangular array of terms, out of which different systems of determinants may be engendered as from the womb of a common parent.

Arthur Cayley published a treatise on geometric transformations using matrices that were not rotated versions of the coefficients being investigated as had previously been done. Instead he defined operations such as addition, subtraction, multiplication, and division as transformations of those matrices and showed the associative and distributive properties held true. Cayley investigated and demonstrated the non-commutative property of matrix multiplication as well as the commutative property of matrix addition. Early matrix theory had limited the use of arrays almost exclusively to determinants and Arthur Cayley's abstract matrix operations were revolutionary. He was instrumental in proposing a matrix concept independent of equation systems. In 1858 Cayley published his *A memoir on the theory of matrices* in which he proposed and demonstrated the Cayley-Hamilton theorem.

An English mathematician named Cullis was the first to use modern bracket notation for matrices in 1913 and he simultaneously demonstrated the first significant use of the notation A = $[a_{i,j}]$ to represent a matrix where $a_{i,j}$ refers to the ith row and the jth column.

The study of determinants sprang from several sources. Number-theoretical problems led Gauss to relate coefficients of quadratic forms, that is, expressions such as $x^2 + xy - 2y^2$, and linear maps in three dimensions to matrices. Eisenstein further developed these notions, including the remark that, in modern parlance, matrix products are non-commutative. Cauchy was the first to prove general statements about determinants, using as definition of the determinant of a matrix A = $[a_{i,j}]$ the following: replace the powers a_j^k by a_{jk} in the polynomial

$$a_1 a_2 \cdots a_n \prod_{i<j} (a_j - a_i),$$

where Π denotes the product of the indicated terms. He also showed, in 1829, that the eigenvalues of symmetric matrices are real. Jacobi studied "functional determinants"—later called Jacobi determinants by Sylvester—which can be used to describe geometric transformations at a local (or infinitesimal) level, Kronecker's *Vorlesungen über die Theorie der Determinanten* and Weierstrass' *Zur Determinantentheorie*, both published in 1903, first treated determinants axiomatically, as opposed to previous more concrete approaches such as the mentioned formula of Cauchy. At that point, determinants were firmly established.

Many theorems were first established for small matrices only, for example the Cayley–Hamilton theorem was proved for 2×2 matrices by Cayley in the aforementioned memoir, and by Hamilton for 4×4 matrices. Frobenius, working on bilinear forms, generalized the theorem to all dimensions (1898). Also at the end of the 19th century the Gauss–Jordan elimination (generalizing a special case now known as Gauss elimination) was established by Jordan. In the early 20th century, matrices attained a central role in linear algebra. partially due to their use in classification of the hypercomplex number systems of the previous century.

The inception of matrix mechanics by Heisenberg, Born and Jordan led to studying matrices with infinitely many rows and columns. Later, von Neumann carried out the mathematical formulation of quantum mechanics, by further developing functional analytic notions such as linear operators on Hilbert spaces, which, very roughly speaking, correspond to Euclidean space, but with an infinity of independent directions.

Other Historical Usages of the Word "Matrix" in Mathematics

The word has been used in unusual ways by at least two authors of historical importance.

Bertrand Russell and Alfred North Whitehead in their *Principia Mathematica* (1910–1913) use the word "matrix" in the context of their Axiom of reducibility. They proposed this axiom as a means to reduce any function to one of lower type, successively, so that at the "bottom" (0 order) the function is identical to its extension:

> "Let us give the name of *matrix* to any function, of however many variables, which does not involve any apparent variables. Then any possible function other than a matrix is derived from a matrix by means of generalization, that is, by considering the proposition which asserts that the function in question is true with all possible values or with some value of one of the arguments, the other argument or arguments remaining undetermined".

For example, a function $\Phi(x, y)$ of two variables x and y can be reduced to a *collection* of functions of a single variable, for example, y, by "considering" the function for all possible values of "individuals" a_i substituted in place of variable x. And then the resulting collection of functions of the single variable y, that is, $\forall a_i: \Phi(a_i, y)$, can be reduced to a "matrix" of values by "considering" the function for all possible values of "individuals" b_i substituted in place of variable y:

$$\forall b_j \forall a_i: \Phi(a_i, b_j).$$

Alfred Tarski in his 1946 *Introduction to Logic* used the word "matrix" synonymously with the notion of truth table as used in mathematical logic.

Glossary

off-diagonal element

An element where i and j differ. Zero for a diagonal matrix.

complex matrix

a matrix containing complex numbers

symmetric sparse matrix

a symmetric sparse matrix

Eigenvalues and Eigenvectors

In linear algebra, an eigenvector or characteristic vector of a linear transformation is a non-zero vector that does not change its direction when that linear transformation is applied to it. More formally, if T is a linear transformation from a vector space V over a field F into itself and v is a vector in V that is not the zero vector, then v is an eigenvector of T if $T(v)$ is a scalar multiple of v. This condition can be written as the equation

$$T(\mathbf{v}) = \lambda \mathbf{v},$$

where λ is a scalar in the field F, known as the eigenvalue, characteristic value, or characteristic root associated with the eigenvector v.

If the vector space V is finite-dimensional, then the linear transformation T can be represented as a square matrix A, and the vector v by a column vector, rendering the above mapping as a matrix multiplication on the left hand side and a scaling of the column vector on the right hand side in the equation

$$A\mathbf{v} = \lambda \mathbf{v}.$$

There is a correspondence between n by n square matrices and linear transformations from an n-dimensional vector space to itself. For this reason, it is equivalent to define eigenvalues and eigenvectors using either the language of matrices or the language of linear transformations.

Geometrically an eigenvector, corresponding to a real nonzero eigenvalue, points in a direction that is stretched by the transformation and the eigenvalue is the factor by which it is stretched. If the eigenvalue is negative, the direction is reversed.

Overview

Eigenvalues and eigenvectors feature prominently in the analysis of linear transformations. The prefix *eigen-* is adopted from the German word *eigen* for "proper", "inherent"; "own", "individual", "special"; "specific", "peculiar", or "characteristic". Originally utilized to study principal axes of the rotational motion of rigid bodies, eigenvalues and eigenvectors have a wide range of applications, for example in stability analysis, vibration analysis, atomic orbitals, facial recognition, and matrix diagonalization.

In essence, an eigenvector v of a linear transformation T is a non-zero vector that, when T is applied to it, does not change direction. Applying T to the eigenvector only scales the eigenvector by the scalar value λ, called an eigenvalue. This condition can be written as the equation

$$T(\mathbf{v}) = \lambda \mathbf{v},$$

referred to as the eigenvalue equation or eigenequation. In general, λ may be any scalar. For example, λ may be negative, in which case the eigenvector reverses direction as part of the scaling, or it may be zero or complex.

In this shear mapping the red arrow changes direction but the blue arrow does not. The blue arrow is an eigenvector of this shear mapping because it doesn't change direction, and since its length is unchanged, its eigenvalue is 1.

The Mona Lisa example pictured at right provides a simple illustration. Each point on the painting can be represented as a vector pointing from the center of the painting to that point. The linear transformation in this example is called a shear mapping. Points in the top half are moved to the right and points in the bottom half are moved to the left proportional to how far they are from the horizontal axis that goes through the middle of the painting. The vectors pointing to each point in the original image are therefore tilted right or left and made longer or shorter by the transformation. Notice that points *along* the horizontal axis do not move at all when this transformation is applied. Therefore, any vector that points directly to the right or left with no vertical component is an eigenvector of this transformation because the mapping does not change its direction. Moreover, these eigenvectors all have an eigenvalue equal to one because the mapping does not change their length, either.

Linear transformations can take many different forms, mapping vectors in a variety of vector spaces, so the eigenvectors can also take many forms. For example, the linear transformation could be a differential operator like $\frac{d}{dx}$, in which case the eigenvectors are functions called eigenfunctions that are scaled by that differential operator, such as

$$\frac{d}{dx} e^{\lambda x} = \lambda e^{\lambda x}.$$

Alternatively, the linear transformation could take the form of an n by n matrix, in which case the eigenvectors are n by 1 matrices that are also referred to as eigenvectors. If the linear transformation is expressed in the form of an n by n matrix A, then the eigenvalue equation above for a linear transformation can be rewritten as the matrix multiplication

$$Av = \lambda v,$$

where the eigenvector v is an n by 1 matrix. For a matrix, eigenvalues and eigenvectors can be used to decompose the matrix, for example by diagonalizing it.

Eigenvalues and eigenvectors give rise to many closely related mathematical concepts, and the prefix *eigen-* is applied liberally when naming them:

- The set of all eigenvectors of a linear transformation, each paired with its corresponding eigenvalue, is called the eigensystem of that transformation.

- The set of all eigenvectors of T corresponding to the same eigenvalue, together with the zero vector, is called an eigenspace or characteristic space of T.

- If the set of eigenvectors of T form a basis of the domain of T, then this basis is called an eigenbasis.

History

Eigenvalues are often introduced in the context of linear algebra or matrix theory. Historically, however, they arose in the study of quadratic forms and differential equations.

In the 18th century Euler studied the rotational motion of a rigid body and discovered the importance of the principal axes. Lagrange realized that the principal axes are the eigenvectors of the inertia matrix. In the early 19th century, Cauchy saw how their work could be used to classify the quadric surfaces, and generalized it to arbitrary dimensions. Cauchy also coined the term *racine caractéristique* (characteristic root) for what is now called *eigenvalue*; his term survives in *characteristic equation*.

Fourier used the work of Laplace and Lagrange to solve the heat equation by separation of variables in his famous 1822 book *Théorie analytique de la chaleur*. Sturm developed Fourier's ideas further and brought them to the attention of Cauchy, who combined them with his own ideas and arrived at the fact that real symmetric matrices have real eigenvalues. This was extended by Hermite in 1855 to what are now called Hermitian matrices. Around the same time, Brioschi proved that the eigenvalues of orthogonal matrices lie on the unit circle, and Clebsch found the corresponding result for skew-symmetric matrices. Finally, Weierstrass clarified an important aspect in the stability theory started by Laplace by realizing that defective matrices can cause instability.

In the meantime, Liouville studied eigenvalue problems similar to those of Sturm; the discipline that grew out of their work is now called *Sturm–Liouville theory*. Schwarz studied the first eigenvalue of Laplace's equation on general domains towards the end of the 19th century, while Poincaré studied Poisson's equation a few years later.

At the start of the 20th century, Hilbert studied the eigenvalues of integral operators by viewing the operators as infinite matrices. He was the first to use the German word *eigen*, which means "own", to denote eigenvalues and eigenvectors in 1904, though he may have been following a related usage by Helmholtz. For some time, the standard term in English was "proper value", but the more distinctive term "eigenvalue" is standard today.

The first numerical algorithm for computing eigenvalues and eigenvectors appeared in 1929, when Von Mises published the power method. One of the most popular methods today, the QR algorithm, was proposed independently by John G.F. Francis and Vera Kublanovskaya in 1961.

Eigenvalues and Eigenvectors of Matrices

Eigenvalues and eigenvectors are often introduced to students in the context of linear algebra courses focused on matrices. Furthermore, linear transformations can be represented using matrices, which is especially common in numerical and computational applications.

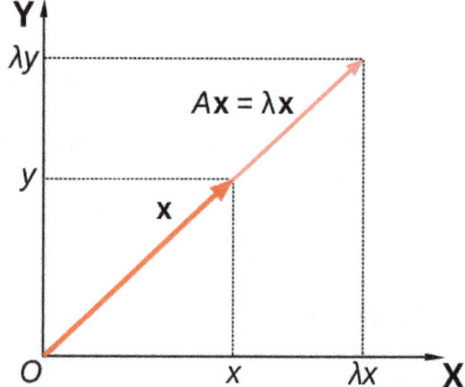

Matrix A acts by stretching the vector x, not changing its direction, so x is an eigenvector of A.

Consider n-dimensional vectors that are formed as a list of n scalars, such as the three-dimensional vectors

$$x = \begin{bmatrix} 1 \\ 3 \\ 4 \end{bmatrix} \quad \text{and} \quad y = \begin{bmatrix} -20 \\ -60 \\ -80 \end{bmatrix}.$$

These vectors are said to be scalar multiples of each other, or parallel or collinear, if there is a scalar λ such that

$$x = \lambda y.$$

In this case $\lambda = -1/20$.

Now consider the linear transformation of n-dimensional vectors defined by an n by n matrix A,

$$Av = w,$$

or

$$\begin{bmatrix} A_{11} & A_{12} & \cdots & A_{1n} \\ A_{21} & A_{22} & \cdots & A_{2n} \\ \vdots & \vdots & \ddots & \vdots \\ A_{n1} & A_{n2} & \cdots & A_{nn} \end{bmatrix} \begin{bmatrix} v_1 \\ v_2 \\ \vdots \\ v_n \end{bmatrix} = \begin{bmatrix} w_1 \\ w_2 \\ \vdots \\ w_n \end{bmatrix}$$

where, for each row,

$$w_i = A_{i1}v_1 + A_{i2}v_2 + \cdots + A_{in}v_n = \sum_{j=1}^{n} A_{ij}v_j.$$

If it occurs that v and w are scalar multiples, that is if

$$Av = \lambda v, \tag{1}$$

then v is an eigenvector of the linear transformation A and the scale factor λ is the eigenvalue corresponding to that eigenvector. Equation (1) is the eigenvalue equation for the matrix A.

Equation (1) can be stated equivalently as

$$(A - \lambda I)v = 0, \tag{2}$$

where I is the n by n identity matrix.

Eigenvalues and the Characteristic Polynomial

Equation (2) has a non-zero solution v if and only if the determinant of the matrix $(A - \lambda I)$ is zero. Therefore, the eigenvalues of A are values of λ that satisfy the equation

$$|A - \lambda I| = 0 \tag{3}$$

Using Leibniz' rule for the determinant, the left hand side of Equation (3) is a polynomial function of the variable λ and the degree of this polynomial is n, the order of the matrix A. Its coefficients depend on the entries of A, except that its term of degree n is always $(-1)^n \lambda^n$. This polynomial is called the *characteristic polynomial* of A. Equation (3) is called the *characteristic equation* or the *secular equation* of A.

The fundamental theorem of algebra implies that the characteristic polynomial of an n by n matrix A, being a polynomial of degree n, can be factored into the product of n linear terms,

$$|A - \lambda I| = (\lambda_1 - \lambda)(\lambda_2 - \lambda)\cdots(\lambda_n - \lambda), \tag{4}$$

where each λ_i may be real but in general is a complex number. The numbers $\lambda_1, \lambda_2, \ldots \lambda_n$, which may not all have distinct values, are roots of the polynomial and are the eigenvalues of A.

As a brief example, which is described in more detail in the examples section later, consider the matrix

$$M = \begin{bmatrix} 2 & 1 \\ 1 & 2 \end{bmatrix}.$$

Taking the determinant of $(M - \lambda I)$, the characteristic polynomial of M is

$$|M - \lambda I| = \begin{vmatrix} 2-\lambda & 1 \\ 1 & 2-\lambda \end{vmatrix} = 3 - 4\lambda + \lambda^2.$$

Setting the characteristic polynomial equal to zero, it has roots at $\lambda = 1$ and $\lambda = 3$, which are the two eigenvalues of M. The eigenvectors corresponding to each eigenvalue can be found by solving for the components of v in the equation $Mv = \lambda v$. In this example, the eigenvectors are any non-zero scalar multiples of

$$v_{\lambda=1} = \begin{bmatrix} 1 \\ -1 \end{bmatrix}, \quad v_{\lambda=3} = \begin{bmatrix} 1 \\ 1 \end{bmatrix}.$$

If the entries of the matrix A are all real numbers, then the coefficients of the characteristic polynomial will also be real numbers, but the eigenvalues may still have non-zero imaginary parts. The entries of the corresponding eigenvectors therefore may also have non-zero imaginary parts. Similarly, the eigenvalues may be irrational numbers even if all the entries of A are rational numbers or even if they are all integers. However, if the entries of A are all algebraic numbers, which include the rationals, the eigenvalues are complex algebraic numbers.

The non-real roots of a real polynomial with real coefficients can be grouped into pairs of complex conjugates, namely with the two members of each pair having imaginary parts that differ only in sign and the same real part. If the degree is odd, then by the intermediate value theorem at least one of the roots is real. Therefore, any real matrix with odd order has at least one real eigenvalue, whereas a real matrix with even order may not have any real eigenvalues. The eigenvectors associated with these complex eigenvalues are also complex and also appear in complex conjugate pairs.

Algebraic Multiplicity

Let λ_i be an eigenvalue of an n by n matrix A. The algebraic multiplicity $\mu_A(\lambda_i)$ of the eigenvalue is its multiplicity as a root of the characteristic polynomial, that is, the largest integer k such that $(\lambda - \lambda_i)^k$ divides evenly that polynomial.

Suppose a matrix A has dimension n and $d \le n$ distinct eigenvalues. Whereas Equation (4) factors the characteristic polynomial of A into the product of n linear terms with some terms potentially repeating, the characteristic polynomial can instead be written as the product d terms each corresponding to a distinct eigenvalue and raised to the power of the algebraic multiplicity,

$$|A - \lambda I| = (\lambda_1 - \lambda)^{\mu_A(\lambda_1)} (\lambda_2 - \lambda)^{\mu_A(\lambda_2)} \cdots (\lambda_d - \lambda)^{\mu_A(\lambda_d)}.$$

If $d = n$ then the right hand side is the product of n linear terms and this is the same as Equation (4). The size of each eigenvalue's algebraic multiplicity is related to the dimension n as

$$1 \le \mu_A(\lambda_i) \le n,$$

$$\mu_A = \sum_{i=1}^{d} \mu_A(\lambda_i) = n.$$

If $\mu_A(\lambda_i) = 1$, then λ_i is said to be a *simple eigenvalue*. If $\mu_A(\lambda_i)$ equals the geometric multiplicity of λ_i, $\gamma_A(\lambda_i)$, defined in the next section, then λ_i is said to be a *semisimple eigenvalue*.

Eigenspaces, Geometric Multiplicity, and the Eigenbasis for Matrices

Given a particular eigenvalue λ of the n by n matrix A, define the set E to be all vectors v that satisfy Equation (2),

$$E = \{v : (A - \lambda I)v = 0\}.$$

On one hand, this set is precisely the kernel or nullspace of the matrix $(A - \lambda I)$. On the other hand, by definition, any non-zero vector that satisfies this condition is an eigenvector of A associated with λ. So, the set E is the union of the zero vector with the set of all eigenvectors of A associated with λ, and E equals the nullspace of $(A - \lambda I)$. E is called the eigenspace or characteristic space of A associated with λ. In general λ is a complex number and the eigenvectors are complex n by 1 matrices. A property of the nullspace is that it is a linear subspace, so E is a linear subspace of C^n.

Because the eigenspace E is a linear subspace, it is closed under addition. That is, if two vectors u and v belong to the set E, written $(u,v) \in E$, then $(u + v) \in E$ or equivalently $A(u + v) = \lambda(u + v)$. This can be checked using the distributive property of matrix multiplication. Similarly, because E is a linear subspace, it is closed under scalar multiplication. That is, if $v \in E$ and α is a complex number, $(\alpha v) \in E$ or equivalently $A(\alpha v) = \lambda(\alpha v)$. This can be checked by noting that multiplication of complex matrices by complex numbers is commutative. As long as $u + v$ and αv are not zero, they are also eigenvectors of A associated with λ.

The dimension of the eigenspace E associated with λ, or equivalently the maximum number of linearly independent eigenvectors associated with λ, is referred to as the eigenvalue's geometric multiplicity $\gamma_A(\lambda)$. Because E is also the nullspace of $(A - \lambda I)$, the geometric multiplicity of λ is the dimension of the nullspace of $(A - \lambda I)$, also called the *nullity* of $(A - \lambda I)$, which relates to the dimension and rank of $(A - \lambda I)$ as

$$\gamma_A(\lambda) = n - \text{rank}(A - \lambda I).$$

Because of the definition of eigenvalues and eigenvectors, an eigenvalue's geometric multiplicity must be at least one, that is, each eigenvalue has at least one associated eigenvector. Furthermore, an eigenvalue's geometric multiplicity cannot exceed its algebraic multiplicity. Additionally, recall that an eigenvalue's algebraic multiplicity cannot exceed n.

$$1 \leq \gamma_A(\lambda) \leq \mu_A(\lambda) \leq n$$

The condition that $\gamma_A(\lambda) \leq \mu_A(\lambda)$ can be proven by considering a particular eigenvalue ξ of A and diagonalizing the first $\gamma_A(\xi)$ columns of A with respect to ξ's eigenvectors, described in a later section. The resulting similar matrix B is block upper triangular, with its top left block being the diagonal matrix $\xi I_{\gamma A(\xi)}$. As a result, the characteristic polynomial of B will have a factor of $(\xi - \lambda)^{\gamma_A(\xi)}$. The other factors of the characteristic polynomial of B are not known, so the algebraic multiplicity of ξ as an eigenvalue of B is no less than the geometric multiplicity of ξ as an eigenvalue of A. The last element of the proof is the property that similar matrices have the same characteristic polynomial.

Suppose A has $d \leq n$ distinct eigenvalues $\lambda_1, \lambda_2, ..., \lambda_d$, where the geometric multiplicity of λ_i is $\gamma_A(\lambda_i)$. The total geometric multiplicity of A,

$$\gamma_A = \sum_{i=1}^{d} \gamma_A(\lambda_i),$$

$$d \le \gamma_A \le n,$$

is the dimension of the union of all the eigenspaces of A's eigenvalues, or equivalently the maximum number of linearly independent eigenvectors of A. If $\gamma_A = n$, then

- The union of the eigenspaces of all of A's eigenvalues is the entire vector space $\mathbb{C}^n$

- A basis of $\mathbb{C}^n$ can be formed from n linearly independent eigenvectors of A; such a basis is called an eigenbasis

- Any vector in $\mathbb{C}^n$ can be written as a linear combination of eigenvectors of A

Additional Properties of Eigenvalues

Let A be an arbitrary n by n matrix of complex numbers with eigenvalues $\lambda_1, \lambda_2, ..., \lambda_n$. Each eigenvalue appears $\mu_A(\lambda_i)$ times in this list, where $\mu_A(\lambda_i)$ is the eigenvalue's algebraic multiplicity. The following are properties of this matrix and its eigenvalues:

- The trace of A, defined as the sum of its diagonal elements, is also the sum of all eigenvalues,

$$\text{tr}(A) = \sum_{i=1}^{n} A_{ii} = \sum_{i=1}^{n} \lambda_i = \lambda_1 + \lambda_2 + \cdots + \lambda_n.$$

- The determinant of A is the product of all its eigenvalues,

$$\det(A) = \prod_{i=1}^{n} \lambda_i = \lambda_1 \lambda_2 \cdots \lambda_n.$$

- The eigenvalues of the kth power of A, i.e. the eigenvalues of A^k, for any positive integer k, are $\lambda_1^k, \lambda_2^k, ..., \lambda_n^k$.

- The matrix A is invertible if and only if every eigenvalue is nonzero.

- If A is invertible, then the eigenvalues of A^{-1} are $1/\lambda_1, 1/\lambda_2, ..., 1/\lambda_n$ and each eigenvalue's geometric multiplicity coincides. Moreover, since the characteristic polynomial of the inverse is the reciprocal polynomial of the original, the eigenvalues share the same algebraic multiplicity.

- If A is equal to its conjugate transpose A^*, or equivalently if A is Hermitian, then every eigenvalue is real. The same is true of any symmetric real matrix.

- If A is not only Hermitian but also positive-definite, positive-semidefinite, negative-definite, or negative-semidefinite, then every eigenvalue is positive, non-negative, negative, or non-positive, respectively.

- If A is unitary, every eigenvalue has absolute value $|\lambda_i| = 1$.

Left and Right Eigenvectors

Many disciplines traditionally represent vectors as matrices with a single column rather than as matrices with a single row. For that reason, the word "eigenvector" in the context of matrices almost always refers to a right eigenvector, namely a *column* vector that *right* multiples the n by n matrix A in the defining equation, Equation (1),

$$Av = \lambda v.$$

The eigenvalue and eigenvector problem can also be defined for *row* vectors that *left* multiply matrix A. In this formulation, the defining equation is

$$uA = \kappa u,$$

where κ is a scalar and u is a 1 by n matrix. Any row vector u satisfying this equation is called a left eigenvector of A and κ is its associated eigenvalue. Taking the conjugate transpose of this equation,

$$A^* u^* = \kappa^* u^*.$$

Comparing this equation to Equation (1), the left eigenvectors of A are the conjugate transpose of the right eigenvectors of A^*. The eigenvalues of the left eigenvectors are the solution of the characteristic polynomial $|A^* - \kappa^* I| = 0$. Because the identity matrix is Hermitian and $|M^*| = |M|^*$ for a square matrix M, the eigenvalues of the left eigenvectors of A are the complex conjugates of the eigenvalues of the right eigenvectors of A. Recall that if A is a real matrix, all of its complex eigenvalues appear in complex conjugate pairs. Therefore, the eigenvalues of the left and right eigenvectors of a real matrix are the same. Similarly, if A is a real matrix, all of its complex eigenvectors also appear in complex conjugate pairs. Therefore, the left eigenvectors simplify to the transpose of the right eigenvectors of A^T if A is real.

Diagonalization and the Eigendecomposition

Suppose the eigenvectors of A form a basis, or equivalently A has n linearly independent eigenvectors $v_1, v_2, ..., v_n$ with associated eigenvalues $\lambda_1, \lambda_2, ..., \lambda_n$. The eigenvalues need not be distinct. Define a square matrix Q whose columns are the n linearly independent eigenvectors of A,

$$Q = \begin{bmatrix} v_1 & v_2 & ... & v_n \end{bmatrix}.$$

Since each column of Q is an eigenvector of A, right multiplying A by Q scales each column of Q by its associated eigenvalue,

$$AQ = \begin{bmatrix} \lambda_1 v_1 & \lambda_2 v_2 & ... & \lambda_n v_n \end{bmatrix}.$$

With this in mind, define a diagonal matrix Λ where each diagonal element Λ_{ii} is the eigenvalue associated with the ith column of Q. Then

$$AQ = Q\Lambda.$$

Because the columns of Q are linearly independent, Q is invertible. Right multiplying both sides of the equation by Q^{-1},

$$A = Q \Lambda Q^{-1},$$

or by instead left multiplying both sides by Q^{-1},

$$Q^{-1} A Q = \Lambda.$$

A can therefore be decomposed into a matrix composed of its eigenvectors, a diagonal matrix with its eigenvalues along the diagonal, and the inverse of the matrix of eigenvectors. This is called the eigendecomposition and it is a similarity transformation. Such a matrix A is said to be *similar* to the diagonal matrix Λ or *diagonalizable*. The matrix Q is the change of basis matrix of the similarity transformation. Essentially, the matrices A and Λ represent the same linear transformation expressed in two different bases. The eigenvectors are used as the basis when representing the linear transformation as Λ.

Conversely, suppose a matrix A is diagonalizable. Let P be a non-singular square matrix such that $P^{-1}AP$ is some diagonal matrix D. Left multiplying both by P, $AP = PD$. Each column of P must therefore be an eigenvector of A whose eigenvalue is the corresponding diagonal element of D. Since the columns of P must be linearly independent for P to be invertible, there exist n linearly independent eigenvectors of A. It then follows that the eigenvectors of A form a basis if and only if A is diagonalizable.

A matrix that is not diagonalizable is said to be defective. For defective matrices, the notion of eigenvectors generalizes to generalized eigenvectors and the diagonal matrix of eigenvalues generalizes to the Jordan normal form. Over an algebraically closed field, any matrix A has a Jordan normal form and therefore admits a basis of generalized eigenvectors and a decomposition into generalized eigenspaces.

Variational Characterization

In the Hermitian case, eigenvalues can be given a variational characterization. The largest eigenvalue of H is the maximum value of the quadratic form $x^{\mathsf{T}} H x / x^{\mathsf{T}} x$. A value of x that realizes that maximum, is an eigenvector.

Matrix Examples

Two-dimensional Matrix Example

Consider the matrix

$$A = \begin{bmatrix} 2 & 1 \\ 1 & 2 \end{bmatrix}.$$

The figure on the right shows the effect of this transformation on point coordinates in the plane. The eigenvectors v of this transformation satisfy Equation (1), and the values of λ for which the determinant of the matrix $(A - \lambda I)$ equals zero are the eigenvalues.

Taking the determinant to find characteristic polynomial of A,

$$| A - \lambda I | = \left| \begin{bmatrix} 2 & 1 \\ 1 & 2 \end{bmatrix} - \lambda \begin{bmatrix} 1 & 0 \\ 0 & 1 \end{bmatrix} \right| = \begin{vmatrix} 2 - \lambda & 1 \\ 1 & 2 - \lambda \end{vmatrix},$$

$$= 3 - 4\lambda + \lambda^2.$$

Setting the characteristic polynomial equal to zero, it has roots at $\lambda = 1$ and $\lambda = 3$, which are the two eigenvalues of A.

For $\lambda = 1$, Equation (2) becomes,

$$(A - I)v_{\lambda=1} = \begin{bmatrix} 1 & 1 \\ 1 & 1 \end{bmatrix} \begin{bmatrix} v_1 \\ v_2 \end{bmatrix} = \begin{bmatrix} 0 \\ 0 \end{bmatrix}.$$

Any non-zero vector with $v_1 = -v_2$ solves this equation. Therefore,

$$v_{\lambda=1} = \begin{bmatrix} 1 \\ -1 \end{bmatrix}$$

is an eigenvector of A corresponding to $\lambda = 1$, as is any scalar multiple of this vector.

For $\lambda = 3$, Equation (2) becomes

$$(A - 3I)v_{\lambda=3} = \begin{bmatrix} -1 & 1 \\ 1 & -1 \end{bmatrix} \begin{bmatrix} v_1 \\ v_2 \end{bmatrix} = \begin{bmatrix} 0 \\ 0 \end{bmatrix}.$$

Any non-zero vector with $v_1 = v_2$ solves this equation. Therefore,

$$v_{\lambda=3} = \begin{bmatrix} 1 \\ 1 \end{bmatrix}$$

is an eigenvector of A corresponding to $\lambda = 3$, as is any scalar multiple of this vector.

Thus, the vectors $v_{\lambda=1}$ and $v_{\lambda=3}$ are eigenvectors of A associated with the eigenvalues $\lambda = 1$ and $\lambda = 3$, respectively.

Three-dimensional Matrix Example

Consider the matrix

$$A = \begin{bmatrix} 2 & 0 & 0 \\ 0 & 3 & 4 \\ 0 & 4 & 9 \end{bmatrix}.$$

The characteristic polynomial of A is

$$|A-\lambda I|=\begin{vmatrix} 2 & 0 & 0 \\ 0 & 3 & 4 \\ 0 & 4 & 9 \end{vmatrix}-\lambda\begin{vmatrix} 1 & 0 & 0 \\ 0 & 1 & 0 \\ 0 & 0 & 1 \end{vmatrix}=\begin{vmatrix} 2-\lambda & 0 & 0 \\ 0 & 3-\lambda & 4 \\ 0 & 4 & 9-\lambda \end{vmatrix},$$

$$=(2-\lambda)\big[(3-\lambda)(9-\lambda)-16\big]=-\lambda^3+14\lambda^2-35\lambda+22.$$

The roots of the characteristic polynomial are 2, 1, and 11, which are the only three eigenvalues of A. These eigenvalues correspond to the eigenvectors $[1\,0\,0]^T$, $[0\,2-1]^T$, and $[0\,1\,2]^T$, or any non-zero multiple thereof.

Three-dimensional Matrix Example with Complex Eigenvalues

Consider the cyclic permutation matrix

$$A=\begin{bmatrix} 0 & 1 & 0 \\ 0 & 0 & 1 \\ 1 & 0 & 0 \end{bmatrix}.$$

This matrix shifts the coordinates of the vector up by one position and moves the first coordinate to the bottom. Its characteristic polynomial is $1-\lambda^3$, whose roots are

$$\lambda_1=1$$

$$\lambda_2=-1/2+i\sqrt{3}/2$$

$$\lambda_3=\lambda_2^*=-1/2-i\sqrt{3}/2$$

where $i=\sqrt{-1}$ is the imaginary unit.

For the real eigenvalue $\lambda_1=1$, any vector with three equal non-zero entries is an eigenvector. For example,

$$A\begin{bmatrix} 5 \\ 5 \\ 5 \end{bmatrix}=\begin{bmatrix} 5 \\ 5 \\ 5 \end{bmatrix}=1\cdot\begin{bmatrix} 5 \\ 5 \\ 5 \end{bmatrix}.$$

For the complex conjugate pair of imaginary eigenvalues, note that

$$\lambda_2\lambda_3=1,\quad \lambda_2^2=\lambda_3,\quad \lambda_3^2=\lambda_2.$$

Then

$$A\begin{bmatrix} 1 \\ \lambda_2 \\ \lambda_3 \end{bmatrix}=\begin{bmatrix} \lambda_2 \\ \lambda_3 \\ 1 \end{bmatrix}=\lambda_2\cdot\begin{bmatrix} 1 \\ \lambda_2 \\ \lambda_3 \end{bmatrix},$$

and

$$A \begin{bmatrix} 1 \\ \lambda_3 \\ \lambda_2 \end{bmatrix} = \begin{bmatrix} \lambda_3 \\ \lambda_2 \\ 1 \end{bmatrix} = \lambda_3 \cdot \begin{bmatrix} 1 \\ \lambda_3 \\ \lambda_2 \end{bmatrix}.$$

Therefore, the other two eigenvectors of A are complex and are $v_{\lambda_2} = [1 \; \lambda_2 \; \lambda_3]^{\mathrm{T}}$ and $v_{\lambda_3} = [1 \; \lambda_3 \; \lambda_2]^{\mathrm{T}}$ with eigenvalues λ_2 and λ_3, respectively. Note that the two complex eigenvectors also appear in a complex conjugate pair,

$$v_{\lambda_2} = v_{\lambda_3}^{*}.$$

Diagonal Matrix Example

Matrices with entries only along the main diagonal are called *diagonal matrices.* The eigenvalues of a diagonal matrix are the diagonal elements themselves. Consider the matrix

$$A = \begin{bmatrix} 1 & 0 & 0 \\ 0 & 2 & 0 \\ 0 & 0 & 3 \end{bmatrix}.$$

The characteristic polynomial of A is

$$|A - \lambda I| = (1 - \lambda)(2 - \lambda)(3 - \lambda),$$

which has the roots $\lambda_1 = 1$, $\lambda_2 = 2$, and $\lambda_3 = 3$. These roots are the diagonal elements as well as the eigenvalues of A.

Each diagonal element corresponds to an eigenvector whose only non-zero component is in the same row as that diagonal element. In the example, the eigenvalues correspond to the eigenvectors,

$$v_{\lambda_1} = \begin{bmatrix} 1 \\ 0 \\ 0 \end{bmatrix}, \quad v_{\lambda_2} = \begin{bmatrix} 0 \\ 1 \\ 0 \end{bmatrix}, \quad v_{\lambda_3} = \begin{bmatrix} 0 \\ 0 \\ 1 \end{bmatrix},$$

respectively, as well as scalar multiples of these vectors.

Triangular Matrix Example

A matrix whose elements above the main diagonal are all zero is called a *lower triangular matrix,* while a matrix whose elements below the main diagonal are all zero is called an *upper triangular matrix.* As with diagonal matrices, the eigenvalues of triangular matrices are the elements of the main diagonal.

Consider the lower triangular matrix,

$$A = \begin{bmatrix} 1 & 0 & 0 \\ 1 & 2 & 0 \\ 2 & 3 & 3 \end{bmatrix}.$$

The characteristic polynomial of A is

$$|A - \lambda I| = (1 - \lambda)(2 - \lambda)(3 - \lambda),$$

which has the roots $\lambda_1 = 1$, $\lambda_2 = 2$, and $\lambda_3 = 3$. These roots are the diagonal elements as well as the eigenvalues of A.

These eigenvalues correspond to the eigenvectors,

$$v_{\lambda_1} = \begin{bmatrix} 1 \\ -1 \\ 1/2 \end{bmatrix}, \quad v_{\lambda_2} = \begin{bmatrix} 0 \\ 1 \\ -3 \end{bmatrix}, \quad v_{\lambda_3} = \begin{bmatrix} 0 \\ 0 \\ 1 \end{bmatrix},$$

respectively, as well as scalar multiples of these vectors.

Matrix with Repeated Eigenvalues Example

As in the previous example, the lower triangular matrix

$$A = \begin{bmatrix} 2 & 0 & 0 & 0 \\ 1 & 2 & 0 & 0 \\ 0 & 1 & 3 & 0 \\ 0 & 0 & 1 & 3 \end{bmatrix},$$

has a characteristic polynomial that is the product of its diagonal elements,

$$|A - \lambda I| = \begin{vmatrix} 2 - \lambda & 0 & 0 & 0 \\ 1 & 2 - \lambda & 0 & 0 \\ 0 & 1 & 3 - \lambda & 0 \\ 0 & 0 & 1 & 3 - \lambda \end{vmatrix} = (2 - \lambda)^2 (3 - \lambda)^2.$$

The roots of this polynomial, and hence the eigenvalues, are 2 and 3. The *algebraic multiplicity* of each eigenvalue is 2; in other words they are both double roots. The sum of the algebraic multiplicities of each distinct eigenvalue is $\mu_A = 4 = n$, the order of the characteristic polynomial and the dimension of A.

On the other hand, the *geometric multiplicity* of the eigenvalue 2 is only 1, because its eigenspace is spanned by just one vector $[0 \ 1 \ -1 \ 1]^T$ and is therefore 1-dimensional. Similarly, the geometric multiplicity of the eigenvalue 3 is 1 because its eigenspace is spanned by just one vector $[0 \ 0 \ 0 \ 1]^T$.

The total geometric multiplicity γ_A is 2, which is the smallest it could be for a matrix with two distinct eigenvalues. Geometric multiplicities are defined in a later section.

Eigenvalues and Eigenfunctions of Differential Operators

The definitions of eigenvalue and eigenvectors of a linear transformation T remains valid even if the underlying vector space is an infinite-dimensional Hilbert or Banach space. A widely used class of linear transformations acting on infinite-dimensional spaces are the differential operators on function spaces. Let D be a linear differential operator on the space C^∞ of infinitely differentiable real functions of a real argument t. The eigenvalue equation for D is the differential equation

$$Df(t) = \lambda f(t)$$

The functions that satisfy this equation are eigenvectors of D and are commonly called eigenfunctions.

Derivative Operator Example

Consider the derivative operator $\frac{d}{dt}$ with eigenvalue equation

$$\frac{d}{dt} f(t) = \lambda f(t).$$

This differential equation can be solved by multiplying both sides by $dt/f(t)$ and integrating. Its solution, the exponential function

$$f(t) = f(0)e^{\lambda t},$$

is the eigenfunction of the derivative operator. Note that in this case the eigenfunction is itself a function of its associated eigenvalue. In particular, note that for $\lambda = 0$ the eigenfunction $f(t)$ is a constant.

General Definition

The concept of eigenvalues and eigenvectors extends naturally to arbitrary linear transformations on arbitrary vector spaces. Let V be any vector space over some field K of scalars, and let T be a linear transformation mapping V into V,

$$T : V \to V.$$

We say that a non-zero vector $v \in V$ is an eigenvector of T if and only if there exists a scalar $\lambda \in K$ such that

$$T(\mathbf{v}) = \lambda \mathbf{v}. \tag{5}$$

This equation is called the eigenvalue equation for T, and the scalar λ is the eigenvalue of T corresponding to the eigenvector v. Note that $T(v)$ is the result of applying the transformation T to the vector v, while λv is the product of the scalar λ with v.

Eigenspaces, Geometric Multiplicity, and the Eigenbasis

Given an eigenvalue λ, consider the set

$$E = \{\mathbf{v} : T(\mathbf{v}) = \lambda \mathbf{v}\},$$

which is the union of the zero vector with the set of all eigenvectors associated with λ. E is called the eigenspace or characteristic space of T associated with λ.

By definition of a linear transformation,

$$T(\mathbf{x} + \mathbf{y}) = T(\mathbf{x}) + T(\mathbf{y}),$$
$$T(\alpha \mathbf{x}) = \alpha T(\mathbf{x}),$$

for $(x,y) \in V$ and $\alpha \in K$. Therefore, if u and v are eigenvectors of T associated with eigenvalue λ, namely $(u,v) \in E$, then

$$T(\mathbf{u} + \mathbf{v}) = \lambda(\mathbf{u} + \mathbf{v}),$$
$$T(\alpha \mathbf{v}) = \lambda(\alpha \mathbf{v}).$$

So, both u + v and αv are either zero or eigenvectors of T associated with λ, namely $(u+v,\alpha v) \in E$, and E is closed under addition and scalar multiplication. The eigenspace E associated with λ is therefore a linear subspace of V. If that subspace has dimension 1, it is sometimes called an eigenline.

The geometric multiplicity $\gamma_T(\lambda)$ of an eigenvalue λ is the dimension of the eigenspace associated with λ, i.e., the maximum number of linearly independent eigenvectors associated with that eigenvalue. By the definition of eigenvalues and eigenvectors, $\gamma_T(\lambda) \geq 1$ because every eigenvalue has at least one eigenvector.

The eigenspaces of T always form a direct sum. As a consequence, eigenvectors of *different* eigenvalues are always linearly independent. Therefore, the sum of the dimensions of the eigenspaces cannot exceed the dimension n of the vector space on which T operates, and there cannot be more than n distinct eigenvalues.

Any subspace spanned by eigenvectors of T is an invariant subspace of T, and the restriction of T to such a subspace is diagonalizable. Moreover, if the entire vector space V can be spanned by the eigenvectors of T, or equivalently if the direct sum of the eigenspaces associated with all the eigenvalues of T is the entire vector space V, then a basis of V called an eigenbasis can be formed from linearly independent eigenvectors of T. When T admits an eigenbasis, T is diagonalizable.

Zero Vector as an Eigenvector

While the definition of an eigenvector used in this chapter excludes the zero vector, it is possible to define eigenvalues and eigenvectors such that the zero vector is an eigenvector.

Consider again the eigenvalue equation, Equation (5). Define an eigenvalue to be any scalar $\lambda \in K$ such that there exists a non-zero vector $v \in V$ satisfying Equation (5). It is important that this version of the definition of an eigenvalue specify that the vector be non-zero, otherwise by this definition the zero vector would allow any scalar in K to be an eigenvalue. Define an eigenvector v associated with the eigenvalue λ to be any vector that, given λ, satisfies Equation (5). Given the eigenvalue, the zero vector is among the vectors that satisfy Equation (5), so the zero vector is included among the eigenvectors by this alternate definition.

Spectral Theory

If λ is an eigenvalue of T, then the operator $(T - \lambda I)$ is not one-to-one, and therefore its inverse $(T - \lambda I)^{-1}$ does not exist. The converse is true for finite-dimensional vector spaces, but not for infinite-dimensional vector spaces. In general, the operator $(T - \lambda I)$ may not have an inverse even if λ is not an eigenvalue.

For this reason, in functional analysis eigenvalues can be generalized to the spectrum of a linear operator T as the set of all scalars λ for which the operator $(T - \lambda I)$ has no bounded inverse. The spectrum of an operator always contains all its eigenvalues but is not limited to them.

Associative Algebras and Representation Theory

One can generalize the algebraic object that is acting on the vector space, replacing a single operator acting on a vector space with an algebra representation – an associative algebra acting on a module. The study of such actions is the field of representation theory.

The representation-theoretical concept of weight is an analog of eigenvalues, while *weight vectors* and *weight spaces* are the analogs of eigenvectors and eigenspaces, respectively.

Dynamic Equations

The simplest difference equations have the form

$$x_t = a_1 x_{t-1} + a_2 x_{t-2} + \cdots + a_k x_{t-k}.$$

The solution of this equation for x in terms of t is found by using its characteristic equation

$$\lambda^k - a_1 \lambda^{k-1} - a_2 \lambda^{k-2} - \cdots - a_{k-1} \lambda - a_k = 0,$$

which can be found by stacking into matrix form a set of equations consisting of the above difference equation and the $k-1$ equations $x_{t-1} = x_{t-1}, \ldots, x_{t-k+1} = x_{t-k+1}$, giving a k-dimensional system of the first order in the stacked variable vector $[x_t, \ldots, x_{t-k+1}]$ in terms of its once-lagged value, and taking the characteristic equation of this system's matrix. This equation gives k characteristic roots $\lambda_1, \ldots, \lambda_k$, for use in the solution equation

$$x_t = c_1 \lambda_1^t + \cdots + c_k \lambda_k^t.$$

A similar procedure is used for solving a differential equation of the form

$$\frac{d^k x}{dt^k} + a_{k-1}\frac{d^{k-1}x}{dt^{k-1}} + \cdots + a_1\frac{dx}{dt} + a_0 x = 0.$$

Calculation

Eigenvalues

The eigenvalues of a matrix A can be determined by finding the roots of the characteristic polynomial. Explicit algebraic formulas for the roots of a polynomial exist only if the degree n is 4 or less. According to the Abel–Ruffini theorem there is no general, explicit and exact algebraic formula for the roots of a polynomial with degree 5 or more.

It turns out that any polynomial with degree n is the characteristic polynomial of some companion matrix of order n. Therefore, for matrices of order 5 or more, the eigenvalues and eigenvectors cannot be obtained by an explicit algebraic formula, and must therefore be computed by approximate numerical methods.

In theory, the coefficients of the characteristic polynomial can be computed exactly, since they are sums of products of matrix elements; and there are algorithms that can find all the roots of a polynomial of arbitrary degree to any required accuracy. However, this approach is not viable in practice because the coefficients would be contaminated by unavoidable round-off errors, and the roots of a polynomial can be an extremely sensitive function of the coefficients (as exemplified by Wilkinson's polynomial).

Efficient, accurate methods to compute eigenvalues and eigenvectors of arbitrary matrices were not known until the advent of the QR algorithm in 1961.Combining the Householder transformation with the LU decomposition results in an algorithm with better convergence than the QR algorithm. For large Hermitian sparse matrices, the Lanczos algorithm is one example of an efficient iterative method to compute eigenvalues and eigenvectors, among several other possibilities.

Eigenvectors

Once the (exact) value of an eigenvalue is known, the corresponding eigenvectors can be found by finding non-zero solutions of the eigenvalue equation, that becomes a system of linear equations with known coefficients. For example, once it is known that 6 is an eigenvalue of the matrix

$$A = \begin{bmatrix} 4 & 1 \\ 6 & 3 \end{bmatrix}$$

we can find its eigenvectors by solving the equation $Av = 6v$, that is

$$\begin{bmatrix} 4 & 1 \\ 6 & 3 \end{bmatrix}\begin{bmatrix} x \\ y \end{bmatrix} = 6 \cdot \begin{bmatrix} x \\ y \end{bmatrix}$$

This matrix equation is equivalent to two linear equations

$$\begin{cases} 4x+y &= 6x \\ 6x+3y &= 6y \end{cases} \quad \text{that is} \quad \begin{cases} -2x+y &= 0 \\ +6x-3y &= 0 \end{cases}$$

Both equations reduce to the single linear equation $y = 2x.$. Therefore, any vector of the form $[a, 2a]'$, for any non-zero real number a, is an eigenvector of A with eigenvalue $\lambda = 6$.

The matrix A above has another eigenvalue $\lambda = 1$. A similar calculation shows that the corresponding eigenvectors are the non-zero solutions of $3x + y = 0$, that is, any vector of the form $[b, -3b]'$, for any non-zero real number b.

Some numeric methods that compute the eigenvalues of a matrix also determine a set of corresponding eigenvectors as a by-product of the computation.

Applications

Eigenvalues of Geometric Transformations

The following table presents some example transformations in the plane along with their 2×2 matrices, eigenvalues, and eigenvectors.

	scaling	unequal scaling	rotation	horizontal shear	hyperbolic rotation
illustration					
matrix	$\begin{bmatrix} k & 0 \\ 0 & k \end{bmatrix}$	$\begin{bmatrix} k_1 & 0 \\ 0 & k_2 \end{bmatrix}$	$\begin{bmatrix} c & -s \\ s & c \end{bmatrix}$ $c = \cos\theta$ $s = \sin\theta$	$\begin{bmatrix} 1 & k \\ 0 & 1 \end{bmatrix}$	$\begin{bmatrix} c & s \\ s & c \end{bmatrix}$ $c = \cosh\varphi$ $s = \sinh\varphi$
characteristic polynomial	$(\lambda - k)^2$	$(\lambda - k_1)(\lambda - k_2)$	$\lambda^2 - 2c\lambda + 1$	$(\lambda - 1)^2$	$\lambda^2 - 2c\lambda + 1$
eigenvalues λ_i	$\lambda_1 = \lambda_2 = k$	$\lambda_1 = k_1$ $\lambda_2 = k_2$	$\lambda_1 = e^{i\theta} = c + si$ $\lambda_1 = \lambda_2 = 1$	$\lambda_1 = \lambda_2 = 1$	$\lambda_1 = e^{\varphi}$ $\lambda_2 = e^{-\varphi},,$
algebraic multipl. $\mu_i = \mu(\lambda_i)$	$\mu_1 = 2$	$\mu_1 = 1$ $\mu_2 = 1$	$\mu_1 = 1$ $\mu_2 = 1$	$\mu_1 = 2$	$\mu_1 = 1$ $\mu_2 = 1$
geometric multipl. $\gamma_i = \gamma(\lambda_i)$	$\gamma_1 = 2$	$\gamma_1 = 1$ $\gamma_2 = 1$	$\gamma_1 = 1$ $\gamma_2 = 1$	$\gamma_1 = 1$	$\gamma_1 = 1$ $\gamma_2 = 1$

| eigenvectors | All non-zero vectors | $u_1 = \begin{bmatrix} 1 \\ 0 \end{bmatrix}$ $u_2 = \begin{bmatrix} 0 \\ 1 \end{bmatrix}$ | $u_1 = \begin{bmatrix} 1 \\ -i \end{bmatrix}$ $u_2 = \begin{bmatrix} 1 \\ +i \end{bmatrix}$ | $u_1 = \begin{bmatrix} 1 \\ 0 \end{bmatrix}$ | $u_1 = \begin{bmatrix} 1 \\ 1 \end{bmatrix}$ $u_2 = \begin{bmatrix} 1 \\ -1 \end{bmatrix}.$ |

Note that the characteristic equation for a rotation is a quadratic equation with discriminant $D = -4(\sin\theta)^2$, which is a negative number whenever θ is not an integer multiple of 180°. Therefore, except for these special cases, the two eigenvalues are complex numbers, $\cos\theta \pm i\sin\theta$; and all eigenvectors have non-real entries. Indeed, except for those special cases, a rotation changes the direction of every nonzero vector in the plane.

A linear transformation that takes a square to a rectangle of the same area (a squeeze mapping) has reciprocal eigenvalues.

Schrödinger Equation

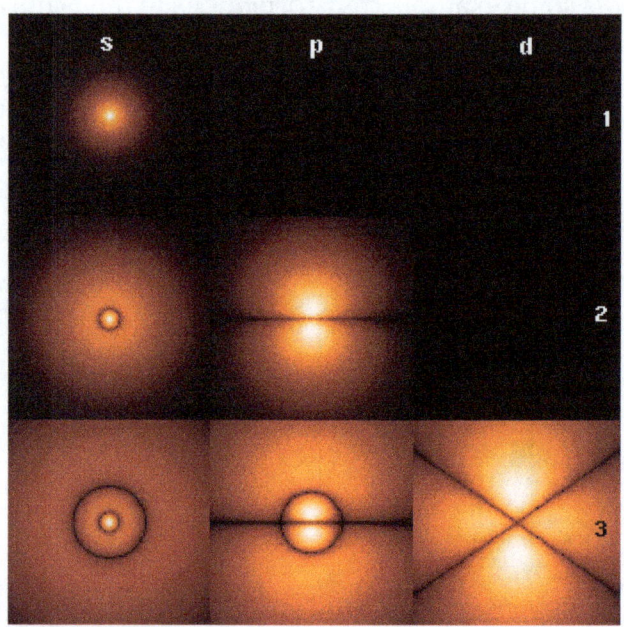

The wavefunctions associated with the bound states of an electron in a hydrogen atom can be seen as the eigenvectors of the hydrogen atom Hamiltonian as well as of the angular momentum operator. They are associated with eigenvalues interpreted as their energies (increasing downward: $n = 1, 2, 3, \dots$) and angular momentum (increasing across: s, p, d, ...). The illustration shows the square of the absolute value of the wavefunctions. Brighter areas correspond to higher probability density for a position measurement. The center of each figure is the atomic nucleus, a proton.

An example of an eigenvalue equation where the transformation T is represented in terms of a differential operator is the time-independent Schrödinger equation in quantum mechanics:

$$H\psi_E = E\psi_E$$

where H, the Hamiltonian, is a second-order differential operator and ψ_E, the wavefunction, is one of its eigenfunctions corresponding to the eigenvalue E, interpreted as its energy.

However, in the case where one is interested only in the bound state solutions of the Schrödinger equation, one looks for ψ_E within the space of square integrable functions. Since this space is a Hilbert space with a well-defined scalar product, one can introduce a basis set in which ψ_E and H can be represented as a one-dimensional array (i.e., a vector) and a matrix respectively. This allows one to represent the Schrödinger equation in a matrix form.

The bra–ket notation is often used in this context. A vector, which represents a state of the system, in the Hilbert space of square integrable functions is represented by $|\Psi_E\rangle$. In this notation, the Schrödinger equation is:

$$H\,|\Psi_E\rangle = E\,|\Psi_E\rangle$$

where $|\Psi_E\rangle$ is an eigenstate of H and E represents the eigenvalue. H is an observable self adjoint operator, the infinite-dimensional analog of Hermitian matrices. As in the matrix case, in the equation above $H\,|\Psi_E\rangle$ is understood to be the vector obtained by application of the transformation H to $|\Psi_E\rangle$.

Molecular Orbitals

In quantum mechanics, and in particular in atomic and molecular physics, within the Hartree–Fock theory, the atomic and molecular orbitals can be defined by the eigenvectors of the Fock operator. The corresponding eigenvalues are interpreted as ionization potentials via Koopmans' theorem. In this case, the term eigenvector is used in a somewhat more general meaning, since the Fock operator is explicitly dependent on the orbitals and their eigenvalues. Thus, if one wants to underline this aspect, one speaks of nonlinear eigenvalue problems. Such equations are usually solved by an iteration procedure, called in this case self-consistent field method. In quantum chemistry, one often represents the Hartree–Fock equation in a non-orthogonal basis set. This particular representation is a generalized eigenvalue problem called Roothaan equations.

Geology and Glaciology

In geology, especially in the study of glacial till, eigenvectors and eigenvalues are used as a method by which a mass of information of a clast fabric's constituents' orientation and dip can be summarized in a 3-D space by six numbers. In the field, a geologist may collect such data for hundreds or thousands of clasts in a soil sample, which can only be compared graphically such as in a Tri-Plot (Sneed and Folk) diagram, or as a Stereonet on a Wulff Net.

The output for the orientation tensor is in the three orthogonal (perpendicular) axes of space. The three eigenvectors are ordered v_1, v_2, v_3 by their eigenvalues $E_1 \geq E_2 \geq E_3$; v_1 then is the primary orientation/dip of clast, v_2 is the secondary and v_3 is the tertiary, in terms of strength. The clast orientation is defined as the direction of the eigenvector, on a compass rose of 360°. Dip is measured as the eigenvalue, the modulus of the tensor: this is valued from 0° (no dip) to 90° (vertical). The relative values of E_1, E_2, and E_3 are dictated by the nature of the sediment's fabric. If $E_1 = E_2 = E_3$, the fabric is said to be isotropic. If $E_1 = E_2 > E_3$, the fabric is said to be planar. If $E_1 > E_2 > E_3$, the fabric is said to be linear.

Principal Component Analysis

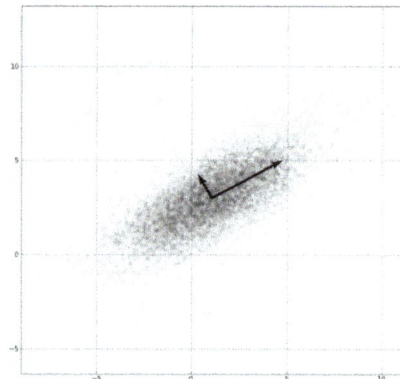

PCA of the multivariate Gaussian distribution centered at $(1,3)$ with a standard deviation of 3 in roughly the $(0.878, 0.478)$ direction and of 1 in the orthogonal direction. The vectors shown are unit eigenvectors of the (symmetric, positive-semidefinite) covariance matrix scaled by the square root of the corresponding eigenvalue. (Just as in the one-dimensional case, the square root is taken because the standard deviation is more readily visualized than the variance.

The eigendecomposition of a symmetric positive semidefinite (PSD) matrix yields an orthogonal basis of eigenvectors, each of which has a nonnegative eigenvalue. The orthogonal decomposition of a PSD matrix is used in multivariate analysis, where the sample covariance matrices are PSD. This orthogonal decomposition is called principal components analysis (PCA) in statistics. PCA studies linear relations among variables. PCA is performed on the covariance matrix or the correlation matrix (in which each variable is scaled to have its sample variance equal to one). For the covariance or correlation matrix, the eigenvectors correspond to principal components and the eigenvalues to the variance explained by the principal components. Principal component analysis of the correlation matrix provides an orthonormal eigen-basis for the space of the observed data: In this basis, the largest eigenvalues correspond to the principal components that are associated with most of the covariability among a number of observed data.

Principal component analysis is used to study large data sets, such as those encountered in bioinformatics, data mining, chemical research, psychology, and in marketing. PCA is popular especially in psychology, in the field of psychometrics. In Q methodology, the eigenvalues of the correlation matrix determine the Q-methodologist's judgment of *practical* significance (which differs from the statistical significance of hypothesis testing; cf. criteria for determining the number of factors). More generally, principal component analysis can be used as a method of factor analysis in structural equation modeling.

Vibration Analysis

Eigenvalue problems occur naturally in the vibration analysis of mechanical structures with many degrees of freedom. The eigenvalues are the natural frequencies (or eigenfrequencies) of vibration, and the eigenvectors are the shapes of these vibrational modes. In particular, undamped vibration is governed by

$$m\ddot{x} + kx = 0$$

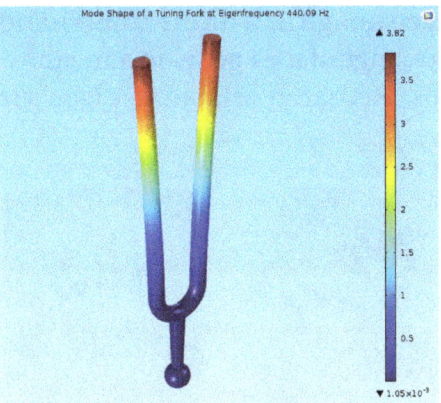

Mode Shape of a Tuning Fork at Eigenfrequency 440.09 Hz

or

$$m\ddot{x} = -kx$$

that is, acceleration is proportional to position (i.e., we expect x to be sinusoidal in time).

In n dimensions, m becomes a mass matrix and k a stiffness matrix. Admissible solutions are then a linear combination of solutions to the generalized eigenvalue problem

$$-kx = \omega^2 mx$$

where ω^2 is the eigenvalue and ω is the (imaginary) angular frequency. Note that the principal vibration modes are different from the principal compliance modes, which are the eigenvectors of k alone. Furthermore, damped vibration, governed by

$$m\ddot{x} + c\dot{x} + kx = 0$$

leads to a so-called quadratic eigenvalue problem,

$$(\omega^2 m + \omega c + k)x = 0.$$

This can be reduced to a generalized eigenvalue problem by clever use of algebra at the cost of solving a larger system.

The orthogonality properties of the eigenvectors allows decoupling of the differential equations so that the system can be represented as linear summation of the eigenvectors. The eigenvalue problem of complex structures is often solved using finite element analysis, but neatly generalize the solution to scalar-valued vibration problems.

Eigenfaces

In image processing, processed images of faces can be seen as vectors whose components are the brightnesses of each pixel. The dimension of this vector space is the number of pixels. The eigenvectors of the covariance matrix associated with a large set of normalized pictures of faces are called eigenfaces; this is an example of principal components analysis. They are very useful for expressing any face image as a linear combination of some of them. In the facial recognition branch of biometrics, eigenfaces provide a means of applying data compression to faces for identification purposes. Research related to eigen vision systems determining hand gestures has also been made.

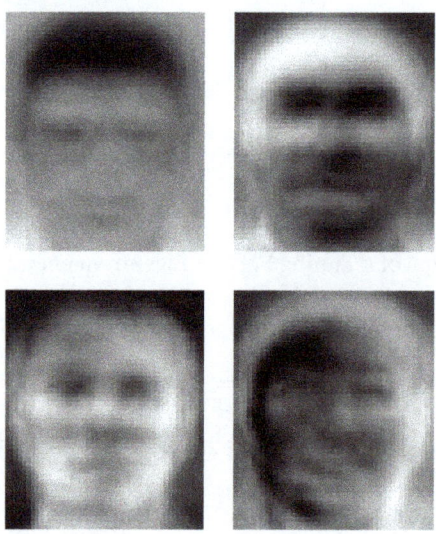

Eigenfaces as examples of eigenvectors

Similar to this concept, eigenvoices represent the general direction of variability in human pronunciations of a particular utterance, such as a word in a language. Based on a linear combination of such eigenvoices, a new voice pronunciation of the word can be constructed. These concepts have been found useful in automatic speech recognition systems for speaker adaptation.

Tensor of Moment of Inertia

In mechanics, the eigenvectors of the moment of inertia tensor define the principal axes of a rigid body. The tensor of moment of inertia is a key quantity required to determine the rotation of a rigid body around its center of mass.

Stress Tensor

In solid mechanics, the stress tensor is symmetric and so can be decomposed into a diagonal tensor with the eigenvalues on the diagonal and eigenvectors as a basis. Because it is diagonal, in this orientation, the stress tensor has no shear components; the components it does have are the principal components.

Graphs

In spectral graph theory, an eigenvalue of a graph is defined as an eigenvalue of the graph's adjacency matrix A, or (increasingly) of the graph's Laplacian matrix due to its Discrete Laplace operator, which is either $T - A$ (sometimes called the *combinatorial Laplacian*) or $I - T^{-1/2} A T^{-1/2}$ (sometimes called the *normalized Laplacian*), where T is a diagonal matrix with T_{ii} equal to the degree of vertex v_i, and in $T^{-1/2}$, the *ith* diagonal entry is

$1/\sqrt{\deg(v_i)}$. The *kth* principal eigenvector of a graph is defined as either the eigenvector corresponding to the *kth* largest or *kth* smallest eigenvalue of the Laplacian. The first principal eigenvector of the graph is also referred to merely as the principal eigenvector.

The principal eigenvector is used to measure the centrality of its vertices. An example is Google's PageRank algorithm. The principal eigenvector of a modified adjacency matrix of the World Wide Web graph gives the page ranks as its components. This vector corresponds to the stationary distribution of the Markov chain represented by the row-normalized adjacency matrix; however, the adjacency matrix must first be modified to ensure a stationary distribution exists. The second smallest eigenvector can be used to partition the graph into clusters, via spectral clustering. Other methods are also available for clustering.

Basic Reproduction Number

The basic reproduction number (R_0) is a fundamental number in the study of how infectious diseases spread. If one infectious person is put into a population of completely susceptible people, then R_0 is the average number of people that one typical infectious person will infect. The generation time of an infection is the time, t_G, from one person becoming infected to the next person becoming infected. In a heterogeneous population, the next generation matrix defines how many people in the population will become infected after time t_G has passed. R_0 is then the largest eigenvalue of the next generation matrix.

References

- Banerjee, Sudipto; Roy, Anindya (2014), Linear Algebra and Matrix Analysis for Statistics, Texts in Statistical Science (1st ed.), Chapman and Hall/CRC, ISBN 978-1420095388

- Dorier, Jean-Luc; Robert, Aline; Robinet, Jacqueline; Rogalsiu, Marc (2000). Dorier, Jean-Luc, ed. The Obstacle of Formalism in Linear Algebra. Springer. pp. 85–124. ISBN 978-0-7923-6539-6. Retrieved 9 July 2014.

- P. K. Jain, Khalil Ahmad (1995). "5.1 Definitions and basic properties of inner product spaces and Hilbert spaces". Functional analysis (2nd ed.). New Age International. p. 203. ISBN 81-224-0801-X.

- Meyer, Carl D. (February 15, 2001), Matrix Analysis and Applied Linear Algebra, Society for Industrial and Applied Mathematics (SIAM), ISBN 978-0-89871-454-8

- Joe D. Hoffman; Steven Frankel (2001). Numerical Methods for Engineers and Scientists, Second Edition,. CRC Press. p. 30. ISBN 978-0-8247-0443-8.

- Thomas S. Shores (2007). Applied Linear Algebra and Matrix Analysis. Springer Science & Business Media. p. 132. ISBN 978-0-387-48947-6.

- Farebrother, R.W. (1988), Linear Least Squares Computations, STATISTICS: Textbooks and Monographs, Marcel Dekker, ISBN 978-0-8247-7661-9 .

- Lipson, Marc; Lipschutz, Seymour (2001), Schaum's outline of theory and problems of linear algebra, New York: McGraw-Hill, pp. 69–80, ISBN 978-0-07-136200-9 .

- Press, WH; Teukolsky, SA; Vetterling, WT; Flannery, BP (2007), "Section 2.2", Numerical Recipes: The Art of Scientific Computing (3rd ed.), New York: Cambridge University Press, ISBN 978-0-521-88068-8

- Lang, Serge (2002), Algebra, Graduate Texts in Mathematics, 211 (Revised third ed.), New York: Springer-Verlag, ISBN 978-0-387-95385-4.

- Roman, Steven (2005), Advanced Linear Algebra, Graduate Texts in Mathematics, 135 (2nd ed.), Berlin, New York: Springer-Verlag, ISBN 978-0-387-24766-3

- Bourbaki, Nicolas (1987), Topological vector spaces, Elements of mathematics, Berlin, New York: Springer-Verlag, ISBN 978-3-540-13627-9

- Braun, Martin (1993), Differential equations and their applications: an introduction to applied mathematics, Berlin, New York: Springer-Verlag, ISBN 978-0-387-97894-9

- Evans, Lawrence C. (1998), Partial differential equations, Providence, R.I.: American Mathematical Society, ISBN 978-0-8218-0772-9

- Gasquet, Claude; Witomski, Patrick (1999), Fourier Analysis and Applications: Filtering, Numerical Computation, Wavelets, Texts in Applied Mathematics, New York: Springer-Verlag, ISBN 0-387-98485-2

- Ifeachor, Emmanuel C.; Jervis, Barrie W. (2001), Digital Signal Processing: A Practical Approach (2nd ed.), Harlow, Essex, England: Prentice-Hall (published 2002), ISBN 0-201-59619-9

- Krantz, Steven G. (1999), A Panorama of Harmonic Analysis, Carus Mathematical Monographs, Washington, DC: Mathematical Association of America, ISBN 0-88385-031-1

4

Vector and Scalar Field Theory

Vectors are geometric objects that have magnitude and direction. Vector fields are used to decipher the direction of fluid throughout space. The chapter strategically encompasses and incorporates the major components like gradient theorem, Stokes' theorem, divergence theorem and Green's theorem, providing a complete understanding.

Vector Field

In vector calculus, a vector field is an assignment of a vector to each point in a subset of space. A vector field in the plane (for instance), can be visualised as: a collection of arrows with a given magnitude and direction, each attached to a point in the plane. Vector fields are often used to model, for example, the speed and direction of a moving fluid throughout space, or the strength and direction of some force, such as the magnetic or gravitational force, as it changes from point to point.

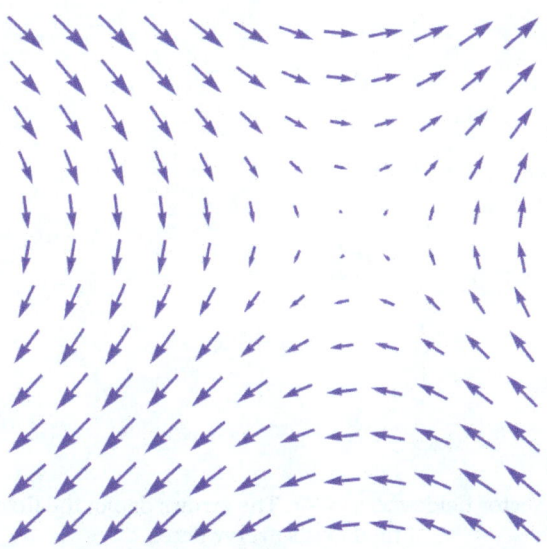

A portion of the vector field ($\sin y$, $\sin x$)

The elements of differential and integral calculus extend naturally to vector fields. When a vector field represents force, the line integral of a vector field represents the work done by a force moving along a path, and under this interpretation conservation of energy is exhibited as a special case of the fundamental theorem of calculus. Vector fields can usefully be thought of as representing the velocity of a moving flow in space, and this physical intuition leads to notions such as the divergence (which represents the rate of change of volume of a flow) and curl (which represents the rotation of a flow).

In coordinates, a vector field on a domain in n-dimensional Euclidean space can be represented as a vector-valued function that associates an n-tuple of real numbers to each point of the domain. This representation of a vector field depends on the coordinate system, and there is a well-defined transformation law in passing from one coordinate system to the other. Vector fields are often discussed on open subsets of Euclidean space, but also make sense on other subsets such as surfaces, where they associate an arrow tangent to the surface at each point (a tangent vector).

More generally, vector fields are defined on differentiable manifolds, which are spaces that look like Euclidean space on small scales, but may have more complicated structure on larger scales. In this setting, a vector field gives a tangent vector at each point of the manifold (that is, a section of the tangent bundle to the manifold). Vector fields are one kind of tensor field.

Definition

Vector Fields on Subsets of Euclidean Space

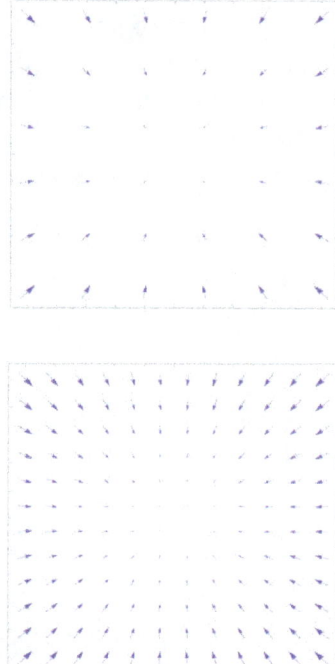

Two representations of the same vector field: v(x, y) = −r. The arrows depict the field at discrete points, however, the field exists everywhere.

Given a subset S in $\mathbb{R}^n$, a vector field is represented by a vector-valued function $V: S \to \mathbb{R}^n$ in standard Cartesian coordinates ($x_1, ..., x_n$). If each component of V is continuous, then V is a continuous vector field, and more generally V is a C^k vector field if each component of V is k times continuously differentiable.

A vector field can be visualized as assigning a vector to individual points within an n-dimensional space.

Given two C^k-vector fields V, W defined on S and a real valued C^k-function f defined on S, the two operations scalar multiplication and vector addition

$$(fV)(p) := f(p)V(p)$$

$$(V + W)(p) := V(p) + W(p)$$

define the module of C^k-vector fields over the ring of C^k-functions.

Coordinate Transformation Law

In physics, a vector is additionally distinguished by how its coordinates change when one measures the same vector with respect to a different background coordinate system. The transformation properties of vectors distinguish a vector as a geometrically distinct entity from a simple list of scalars, or from a covector.

Thus, suppose that $(x_1,...,x_n)$ is a choice of Cartesian coordinates, in terms of which the components of the vector V are

$$V_x = (V_{1,x},...,V_{n,x})$$

and suppose that $(y_1,...,y_n)$ are n functions of the x_i defining a different coordinate system. Then the components of the vector V in the new coordinates are required to satisfy the transformation law

$$V_{i,y} = \sum_{j=1}^{n} \frac{\partial y_i}{\partial x_j} V_{j,x}. \tag{1}$$

Such a transformation law is called contravariant. A similar transformation law characterizes vector fields in physics: specifically, a vector field is a specification of n functions in each coordinate system subject to the transformation law (1) relating the different coordinate systems.

Vector fields are thus contrasted with scalar fields, which associate a number or *scalar* to every point in space, and are also contrasted with simple lists of scalar fields, which do not transform under coordinate changes.

Vector Fields on Manifolds

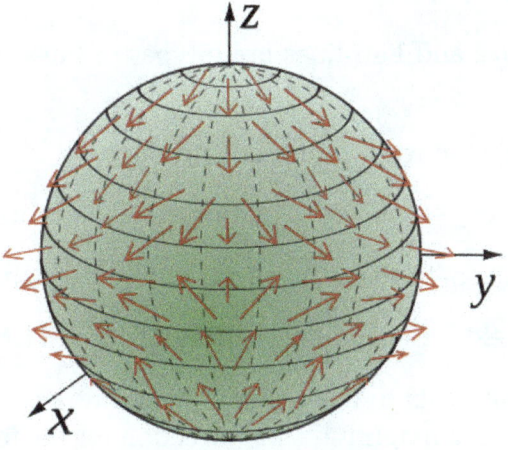

A vector field on a sphere

Given a differentiable manifold M, a vector field on M is an assignment of a tangent vector to each point in M. More precisely, a vector field F is a mapping from M into the tangent bundle TM so that $p°F$ is the identity mapping where p denotes the projection from TM to M. In other words, a vector field is a section of the tangent bundle.

If the manifold M is smooth or analytic—that is, the change of coordinates is smooth (analytic)—then one can make sense of the notion of smooth (analytic) vector fields. The collection of all smooth vector fields on a smooth manifold M is often denoted by $\Gamma(TM)$ or $C^\infty(M,TM)$ (especially when thinking of vector fields as sections); the collection of all smooth vector fields is also denoted by $\mathfrak{X}(M)$ (a fraktur "X").

Examples

The flow field around an airplane is a vector field in $\mathbf{R}^3$, here visualized by bubbles that follow the streamlines showing a wingtip vortex.

- A vector field for the movement of air on Earth will associate for every point on the surface of the Earth a vector with the wind speed and direction for that point. This can be drawn using arrows to represent the wind; the length (magnitude) of the arrow will be an indication of the wind speed. A "high" on the usual barometric pressure map would then act as a source (arrows pointing away), and a "low" would be a sink (arrows pointing towards), since air tends to move from high pressure areas to low pressure areas.

- Velocity field of a moving fluid. In this case, a velocity vector is associated to each point in the fluid.

- Streamlines, Streaklines and Pathlines are 3 types of lines that can be made from vector fields. They are :

  streaklines — as revealed in wind tunnels using smoke.

  streamlines (or fieldlines)— as a line depicting the instantaneous field at a given time.

  pathlines — showing the path that a given particle (of zero mass) would follow.

- Magnetic fields. The fieldlines can be revealed using small iron filings.

- Maxwell's equations allow us to use a given set of initial conditions to deduce, for every point in Euclidean space, a magnitude and direction for the force experienced by a charged test particle at that point; the resulting vector field is the electromagnetic field.

- A gravitational field generated by any massive object is also a vector field. For example, the gravitational field vectors for a spherically symmetric body would all point towards the sphere's center with the magnitude of the vectors reducing as radial distance from the body increases.

Gradient Field in Euclidean Spaces

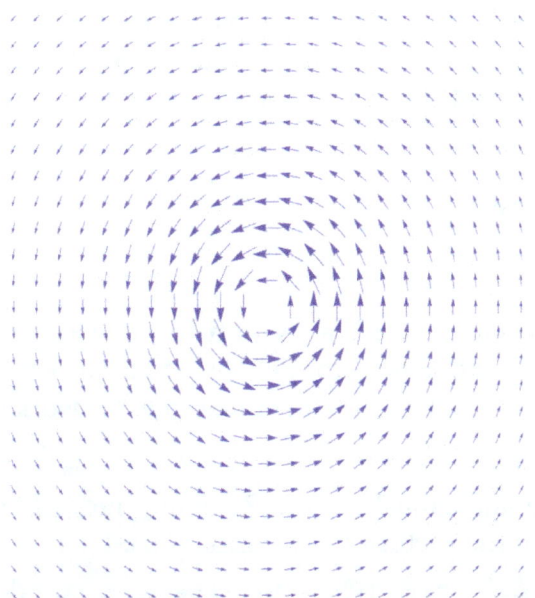

A vector field that has circulation about a point cannot be written as the gradient of a function.

Vector fields can be constructed out of scalar fields using the gradient operator (denoted by the del: ∇).

A vector field V defined on an open set S is called a gradient field or a conservative field if there exists a real-valued function (a scalar field) f on S such that

$$V = \nabla f = \left(\frac{\partial f}{\partial x_1}, \frac{\partial f}{\partial x_2}, \frac{\partial f}{\partial x_3}, \ldots, \frac{\partial f}{\partial x_n} \right).$$

The associated flow is called the gradient flow, and is used in the method of gradient descent.

The path integral along any closed curve γ ($\gamma(0) = \gamma(1)$) in a conservative field is zero:

$$\oint_\gamma V(\mathbf{x}) \cdot d\mathbf{x} = \oint_\gamma \nabla f(\mathbf{x}) \cdot d\mathbf{x} = f(\gamma(1)) - f(\gamma(0)).$$

where the angular brackets and comma: $\langle , \rangle$ denotes the inner product of two vectors (strictly speaking – the integrand $V(x)$ is a 1-form rather than a vector in the elementary sense).

Central Field in Euclidean Spaces

A C^∞-vector field over $\mathbb{R}^n \setminus \{0\}$ is called a central field if

$$V(T(p)) = T(V(p)) \qquad (T \in O(n, \mathbf{R}))$$

where O(n, R) is the orthogonal group. We say central fields are invariant under orthogonal transformations around o.

The point o is called the center of the field.

Since orthogonal transformations are actually rotations and reflections, the invariance conditions mean that vectors of a central field are always directed towards, or away from, o; this is an alternate (and simpler) definition. A central field is always a gradient field, since defining it on one semiaxis and integrating gives an antigradient.

Operations on Vector Fields

Line Integral

A common technique in physics is to integrate a vector field along a curve, i.e. to determine its line integral. Given a particle in a gravitational vector field, where each vector represents the force acting on the particle at a given point in space, the line integral is the work done on the particle when it travels along a certain path.

The line integral is constructed analogously to the Riemann integral and it exists if the curve is rectifiable (has finite length) and the vector field is continuous.

Given a vector field V and a curve γ parametrized by [a, b] (where a and b are real) the line integral is defined as

$$\int_{\gamma} V(\mathbf{x}) \cdot \mathrm{d}\mathbf{x} = \int_{a}^{b} V(\gamma(t)) \cdot \dot{\gamma}(t) \, \mathrm{d}t.$$

Divergence

The divergence of a vector field on Euclidean space is a function (or scalar field). In three-dimensions, the divergence is defined by

$$\operatorname{div} \mathbf{F} = \nabla \cdot \mathbf{F} = \frac{\partial F_1}{\partial x} + \frac{\partial F_2}{\partial y} + \frac{\partial F_3}{\partial z},$$

with the obvious generalization to arbitrary dimensions. The divergence at a point represents the degree to which a small volume around the point is a source or a sink for the vector flow, a result which is made precise by the divergence theorem.

The divergence can also be defined on a Riemannian manifold, that is, a manifold with a Riemannian metric that measures the length of vectors.

Curl in Three-dimensions

The curl is an operation which takes a vector field and produces another vector field. The curl is defined only in three-dimensions, but some properties of the curl can be captured in higher dimensions with the exterior derivative. In three-dimensions, it is defined by

$$\text{curl}\,\mathbf{F} = \nabla \times \mathbf{F} = \left(\frac{\partial F_3}{\partial y} - \frac{\partial F_2}{\partial z} \right)\mathbf{e}_1 - \left(\frac{\partial F_3}{\partial x} - \frac{\partial F_1}{\partial z} \right)\mathbf{e}_2 + \left(\frac{\partial F_2}{\partial x} - \frac{\partial F_1}{\partial y} \right)\mathbf{e}_3.$$

The curl measures the density of the angular momentum of the vector flow at a point, that is, the amount to which the flow circulates around a fixed axis. This intuitive description is made precise by Stokes' theorem.

Index of a Vector Field

The index of a vector field is an integer that helps to describe the behaviour of a vector field around an isolated zero (i.e., an isolated singularity of the field). In the plane, the index takes the value -1 at a saddle singularity but +1 at a source or sink singularity.

Let the dimension of the manifold on which the vector field is defined be n. Take a small sphere S around the zero so that no other zeros lie in the interior of S. A map from this sphere to a unit sphere of dimensions $n-1$ can be constructed by dividing each vector on this sphere by its length to form a unit length vector, which is a point on the unit sphere S^{n-1}. This defines a continuous map from S to S^{n-1}. The index of the vector field at the point is the degree of this map. It can be shown that this integer does not depend on the choice of S, and therefore depends only on the vector field itself.

The index of the vector field as a whole is defined when it has just a finite number of zeroes. In this case, all zeroes are isolated, and the index of the vector field is defined to be the sum of the indices at all zeroes.

The index is not defined at any non-singular point (i.e., a point where the vector is non-zero). it is equal to +1 around a source, and more generally equal to $(-1)^k$ around a saddle that has k contracting dimensions and n-k expanding dimensions. For an ordinary (2-dimensional) sphere in three-dimensional space, it can be shown that the index of any vector field on the sphere must be 2. This shows that every such vector field must have a zero. This implies the hairy ball theorem, which states that if a vector in R^3 is assigned to each point of the unit sphere S^2 in a continuous manner, then it is impossible to "comb the hairs flat", i.e., to choose the vectors in a continuous way such that they are all non-zero and tangent to S^2.

For a vector field on a compact manifold with a finite number of zeroes, the Poincaré-Hopf theorem states that the index of the vector field is equal to the Euler characteristic of the manifold.

History

Vector fields arose originally in classical field theory in 19th century physics, specifically in magnetism. They were formalized by Michael Faraday, in his concept of *lines of force,* who emphasized that the field *itself* should be an object of study, which it has become throughout physics in the form of field theory.

In addition to the magnetic field, other phenomena that were modeled as vector fields by Faraday include the electrical field and light field.

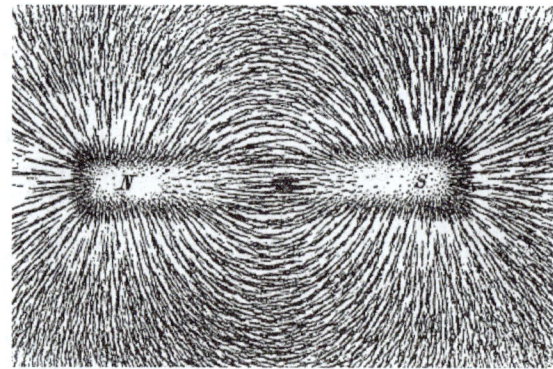

Magnetic field lines of an iron bar (magnetic dipole)

Flow Curves

Consider the flow of a fluid through a region of space. At any given time, any point of the fluid has a particular velocity associated with it; thus there is a vector field associated to any flow. The converse is also true: it is possible to associate a flow to a vector field having that vector field as its velocity.

Given a vector field V defined on S, one defines curves $\gamma(t)$ on S such that for each t in an interval I

$$\gamma'(t) = V(\gamma(t)).$$

By the Picard–Lindelöf theorem, if V is Lipschitz continuous there is a *unique C^1-curve* γ_x for each point x in S so that, for some $\varepsilon > 0$,

$$\gamma_x(0) = x$$

$$\gamma_x'(t) = V(\gamma_x(t)) \qquad (t \in (-\varepsilon, +\varepsilon) \subset \mathbf{R}).$$

The curves γ_x are called integral curves or trajectories (or less commonly, flow lines) of the vector field V and partition S into equivalence classes. It is not always possible to extend the interval $(-\varepsilon, +\varepsilon)$ to the whole real number line. The flow may for example reach the edge of S in a finite time. In two or three dimensions one can visualize the vector field as giving rise to a flow on S. If we drop a particle into this flow at a point p it will move along the curve γ_p in the flow depending on the initial point p. If p is a stationary point of V (i.e., the vector field is equal to the zero vector at the point p), then the particle will remain at p.

Typical applications are streamline in fluid, geodesic flow, and one-parameter subgroups and the exponential map in Lie groups.

Complete Vector Fields

By definition, a vector field is called complete if every one of its flow curves exist for all time. In particular, compactly supported vector fields on a manifold are complete. If X is a complete vector field on M, then the one-parameter group of diffeomorphisms generated by the flow along X exists for all time. On a compact manifold without boundary, every smooth vector field is complete. An example of an incomplete vector field V on the real line R is given by $V(x) = x^2$. For, the differential equation $dx/dt = x^2$, with initial condition $x(0) = x_0$, has as its unique solution $x(t) = 1/(x_0 - t)$ if $x_0 \neq 0$ (and $x(t) = 0$ for all t if $x_0 = 0$). Hence for $x_0 \neq 0$, $x(t)$ is undefined at $t = x_0$ so cannot be defined for all values of t.

Difference between Scalar and Vector Field

The difference between a scalar and vector field is not that "a scalar is just one number while a vector is several numbers". The difference is in: how their coordinates respond to coordinate transformations. A scalar *is* a coordinate whereas a vector *can be described* by coordinates, but it *is not* the collection of its coordinates.

Example 1

This example is about 2-dimensional Euclidean space (R²) where we examine Euclidean (x, y) and polar (r, θ) coordinates (which are undefined at the origin). Thus $x = r \cos \theta$ and $y = r \sin \theta$ and also $r^2 = x^2 + y^2$, $\cos \theta = x/(x^2 + y^2)^{1/2}$ and $\sin \theta = y/(x^2 + y^2)^{1/2}$. Suppose we have a scalar field which is given by the constant function 1, and a vector field which attaches a vector in the r-direction with length 1 to each point. More precisely, they are given by the functions

$$s_{\text{polar}} : (r, \theta) \mapsto 1, \quad v_{\text{polar}} : (r, \theta) \mapsto (1, \theta).$$

Let us convert these fields to Euclidean coordinates. The vector of length 1 in the r-direction has the x coordinate $\cos \theta$ and the y coordinate $\sin \theta$. Thus in Euclidean coordinates the same fields are described by the functions

$$s_{\text{Euclidean}} : (x, y) \mapsto 1,$$

$$v_{\text{Euclidean}} : (x, y) \mapsto (\cos \theta, \sin \theta) = \left(\frac{x}{\sqrt{x^2 + y^2}}, \frac{y}{\sqrt{x^2 + y^2}} \right).$$

We see that while the scalar field remains the same, the vector field now looks different. The same holds even in the 1-dimensional case, as illustrated by the next example.

Example 2

Consider the 1-dimensional Euclidean space R with its standard Euclidean coordinate x. Suppose we have a scalar field and a vector field which are both given in the x coordinate by the constant function 1,

$$s_{\text{Euclidean}} : x \mapsto 1, \quad v_{\text{Euclidean}} : x \mapsto 1.$$

Thus, we have a scalar field which has the value 1 everywhere and a vector field which attaches a vector in the x-direction with magnitude 1 unit of x to each point.

Now consider the coordinate $\xi := 2x$. If x changes one unit then ξ changes 2 units. But since we wish the integral of v along a path to be independent of coordinate, this means v*dx=v'*dξ. So from x increase by 1 unit, ξ increases by 1/2 unit, so v' must be 2. Thus this vector field has a magnitude of 2 in units of ξ. Therefore, in the ξ coordinate the scalar field and the vector field are described by the functions

$$s_{unusual} : \xi \mapsto 1, \quad v_{unusual} : \xi \mapsto 2$$

which are different.

f-relatedness

Given a smooth function between manifolds, $f: M \rightarrow N$, the derivative is an induced map on tangent bundles, $f_*: TM \rightarrow TN$. Given vector fields $V: M \rightarrow TM$ and $W: N \rightarrow TN$, we say that W is f-related to V if the equation $W \circ f = f_* \circ V$ holds.

If V_i is f-related to W_i, $i = 1, 2$, then the Lie bracket $[V_1, V_2]$ is f-related to $[W_1, W_2]$.

Generalizations

Replacing vectors by p-vectors (pth exterior power of vectors) yields p-vector fields; taking the dual space and exterior powers yields differential k-forms, and combining these yields general tensor fields.

Algebraically, vector fields can be characterized as derivations of the algebra of smooth functions on the manifold, which leads to defining a vector field on a commutative algebra as a derivation on the algebra, which is developed in the theory of differential calculus over commutative algebras.

Scalar Field Theory

In theoretical physics, scalar field theory can refer to a classical or quantum theory of scalar fields. A scalar field is invariant under any Lorentz transformation.

The only fundamental scalar quantum field that has been observed in nature is the Higgs field. However, scalar quantum fields feature in the effective field theory descriptions of many physical phenomena. An example is the pion, which is actually a pseudoscalar.

Since they do not involve polarization complications, scalar fields are often the easiest to appreciate second quantization through. For this reason, scalar field theories are often used for purposes of introduction of novel concepts and techniques.

The signature of the metric employed below is (+, −, −, −).

Linear (Free) Theory

The most basic scalar field theory is the linear theory. Through the Fourier decomposition of the fields, it represents the normal modes of an infinity of coupled oscillators. The action for the free relativistic scalar field theory is then

$$S = \int d^{D-1}x \, dt \, \mathcal{L}$$

$$= \int d^{D-1}x \, dt \left[\frac{1}{2} \eta^{\mu\nu} \partial_\mu \phi \partial_\nu \phi - \frac{1}{2} m^2 \phi^2 \right]$$

$$= \int d^{D-1}x \, dt \left[\frac{1}{2}(\partial_t \phi)^2 - \frac{1}{2} \delta^{ij} \partial_i \phi \partial_j \phi - \frac{1}{2} m^2 \phi^2 \right],$$

where $\mathcal{L}$ is known as a Lagrangian density; $d^{4-1}x \equiv dx \cdot dy \cdot dz \equiv dx^1 \cdot dx^2 \cdot dx^3$ for the three spatial coordinates; δ^{ij} is the Kronecker delta function; and $\partial_\rho = \partial/\partial x^\rho$ for the ρ-th coordinate x^ρ.

This is an example of a quadratic action, since each of the terms is quadratic in the field, φ. The term proportional to m^2 is sometimes known as a mass term, due to its subsequent interpretation, in the quantized version of this theory, in terms of particle mass.

The equation of motion for this theory is obtained by extremizing the action above. It takes the following form, linear in φ,

$$\eta^{\mu\nu} \partial_\mu \partial_\nu \phi + m^2 \phi = \partial_t^2 \phi - \nabla^2 \phi + m^2 \phi = 0 ,$$

where ∇^2 is the Laplace operator. This is the Klein–Gordon equation, with the interpretation as a classical field equation, rather than as a quantum-mechanical wave equation.

Nonlinear (Interacting) Theory

The most common generalization of the linear theory above is to add a scalar potential $V(\Phi)$ to the Lagrangian, where typically, in addition to a mass term, V is a polynomial in Φ. Such a theory is sometimes said to be interacting, because the Euler-Lagrange equation is now nonlinear, implying a self-interaction. The action for the most general such theory is

$$S = \int d^{D-1}x \, dt \, \mathcal{L}$$

$$= \int d^{D-1}x \, dt \left[\frac{1}{2} \eta^{\mu\nu} \partial_\mu \phi \partial_\nu \phi - V(\phi) \right]$$

$$= \int d^{D-1}x \, dt \left[\frac{1}{2}(\partial_t \phi)^2 - \frac{1}{2} \delta^{ij} \partial_i \phi \partial_j \phi - \frac{1}{2} m^2 \phi^2 - \sum_{n=3}^\infty \frac{1}{n!} g_n \phi^n \right]$$

The $n!$ factors in the expansion are introduced because they are useful in the Feynman diagram expansion of the quantum theory, as described below.

The corresponding Euler-Lagrange equation of motion is now

$$\eta^{\mu\nu}\partial_{\mu}\partial_{\nu}\phi + V'(\phi) = \partial_t^2\phi - \nabla^2\phi + V'(\phi) = 0.$$

Dimensional Analysis and Scaling

Physical quantities in these scalar field theories may have dimensions of length, time or mass, or some combination of the three.

However, in a relativistic theory, any quantity t, with dimensions of time, can be readily converted into a *length*, $l = ct$, by using the velocity of light, c. Similarly, any length l is equivalent to an inverse mass, $\hbar/mc$, using Planck's constant, $\hbar$. In natural units, one thinks of a time as a length, or either time or length as an inverse mass.

In short, one can think of the dimensions of any physical quantity as defined in terms of *just one* independent dimension, rather than in terms of all three. This is most often termed the mass dimension of the quantity. Knowing the dimensions of each quantity, allows one to *uniquely restore* conventional dimensions from a natural units expression in terms of this mass dimension, by simply reinserting the requisite powers of $\hbar$ and c required for dimensional consistency.

One conceivable objection is that this theory is classical, and therefore it is not obvious how Planck's constant should be a part of the theory at all. If desired, one could indeed recast the theory without mass dimensions at all: However, this would be at the expense of slightly obscuring the connection with the quantum scalar field. Given that one has dimensions of mass, Planck's constant is thought of here as an essentially *arbitrary fixed reference quantity of action* (not necessarily connected to quantization), hence with dimensions appropriate to convert between mass and inverse length.

Scaling Dimension

The classical scaling dimension, or mass dimension, Δ, of φ describes the transformation of the field under a rescaling of coordinates:

$$x \rightarrow \lambda x$$

$$\phi \rightarrow \lambda^{-\Delta}\phi.$$

The units of action are the same as the units of $\hbar$, and so the action itself has zero mass dimension. This fixes the scaling dimension of the field φ to be

$$\Delta = \frac{D-2}{2}.$$

Scale Invariance

There is a specific sense in which some scalar field theories are scale-invariant. While the actions above are all constructed to have zero mass dimension, not all actions are invariant under the scaling transformation

$$x \rightarrow \lambda x$$

$$\phi \to \lambda^{-\Delta}\phi \ .$$

The reason that not all actions are invariant is that one usually thinks of the parameters m and g_n as fixed quantities, which are not rescaled under the transformation above. The condition for a scalar field theory to be scale invariant is then quite obvious: all of the parameters appearing in the action should be dimensionless quantities. In other words, a scale invariant theory is one without any fixed length scale (or equivalently, mass scale) in the theory.

For a scalar field theory with D spacetime dimensions, the only dimensionless parameter g_n satisfies $n = {}^{2D}/_{(D-2)}$. For example, in $D = 4$, only g_4 is classically dimensionless, and so the only classically scale-invariant scalar field theory in $D = 4$ is the massless φ^4 theory.

Classical scale invariance, however, normally does not imply quantum scale invariance, because of the renormalization group involved.

Conformal Invariance

A transformation

$$x \to \tilde{x}(x)$$

is said to be conformal if the transformation satisfies

$$\frac{\partial \widetilde{x^\mu}}{\partial x^\rho}\frac{\partial \widetilde{x^\nu}}{\partial x^\sigma}\eta_{\mu\nu} = \lambda^2(x)\eta_{\rho\sigma}$$

for some function $\lambda(x)$.

The conformal group contains as subgroups the isometries of the metric $\eta_{\mu\nu}$ (the Poincaré group) and also the scaling transformations (or dilatations) considered above. In fact, the scale-invariant the-ories in the previous section are also conformally-invariant.

φ^4 Theory

Massive φ^4 theory illustrates a number of interesting phenomena in scalar field theory.

The Lagrangian density is

$$\mathcal{L} = \frac{1}{2}(\partial_t\phi)^2 - \frac{1}{2}\delta^{ij}\partial_i\phi\partial_j\phi - \frac{1}{2}m^2\phi^2 - \frac{g}{4!}\phi^4 .$$

Spontaneous Symmetry Breaking

This Lagrangian has a Z_2 symmetry under the transformation $\varphi \to -\varphi$. This is an example of an internal symmetry, in contrast to a space-time symmetry.

If m^2 is positive, the potential

$$V(\phi) = \frac{1}{2}m^2\phi^2 + \frac{g}{4!}\phi^4$$

has a single minimum, at the origin. The solution $\varphi=0$ is clearly invariant under the Z_2 symmetry.

Conversely, if m^2 is negative, then one can readily see that the potential

$$V(\phi) = \frac{1}{2}m^2\phi^2 + \frac{g}{4!}\phi^4$$

has two minima. This is known as a *double well potential*, and the lowest energy states (known as the vacua, in quantum field theoretical language) in such a theory are not invariant under the Z_2 symmetry of the action (in fact it maps each of the two vacua into the other). In this case, the Z_2 symmetry is said to be *spontaneously broken*.

Kink Solutions

The φ^4 theory with a negative m^2 also has a kink solution, which is a canonical example of a soliton. Such a solution is of the form

$$\phi(\vec{x},t) = \pm\frac{m}{2\sqrt{\dfrac{g}{4!}}}\tanh\left[\frac{m(x-x_0)}{\sqrt{2}}\right]$$

where x is one of the spatial variables (φ is taken to be independent of t, and the remaining spatial variables). The solution interpolates between the two different vacua of the double well potential. It is not possible to deform the kink into a constant solution without passing through a solution of infinite energy, and for this reason the kink is said to be stable. For $D>2$ (i.e., theories with more than one spatial dimension), this solution is called a domain wall.

Another well-known example of a scalar field theory with kink solutions is the sine-Gordon theory.

Complex Scalar Field Theory

In a complex scalar field theory, the scalar field takes values in the complex numbers, rather than the real numbers. The action considered normally takes the form

$$\mathcal{S} = \int d^{D-1}x dt \mathcal{L} = \int d^{D-1}x dt \left[\eta^{\mu\nu}\partial_\mu\phi^*\partial_\nu\phi - V(|\phi|^2)\right]$$

This has a U(1), equivalently O(2) symmetry, whose action on the space of fields rotates $\phi \rightarrow e^{i\alpha}\phi$, for some real phase angle α.

As for the real scalar field, spontaneous symmetry breaking is found if m^2 is negative. This gives rise to Goldstone's Mexican hat potential which is a rotation of the double-well potential of a real scalar field by 2π radians about the $V(\phi)$ axis. The symmetry breaking takes place in one higher dimension, i.e. the choice of vacuum breaks a continuous $U(1)$ symmetry instead of a discrete one.

The two components of the scalar field are reconfigured as a massive mode and a massless Goldstone boson.

O(N) Theory

One can express the complex scalar field theory in terms of two real fields, $\varphi^1 = \text{Re } \varphi$ and $\varphi^2 = \text{Im } \varphi$, which transform in the vector representation of the $U(1) = O(2)$ internal symmetry. Although such fields transform as a vector under the *internal symmetry*, they are still Lorentz scalars.

This can be generalised to a theory of N scalar fields transforming in the vector representation of the $O(N)$ symmetry. The Lagrangian for an $O(N)$-invariant scalar field theory is typically of the form

$$\mathcal{L} = \frac{1}{2}\eta^{\mu\nu}\partial_\mu\phi\cdot\partial_\nu\phi - V(\phi\cdot\phi)$$

using an appropriate $O(N)$-invariant inner product.

Quantum Scalar Field Theory

In quantum field theory, the fields, and all observables constructed from them, are replaced by quantum operators on a Hilbert space. This Hilbert space is built on a vacuum state, and dynamics are governed by a quantum Hamiltonian, a positive-definite operator which annihilates the vacuum. A construction of a quantum scalar field theory is detailed in the canonical quantization, which relies on canonical commutation relations among the fields. Essentially, the infinity of classical oscillators repackaged in the scalar field as its (decoupled) normal modes, above, are now quantized in the standard manner, so the respective quantum operator field describes an infinity of quantum harmonic oscillators acting on a respective Fock space.

In brief, the basic variables are the quantum field φ and its canonical momentum π. Both these operator-valued fields are Hermitian. At spatial points $\vec{x}$, $\vec{y}$, at equal times, their canonical commutation relations are given by

$$\left[\phi(\vec{x}),\phi(\vec{y})\right] = \left[\pi(\vec{x}),\pi(\vec{y})\right] = 0,$$
$$\left[\phi(\vec{x}),\pi(\vec{y})\right] = i\delta(\vec{x}-\vec{y}),$$

while the free Hamiltonian is, similarly to above,

$$H = \int d^3x\left[\frac{1}{2}\pi^2 + \frac{1}{2}(\nabla\phi)^2 + \frac{m^2}{2}\phi^2\right].$$

A spatial Fourier transform leads to momentum space fields

$$\tilde{\phi}\left(\vec{k}\right)=\int d^3x e^{-i\vec{k}\cdot\vec{x}}\phi\left(\vec{x}\right),$$

$$\tilde{\pi}\left(\vec{k}\right)=\int d^3x e^{-i\vec{k}\cdot\vec{x}}\pi\left(\vec{x}\right)$$

which resolve to annihilation and creation operators

$$a\left(\vec{k}\right)=\left(E\tilde{\phi}\left(\vec{k}\right)+i\tilde{\pi}\left(\vec{k}\right)\right),$$

$$a^\dagger\left(\vec{k}\right)=\left(E\tilde{\phi}\left(\vec{k}\right)-i\tilde{\pi}\left(\vec{k}\right)\right),$$

where $E=\sqrt{k^2+m^2}$

These operators satisfy the commutation relations

$$\left[a\left(\vec{k}_1\right),a\left(\vec{k}_2\right)\right]=\left[a^\dagger\left(\vec{k}_1\right),a^\dagger\left(\vec{k}_2\right)\right]=0,$$

$$\left[a\left(\vec{k}_1\right),a^\dagger\left(\vec{k}_2\right)\right]=(2\pi)^3 2E\delta\left(\vec{k}_1-\vec{k}_2\right).$$

The state $|0\rangle$ annihilated by all of the operators a is identified as the *bare vacuum*, and a particle with momentum $k\to$ is created by applying $a^\dagger\left(\vec{k}\right)$ to the vacuum.

Applying all possible combinations of creation operators to the vacuum constructs the relevant Hilbert space: This construction is called Fock space. The vacuum is annihilated by the Hamiltonian

$$H=\int\frac{d^3k}{(2\pi)^3}\frac{1}{2}a^\dagger\left(\vec{k}\right)a\left(\vec{k}\right),$$

where the zero-point energy has been removed by Wick ordering.

Interactions can be included by adding an interaction Hamiltonian. For a φ^4 theory, this corresponds to adding a Wick ordered term $g:\varphi^4:/4!$ to the Hamiltonian, and integrating over x. Scattering amplitudes may be calculated from this Hamiltonian in the interaction picture. These are constructed in perturbation theory by means of the Dyson series, which gives the time-ordered products, or n-particle Green's functions $\langle 0|\mathcal{T}\{\phi(x_1)\cdots\phi(x_n)\}|0\rangle$ as described in the Dyson series. The Green's functions may also be obtained from a generating function that is constructed as a solution to the Schwinger–Dyson equation.

Feynman Path Integral

The Feynman diagram expansion may be obtained also from the Feynman path integral formulation. The time ordered vacuum expectation values of polynomials in φ, known as the n-particle Green's functions, are constructed by integrating over all possible fields, normalized by the vacuum expectation value with no external fields,

$$\langle 0 | T\{\phi(x_1)\cdots\phi(x_n)\} | 0 \rangle = \frac{\int \mathcal{D}\phi \, \phi(x_1)\cdots\phi(x_n) e^{i\int d^4x \left(\frac{1}{2}\partial^\mu\phi\partial_\mu\phi - \frac{m^2}{2}\phi^2 - \frac{g}{4!}\phi^4\right)}}{\int \mathcal{D}\phi \, e^{i\int d^4x \left(\frac{1}{2}\partial^\mu\phi\partial_\mu\phi - \frac{m^2}{2}\phi^2 - \frac{g}{4!}\phi^4\right)}}.$$

All of these Green's functions may be obtained by expanding the exponential in $J(x)\phi(x)$ in the generating function

$$Z[J] = \int \mathcal{D}\phi \, e^{i\int d^4x \left(\frac{1}{2}\partial^\mu\phi\partial_\mu\phi - \frac{m^2}{2}\phi^2 - \frac{g}{4!}\phi^4 + J\phi\right)} = Z[0]\sum_{n=0}^{\infty} \frac{i^n}{n!} J(x_1)\cdots J(x_n)\langle 0 | T\{\phi(x_1)\cdots\phi(x_n)\} | 0 \rangle.$$

A Wick rotation may be applied to make time imaginary. Changing the signature to (++++) then turns the Feynman integral into a statistical mechanics partition function in Euclidean space,

$$Z[J] = \int \mathcal{D}\phi \, e^{-\int d^4x \left[\frac{1}{2}(\nabla\phi)^2 + \frac{m^2}{2}\phi^2 + \frac{g}{4!}\phi^4 + J\phi\right]}.$$

Normally, this is applied to the scattering of particles with fixed momenta, in which case, a Fourier transform is useful, giving instead

$$\tilde{Z}[\tilde{J}] = \int \mathcal{D}\tilde{\phi} \, e^{-\int d^4p \left[\frac{1}{2}\left(p^2 + m^2\right)\tilde{\phi}^2 + \frac{\lambda}{4!}\tilde{\phi}^4 - \tilde{J}\tilde{\phi}\right]}.$$

The standard trick to evaluate this functional integral is to write it as a product of exponential factors, schematically,

$$\tilde{Z}[\tilde{J}] \sim \int \mathcal{D}\tilde{\phi} \prod_p \left[e^{-\frac{1}{2}\left(p^2 + m^2\right)\tilde{\phi}^2} e^{-\frac{g}{4!}\tilde{\phi}^4} e^{\tilde{J}\tilde{\phi}} \right].$$

The second two exponential factors can be expanded as power series, and the combinatorics of this expansion can be represented graphically through Feynman diagrams.

The integral with $\lambda = 0$ can be treated as a product of infinitely many elementary Gaussian integrals: the result may be expressed as a sum of Feynman diagrams, calculated using the following Feynman rules:

- Each field $\sim\varphi(p)$ in the n-point Euclidean Green's function is represented by an external line (half-edge) in the graph, and associated with momentum p.

- Each vertex is represented by a factor $-g$.

- At a given order g^k, all diagrams with n external lines and k vertices are constructed such that the momenta flowing into each vertex is zero. Each internal line is represented by a propagator $1/(q^2 + m^2)$, where q is the momentum flowing through that line.

- Any unconstrained momenta are integrated over all values.

- The result is divided by a symmetry factor, which is the number of ways the lines and vertices of the graph can be rearranged without changing its connectivity.

- Do not include graphs containing "vacuum bubbles", connected subgraphs with no external lines.

The last rule takes into account the effect of dividing by $\tilde{Z}[0]$. The Minkowski-space Feynman rules are similar, except that each vertex is represented by $-ig$, while each internal line is represented by a propagator $i/(q^2-m^2+i\varepsilon)$, where the ε term represents the small Wick rotation needed to make the Minkowski-space Gaussian integral converge.

Renormalization

The integrals over unconstrained momenta, called "loop integrals", in the Feynman graphs typically diverge. This is normally handled by renormalization, which is a procedure of adding divergent counter-terms to the Lagrangian in such a way that the diagrams constructed from the original Lagrangian and counter-terms is finite. A renormalization scale must be introduced in the process, and the coupling constant and mass become dependent upon it.

The dependence of a coupling constant g on the scale λ is encoded by a beta function, $\beta(g)$, defined by

$$\beta(g) = \lambda \frac{\partial g}{\partial \lambda} \ .$$

This dependence on the energy scale is known as "the running of the coupling parameter", and theory of this systematic scale-dependence in quantum field theory is described by the renormalization group.

Beta-functions are usually computed in an approximation scheme, most commonly perturbation theory, where one assumes that the coupling constant is small. One can then make an expansion in powers of the coupling parameters and truncate the higher-order terms (also known as higher loop contributions, due to the number of loops in the corresponding Feynman graphs).

The β-function at one loop (the first perturbative contribution) for the φ^4 theory is

$$\beta(g) = \frac{3}{16\pi^2} g^2 + O\left(g^3\right) \ .$$

The fact that the sign in front of the lowest-order term is positive suggests that the coupling constant increases with energy. If this behavior persisted at large couplings, this would indicate the presence of a Landau pole at finite energy, arising from quantum triviality. However, the question can only be answered non-perturbatively, since it involves strong coupling.

A quantum field theory is said to be *trivial* when the renormalized coupling, computed through its beta function, goes to zero when the ultraviolet cutoff is removed. Consequently, the propagator becomes that of a free particle and the field is no longer interacting.

For a φ^4 interaction, Michael Aizenman proved that the theory is indeed trivial, for space-time dimension $D \geq 5$.

For $D = 4$, the triviality has yet to be proven rigorously, but lattice computations have provided strong evidence for this. This fact is important as quantum triviality can be used to bound or even *predict* parameters such as the Higgs boson mass. This can also lead to a predictable Higgs mass in asymptotic safety scenarios.

Multivariable Calculus

Multivariable calculus (also known as multivariate calculus) is the extension of calculus in one variable to calculus with functions of several variables: the differentiation and integration of functions involving multiple variables, rather than just one.

Typical Operations

Limits and Continuity

A study of limits and continuity in multivariable calculus yields many counter-intuitive results not demonstrated by single-variable functions. For example, there are scalar functions of two variables with points in their domain which give a particular limit when approached along any arbitrary line, yet give a different limit when approached along a parabola. For example, the function

$$f(x,y) = \frac{x^2 y}{x^4 + y^2}$$

approaches zero along any line through the origin. However, when the origin is approached along a parabola $y = x^2$, it has a limit of 0.5. Since taking different paths toward the same point yields different values for the limit, the limit does not exist.

Continuity in each argument is not sufficient for multivariate continuity For instance, in the case of a real-valued function with two real-valued parameters, $f(x, y)$, continuity of in f for fixed x and continuity of y in f for fixed y does not imply continuity of f. As an example, consider

$$f(x,y) = \begin{cases} \dfrac{y}{x} - y & \text{if } 1 \geq x > y \geq 0 \\[2mm] \dfrac{x}{y} - x & \text{if } 1 \geq y > x \geq 0 \\[2mm] 1 - x & \text{if } x = y > 0 \\[2mm] 0 & \text{else.} \end{cases}$$

It is easy to verify that all real-valued functions (with one real-valued argument) that are given by $f_y(x) := f(x, y)$ are continuous in x (for any fixed y). Similarly, all f_x are continuous as f is symmetric with regards to x and y. However, f itself is not continuous as can be seen by considering the sequence $f\left(\dfrac{1}{n},\dfrac{1}{n}\right)$ (for natural n) which should converge to $f(0,0)=0$ if f was continuous.

However, $\lim\limits_{n\to\infty} f\left(\dfrac{1}{n},\dfrac{1}{n}\right)=1$. Thus, function is not continuous at $(0,0)$.

Partial Differentiation

The partial derivative generalizes the notion of the derivative to higher dimensions. A partial derivative of a multivariable function is a derivative with respect to one variable with all other variables held constant.

Partial derivatives may be combined in interesting ways to create more complicated expressions of the derivative. In vector calculus, the del operator (∇) is used to define the concepts of gradient, divergence, and curl in terms of partial derivatives. A matrix of partial derivatives, the Jacobian matrix, may be used to represent the derivative of a function between two spaces of arbitrary dimension. The derivative can thus be understood as a linear transformation which directly varies from point to point in the domain of the function.

Differential equations containing partial derivatives are called partial differential equations or PDEs. These equations are generally more difficult to solve than ordinary differential equations, which contain derivatives with respect to only one variable.

Multiple Integration

The multiple integral expands the concept of the integral to functions of any number of variables. Double and triple integrals may be used to calculate areas and volumes of regions in the plane and in space. Fubini's theorem guarantees that a multiple integral may be evaluated as a *repeated integral* or *iterated integral* as long as the integrand is continuous throughout the domain of integration.

The surface integral and the line integral are used to integrate over curved manifolds such as surfaces and curves.

Fundamental Theorem of Calculus in Multiple Dimensions

In single-variable calculus, the fundamental theorem of calculus establishes a link between the derivative and the integral. The link between the derivative and the integral in multivariable calculus is embodied by the integral theorems of vector calculus:

- Gradient theorem

- Stokes' theorem

- Divergence theorem

- Green's theorem.

In a more advanced study of multivariable calculus, it is seen that these four theorems are specific incarnations of a more general theorem, the generalized Stokes' theorem, which applies to the integration of differential forms over manifolds.

Applications and Uses

Techniques of multivariable calculus are used to study many objects of interest in the material world. In particular,

		Domain/ Codomain	Applicable techniques
Curves		$f : \mathbb{R} \to \mathbb{R}^n$	Lengths of curves, line integrals, and curvature.
Surfaces		$f : \mathbb{R}^2 \to \mathbb{R}^n$	Areas of surfaces, surface integrals, flux through surfaces, and curvature.
Scalar fields		$f : \mathbb{R}^n \to \mathbb{R}$	Maxima and minima, Lagrange multipliers, directional derivatives.
Vector fields		$f : \mathbb{R}^m \to \mathbb{R}^n$	Any of the operations of vector calculus including gradient, divergence, and curl.

Multivariable calculus can be applied to analyze deterministic systems that have multiple degrees of freedom. Functions with independent variables corresponding to each of the degrees of freedom are often used to model these systems, and multivariable calculus provides tools for characterizing the system dynamics.

Multivariate calculus is used in the optimal control of continuous time dynamic systems. It is used in regression analysis to derive formulas for estimating relationships among various sets of empirical data.

Multivariable calculus is used in many fields of natural and social science and engineering to model and study high-dimensional systems that exhibit deterministic behavior. In economics, for example, consumer choice over a variety of goods, and producer choice over various inputs to use and outputs to produce, are modeled with multivariate calculus. Quantitative analysts in finance also often use multivariate calculus to predict future trends in the stock market.

Non-deterministic, or stochastic systems can be studied using a different kind of mathematics, such as stochastic calculus.

Gradient Theorem

The gradient theorem, also known as the fundamental theorem of calculus for line integrals, says that a line integral through a gradient field can be evaluated by evaluating the original scalar field at the endpoints of the curve.

Let $\varphi : U \subseteq \mathbb{R}^n \to \mathbb{R}$ and γ a curve from p to q. Then

$$\varphi(\mathbf{q}) - \varphi(\mathbf{p}) = \int_{\gamma[\mathbf{p},\mathbf{q}]} \nabla\varphi(\mathbf{r}) \cdot d\mathbf{r}.$$

It is a generalization of the fundamental theorem of calculus to any curve in a plane or space (generally n-dimensional) rather than just the real line.

The gradient theorem implies that line integrals through gradient fields are path independent. In physics this theorem is one of the ways of defining a *conservative* force. By placing ϕ as potential, ∇ϕ is a conservative field. Work done by conservative forces does not depend on the path followed by the object, but only the end points, as the above equation shows.

The gradient theorem also has an interesting converse: any path-independent vector field can be expressed as the gradient of a scalar field. Just like the gradient theorem itself, this converse has many striking consequences and applications in both pure and applied mathematics.

Proof

If ϕ is a differentiable function from some open subset U (of R^n) to R, and if r is a differentiable function from some closed interval [a,b] to U, then by the multivariate chain rule, the composite function ϕ ∘ r is differentiable on (a, b) and

$$\frac{d}{dt}(\varphi \circ \mathbf{r})(t) = \nabla\varphi(\mathbf{r}(t)) \cdot \mathbf{r}'(t)$$

for all t in (a, b). Here the · denotes the usual inner product.

Now suppose the domain U of ϕ contains the differentiable curve γ with endpoints p and q, (oriented in the direction from p to q). If r parametrizes γ for t in $[a, b]$, then the above shows that

$$\int_\gamma \nabla\varphi(\mathbf{u})\cdot d\mathbf{u} = \int_a^b \nabla\varphi(\mathbf{r}(t))\cdot \mathbf{r}'(t)dt$$

$$= \int_a^b \frac{d}{dt}\varphi(\mathbf{r}(t))dt = \varphi(\mathbf{r}(b)) - \varphi(\mathbf{r}(a)) = \varphi(\mathbf{q}) - \varphi(\mathbf{p})$$

where the definition of the line integral is used in the first equality, and the fundamental theorem of calculus is used in the third equality.

Examples

Example 1

Suppose $\gamma \subset \mathbf{R}^2$ is the circular arc oriented counterclockwise from $(5, 0)$ to $(-4, 3)$. Using the definition of a line integral,

$$\int_\gamma ydx + xdy = \int_0^{\pi-\tan^{-1}\left(\frac{3}{4}\right)} ((5\sin t)(-5\sin t) + (5\cos t)(5\cos t))dt$$

$$= \int_0^{\pi-\tan^{-1}\left(\frac{3}{4}\right)} 25\left(-\sin^2 t + \cos^2 t\right)dt$$

$$= \int_0^{\pi-\tan^{-1}\left(\frac{3}{4}\right)} 25\cos(2t)dt$$

$$= \frac{25}{2}\sin(2t)\Big|_0^{\pi-\tan^{-1}\left(\frac{3}{4}\right)}$$

$$= \frac{25}{2}\sin\left(2\pi - 2\tan^{-1}\left(\frac{3}{4}\right)\right)$$

$$= -\frac{25}{2}\sin\left(2\tan^{-1}\left(\frac{3}{4}\right)\right)$$

$$= -\frac{25\left(\frac{3}{4}\right)}{\left(\frac{3}{4}\right)^2 + 1} = -12.$$

Notice all of the painstaking computations involved in directly calculating the integral. Instead, since the function $f(x, y) = xy$ is differentiable on all of $\mathbf{R}^2$, we can simply use the gradient theorem to say

$$\int_{\gamma} ydx + xdy = \int_{\gamma} \nabla(xy) \cdot (dx, dy) = xy \mid_{(5,0)}^{(-4,3)} = -4 \cdot 3 - 5 \cdot 0 = -12.$$

Notice that either way gives the same answer, but using the latter method, most of the work is already done in the proof of the gradient theorem.

Example 2

For a more abstract example, suppose $\gamma \subset R^n$ has endpoints p, q, with orientation from p to q. For u in R^n, let $|u|$ denote the Euclidean norm of u. If $\alpha \geq 1$ is a real number, then

$$\int_{\gamma} |\mathbf{x}|^{\alpha-1} \, \mathbf{x} \cdot d\mathbf{x} = \frac{1}{\alpha+1} \int_{\gamma} (\alpha+1) \, |\mathbf{x}|^{(\alpha+1)-2} \, \mathbf{x} \cdot d\mathbf{x}$$

$$= \frac{1}{\alpha+1} \int_{\gamma} \nabla |\mathbf{x}|^{\alpha+1} \cdot d\mathbf{x} = \frac{|\mathbf{q}|^{\alpha+1} - |\mathbf{p}|^{\alpha+1}}{\alpha+1}$$

Here the final equality follows by the gradient theorem, since the function $f(x) = |x|^{\alpha+1}$ is differentiable on R^n if $\alpha \geq 1$.

If $\alpha < 1$ then this equality will still hold in most cases, but caution must be taken if γ passes through or encloses the origin, because the integrand vector field $|x|^{\alpha-1}x$ will fail to be defined there. However, the case $\alpha = -1$ is somewhat different; in this case, the integrand becomes $|x|^{-2}x = \nabla(\log|x|)$, so that the final equality becomes $\log|q| - \log|p|$.

Note that if $n = 1$, then this example is simply a slight variant of the familiar Power rule from single-variable calculus.

Example 3

Suppose there are n point charges arranged in three-dimensional space, and the i-th point charge has charge Q_i and is located at position p_i in R^3. We would like to calculate the work done on a particle of charge q as it travels from a point a to a point b in R^3. Using Coulomb's law, we can easily determine that the force on the particle at position r will be

$$\mathbf{F}(\mathbf{r}) = kq \sum_{i=1}^{n} \frac{Q_i(\mathbf{r} - \mathbf{p}_i)}{|\mathbf{r} - \mathbf{p}_i|^3}$$

Here $|u|$ denotes the Euclidean norm of the vector u in R^3, and $k = 1/(4\pi\varepsilon_o)$, where ε_o is the Vacuum permittivity.

Let $\gamma \subset R^3 - \{p_1, ..., p_n\}$ be an arbitrary differentiable curve from a to b. Then the work done on the particle is

$$W = \int_{\gamma} \mathbf{F}(\mathbf{r}) \cdot d\mathbf{r} = \int_{\gamma} \left(kq \sum_{i=1}^{n} \frac{Q_i(\mathbf{r} - \mathbf{p}_i)}{|\mathbf{r} - \mathbf{p}_i|^3} \right) \cdot d\mathbf{r} = kq \sum_{i=1}^{n} \left(Q_i \int_{\gamma} \frac{\mathbf{r} - \mathbf{p}_i}{|\mathbf{r} - \mathbf{p}_i|^3} \cdot d\mathbf{r} \right)$$

Now for each i, direct computation shows that

$$\frac{\mathbf{r}-\mathbf{p}_i}{|\mathbf{r}-\mathbf{p}_i|^3} = -\nabla\frac{1}{|\mathbf{r}-\mathbf{p}_i|}.$$

Thus, continuing from above and using the gradient theorem,

$$W = -kq\sum_{i=1}^{n}\left(Q_i\int_\gamma \nabla\frac{1}{|\mathbf{r}-\mathbf{p}_i|}\cdot d\mathbf{r}\right) = kq\sum_{i=1}^{n}Q_i\left(\frac{1}{|\mathbf{a}-\mathbf{p}_i|}-\frac{1}{|\mathbf{b}-\mathbf{p}_i|}\right)$$

We are finished. Of course, we could have easily completed this calculation using the powerful language of electrostatic potential or electrostatic potential energy (with the familiar formulas $W = -\Delta U = -q\Delta V$). However, we have not yet *defined* potential or potential energy, because the *converse* of the gradient theorem is required to prove that these are well-defined, differentiable functions and that these formulas hold. Thus, we have solved this problem using only Coulomb's Law, the definition of work, and the gradient theorem.

Converse of the Gradient Theorem

The gradient theorem states that if the vector field F is the gradient of some scalar-valued function (i.e., if F is conservative), then F is a path-independent vector field (i.e., the integral of F over some piecewise-differentiable curve is dependent only on end points). This theorem has a powerful converse:

If F is a path-independent vector field, then F is the gradient of some scalar-valued function.

It is straightforward to show that a vector field is path-independent if and only if the integral of the vector field over every closed loop in its domain is zero. Thus the converse can alternatively be stated as follows: If the integral of F over every closed loop in the domain of F is zero, then F is the gradient of some scalar-valued function.

Proof of the Converse

Example of the Converse Principle

To illustrate the power of this converse principle, we cite an example that has significant physical consequences. In classical electromagnetism, the electric force is a path-independent force ; i.e. the work done on a particle that has returned to its original position within an electric field is zero (assuming that no changing magnetic fields are present).

Therefore, the above theorem implies that the electric force field $\mathbf{F}_e : S \to \mathbf{R}^3$ is conservative (here S is some open, path-connected subset of $\mathbf{R}^3$ that contains a charge distribution). Following the ideas of the above proof, we can set some reference point a in S, and define a function $U_e : S \to \mathbf{R}$ by

$$U_e(\mathbf{r}) := -\int_{\gamma[\mathbf{a},\mathbf{r}]} \mathbf{F}_e(\mathbf{u})\cdot d\mathbf{u}$$

Using the above proof, we know U_e is well-defined and differentiable, and $\mathbf{F}_e = -\nabla U_e$ (from this formula we can use the gradient theorem to easily derive the well-known formula for calculating work done by conservative forces: $W = -\Delta U$). This function U_e is often referred to as the electrostatic potential energy of the system of charges in S (with reference to the zero-of-potential a). In many cases, the domain S is assumed to be unbounded and the reference point a is taken to be "infinity," which can be made rigorous using limiting techniques. This function U_e is an indispensable tool used in the analysis of many physical systems.

Generalizations

Many of the critical theorems of vector calculus generalize elegantly to statements about the integration of differential forms on manifolds. In the language of differential forms and exterior derivatives, the gradient theorem states that

$$\int_{\partial\gamma} \phi = \int_{\gamma} d\phi$$

for any o-form, ϕ, defined on some differentiable curve $\gamma \subset R^n$ (here the integral of ϕ over the boundary of the γ is understood to be the evaluation of ϕ at the endpoints of γ).

Notice the striking similarity between this statement and the generalized version of Stokes' theorem, which says that the integral of any compactly supported differential form ω over the boundary of some orientable manifold Ω is equal to the integral of its exterior derivative $d\omega$ over the whole of Ω, i.e.,

$$\int_{\partial\Omega} \omega = \int_{\Omega} d\omega$$

This powerful statement is a generalization of the gradient theorem from 1-forms defined on one-dimensional manifolds to differential forms defined on manifolds of arbitrary dimension.

The converse statement of the gradient theorem also has a powerful generalization in terms of differential forms on manifolds. In particular, suppose ω is a form defined on a contractible domain, and the integral of ω over any closed manifold is zero. Then there exists a form ψ such that $\omega = d\psi$. Thus, on a contractible domain, every closed form is exact. This result is summarized by the Poincaré lemma.

Stokes' Theorem

In vector calculus, and more generally differential geometry, Stokes' theorem (also called the generalized Stokes' theorem) is a statement about the integration of differential forms on manifolds, which both simplifies and generalizes several theorems from vector calculus. Stokes' theorem says that the integral of a differential form ω over the boundary of some orientable manifold Ω is equal to the integral of its exterior derivative $d\omega$ over the whole of Ω, i.e.,

$$\int_{\partial\Omega} \omega = \int_{\Omega} d\omega.$$

This modern form of Stokes' theorem is a vast generalization of a classical result. Lord Kelvin communicated it to George Stokes in a letter dated July 2, 1850. Stokes set the theorem as a question on the 1854 Smith's Prize exam, which led to the result bearing his name, even though it was actually first published by Hermann Hankel in 1861. This classical Kelvin–Stokes theorem relates the surface integral of the curl of a vector field F over a surface Σ in Euclidean three-space to the line integral of the vector field over its boundary $\partial\Sigma$:

$$\iint_{\Sigma} \nabla \times \mathbf{F} \cdot d\Sigma = \oint_{\partial\Sigma} \mathbf{F} \cdot d\mathbf{r}.$$

This classical statement, along with the classical divergence theorem, fundamental theorem of calculus, and Green's theorem are simply special cases of the general formulation stated above.

Introduction

The fundamental theorem of calculus states that the integral of a function f over the interval $[a, b]$ can be calculated by finding an antiderivative F of f:

$$\int_a^b f(x)dx = F(b) - F(a).$$

Stokes' theorem is a vast generalization of this theorem in the following sense.

- By the choice of F, $\frac{dF}{dx} = f(x)$. In the parlance of differential forms, this is saying that $f(x)\,dx$ is the exterior derivative of the 0-form, i.e. function, F: in other words, that $dF = f\,dx$. The general Stokes theorem applies to higher differential forms ω instead of just 0-forms such as F.

- A closed interval $[a, b]$ is a simple example of a one-dimensional manifold with boundary. Its boundary is the set consisting of the two points a and b. Integrating f over the interval may be generalized to integrating forms on a higher-dimensional manifold. Two technical conditions are needed: the manifold has to be orientable, and the form has to be compactly supported in order to give a well-defined integral.

- The two points a and b form the boundary of the closed interval. More generally, Stokes' theorem applies to oriented manifolds M with boundary. The boundary ∂M of M is itself a manifold and inherits a natural orientation from that of the manifold. For example, the natural orientation of the interval gives an orientation of the two boundary points. Intuitively, a inherits the opposite orientation as b, as they are at opposite ends of the interval. So, "integrating" F over two boundary points a, b is taking the difference $F(b) - F(a)$.

In even simpler terms, one can consider that points can be thought of as the boundaries of curves, that is as 0-dimensional boundaries of 1-dimensional manifolds. So, just as one can find the value of an integral ($f\,dx = dF$) over a 1-dimensional manifold ($[a, b]$) by considering the anti-derivative (F) at the 0-dimensional boundaries ($\{a, b\}$), one can generalize the fundamental theorem of calculus, with a few additional caveats, to deal with the value of integrals ($d\omega$) over n-dimensional manifolds (Ω) by considering the anti-derivative (ω) at the $(n - 1)$-dimensional boundaries ($\partial\Omega$) of the manifold.

So the fundamental theorem reads:

$$\int_{[a,b]} f(x)dx = \int_{[a,b]} dF = \int_{\{a\}^- \cup \{b\}^+} F = F(b) - F(a).$$

Formulation for Smooth Manifolds with Boundary

Let Ω be an oriented smooth manifold with boundary of dimension n and let α be a smooth n-differential form that is compactly supported on Ω. First, suppose that α is compactly supported in the domain of a single, oriented coordinate chart $\{U, \phi\}$. In this case, we define the integral of α over Ω as

$$\int_\Omega \alpha = \int_{\varphi(U)} \left(\varphi^{-1}\right)^* \alpha,$$

i.e., via the pullback of α to $\mathbf{R}^n$.

More generally, the integral of α over Ω is defined as follows: Let $\{\psi_i\}$ be a partition of unity associated with a locally finite cover $\{U_i, \varphi_i\}$ of (consistently oriented) coordinate charts, then define the integral

$$\int_\Omega \alpha \equiv \sum_i \int_{U_i} \psi_i \alpha,$$

where each term in the sum is evaluated by pulling back to $\mathbf{R}^n$ as described above. This quantity is well-defined; that is, it does not depend on the choice of the coordinate charts, nor the partition of unity.

Stokes' theorem reads: If ω is an $(n-1)$-form with compact support on Ω and $\partial\Omega$ denotes the boundary of Ω with its induced orientation, then

$$\int_\Omega d\omega = \int_{\partial\Omega} \omega \left(= \oint_{\partial\Omega} \omega\right).$$

Here d is the exterior derivative, which is defined using the manifold structure only. On the r.h.s., a circle is sometimes used within the integral sign to stress the fact that the $(n-1)$-manifold $\partial\Omega$ is *closed*. The r.h.s. of the equation is often used to formulate *integral* laws; the l.h.s. then leads to equivalent *differential* formulations.

The theorem is often used in situations where Ω is an embedded oriented submanifold of some bigger manifold on which the form ω is defined.

Topological Preliminaries; Integration Over Chains

Let M be a smooth manifold. A (smooth) singular k-simplex in M is defined as a smooth map from the standard simplex in $\mathbf{R}^k$ to M. The group $C_k(M, \mathbf{Z})$ of singular k-chains on M is defined to be the free abelian group on the set of singular k-simplices in M. These groups, together with the boundary map, ∂, define a chain complex. The corresponding homology (resp. cohomology) group is isomorphic to the usual singular homology group $H_k(M, Z)$ (resp. the singular cohomology group $H^k(M, Z)$), defined using continuous rather than smooth simplices in M.

On the other hand, the differential forms, with exterior derivative, d, as the connecting map, form a cochain complex, which defines the de Rham cohomology groups $H_{dR}^k(M, \mathbf{R})$.

Differential k-forms can be integrated over a k-simplex in a natural way, by pulling back to $\mathbf{R}^k$. Extending by linearity allows one to integrate over chains. This gives a linear map from the space of k-forms to the k-th group of singular cochains, $C^*(M, Z)$, the linear functionals on $C_k(M, Z)$. In other words, a k-form ω defines a functional

$$I(\omega)(c) = \oint_c \omega$$

on the k-chains. Stokes' theorem says that this is a chain map from de Rham cohomology to singular cohomology with real coefficients; the exterior derivative, d, behaves like the *dual* of ∂ on forms. This gives a homomorphism from de Rham cohomology to singular cohomology. On the level of forms, this means:

1. closed forms, i.e., $d\omega = 0$, have zero integral over *boundaries*, i.e. over manifolds that can be written as $\partial \sum_c M_c,$, and

2. exact forms, i.e., $\omega = d\sigma$, have zero integral over *cycles*, i.e. if the boundaries sum up to the empty set: $\sum_c M_c = \varnothing$.

De Rham's theorem shows that this homomorphism is in fact an isomorphism. So the converse to 1 and 2 above hold true. In other words, if $\{c_i\}$ are cycles generating the k-th homology group, then for any corresponding real numbers, $\{a_i\}$, there exist a closed form, ω, such that

$$\oint_{c_i} \omega = a_i,$$

and this form is unique up to exact forms.

Underlying Principle

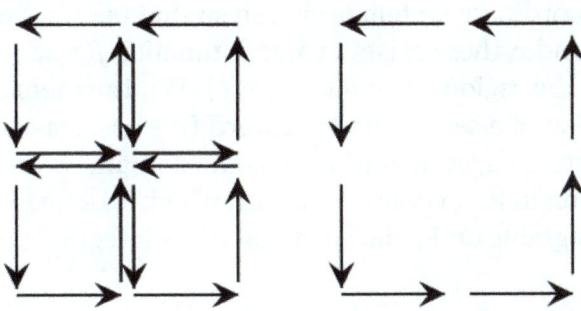

To simplify these topological arguments, it is worthwhile to examine the underlying principle by considering an example for $d = 2$ dimensions. The essential idea can be understood by the diagram on the left, which shows that, in an oriented tiling of a manifold, the interior paths are traversed in opposite directions; their contributions to the path integral thus cancel each other pairwise. As a consequence, only the contribution from the boundary remains. It thus suffices to prove Stokes' theorem for sufficiently fine tilings (or, equivalently, simplices), which usually is not difficult.

Generalization to Rough Sets

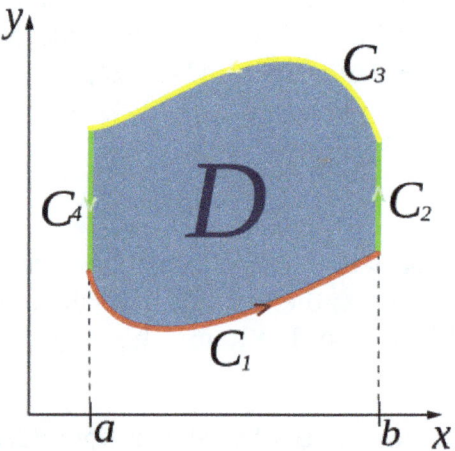

A region (here called D instead of Ω) with piecewise smooth boundary.
This is a manifold with corners, so its boundary is not a smooth manifold.

The formulation above, in which Ω is a smooth manifold with boundary, does not suffice in many applications. For example, if the domain of integration is defined as the plane region between two x-coordinates and the graphs of two functions, it will often happen that the domain has corners. In such a case, the corner points mean that Ω is not a smooth manifold with boundary, and so the statement of Stokes' theorem given above does not apply. Nevertheless, it is possible to check that the conclusion of Stokes' theorem is still true. This is because Ω and its boundary are well-behaved away from a small set of points (a measure zero set).

A version of Stokes' theorem that extends to rough domains was proved by Whitney. Assume that D is a connected bounded open subset of of $\mathbb{R}^n$. Call D a *standard domain* if it satisfies the following property: There exists a subset P of ∂D, open in ∂D, whose complement in ∂D has Hausdorff $(n-1)$-measure zero; and such that every point of P has a *generalized normal vector*. This is a vector $v(x)$ such that, if a coordinate system is chosen so that $v(x)$ is the first basis vector, then, in an open neighborhood around x, there exists a smooth function $f(x_2, ..., x_n)$ such that P is the graph $\{ x_1 = f(x_2, ..., x_n) \}$ and D is the region $\{ x_1 < f(x_2, ..., x_n) \}$. Whitney remarks that the boundary of a standard domain is the union of a set of zero Hausdorff $(n-1)$-measure and a finite or countable union of smooth $(n-1)$-manifolds, each of which has the domain on only one side. He then proves that if D is a standard domain in $\mathbb{R}^n$, ω is a $(n-1)$-form which is defined, continuous, and bounded on $D \cup P$, smooth on D, integrable on P, and such that $d\omega$ is integrable on D, then Stokes' theorem holds, that is,

$$\int_P \omega = \int_D d\omega.$$

The study of measure-theoretic properties of rough sets leads to geometric measure theory. Even more general versions of Stokes' theorem have been proved by Federer and by Harrison.

Special Cases

The general form of the Stokes theorem using differential forms is more powerful and easier to use than the special cases. The traditional versions can be formulated using Cartesian coordinates without the machinery of differential geometry, and thus are more accessible. Further, they are older and their names are more familiar as a result. The traditional forms are often considered more convenient by practicing scientists and engineers but the non-naturalness of the traditional formulation becomes apparent when using other coordinate systems, even familiar ones like spherical or cylindrical coordinates. There is potential for confusion in the way names are applied, and the use of dual formulations.

Kelvin–Stokes Theorem

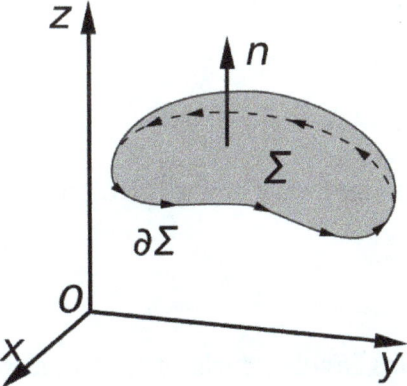

An illustration of the Kelvin–Stokes theorem, with surface Σ, its boundary ∂Σ and the "normal" vector n.

This is a (dualized) 1+1 dimensional case, for a 1-form (dualized because it is a statement about vector fields). This special case is often just referred to as the *Stokes' theorem* in many introductory university vector calculus courses and as used in physics and engineering. It is also sometimes known as the curl theorem.

The classical Kelvin–Stokes theorem:

$$\int_{\Sigma} \nabla \times F \cdot d\Sigma = \oint_{\partial\Sigma} F \cdot d\mathbf{r},$$

which relates the surface integral of the curl of a vector field over a surface Σ in Euclidean three-space to the line integral of the vector field over its boundary, is a special case of the general Stokes theorem (with $n = 2$) once we identify a vector field with a 1 form using the metric on Euclidean three-space. The curve of the line integral, ∂Σ, must have positive orientation, meaning that dr points counterclockwise when the surface normal, $d\Sigma$, points toward the viewer, following the right-hand rule.

One consequence of the Kelvin–Stokes theorem is that the field lines of a vector field with zero curl cannot be closed contours. The formula can be rewritten as:

$$\iint_{\Sigma} \left[\left(\frac{\partial R}{\partial y} - \frac{\partial Q}{\partial z} \right) dy\,dz + \left(\frac{\partial P}{\partial z} - \frac{\partial R}{\partial x} \right) dz\,dx + \left(\frac{\partial Q}{\partial x} - \frac{\partial P}{\partial y} \right) dx\,dy \right]$$

$$= \oint_{\partial \Sigma} (P\,dx + Q\,dy + R\,dz),$$

where P, Q and R are the components of F.

These variants are rarely used:

$$\int_{\Sigma} \big(g(\nabla \times F) + (\nabla g) \times F\big) \cdot d\Sigma = \oint_{\partial \Sigma} gF \cdot d\mathbf{r},$$

$$\int_{\Sigma} \big(F(\nabla \cdot \mathbf{G}) - \mathbf{G}(\nabla \cdot F) + (\mathbf{G} \cdot \nabla)F - (F \cdot \nabla)\mathbf{G}\big) \cdot d\Sigma = \oint_{\partial \Sigma} (F \times \mathbf{G}) \cdot d\mathbf{r},$$

$$\int_{\Sigma} \nabla(F \cdot d\Sigma) - (\nabla \cdot F)d\Sigma = \oint_{\partial \Sigma} d\mathbf{r} \times F.$$

Green's Theorem

Green's theorem is immediately recognizable as the third integrand of both sides in the integral in terms of P, Q, and R cited above.

In Electromagnetism

Two of the four Maxwell equations involve curls of 3-D vector fields and their differential and integral forms are related by the Kelvin–Stokes theorem. Caution must be taken to avoid cases with moving boundaries: the partial time derivatives are intended to exclude such cases. If moving boundaries are included, interchange of integration and differentiation introduces terms related to boundary motion not included in the results below:

Name	Differential form	Integral form (using Kelvin–Stokes theorem plus relativistic invariance, $\int \dfrac{\partial}{\partial t}\ldots \to \dfrac{d}{dt}\int\ldots)$
Maxwell–Faraday equation Faraday's law of induction:	$\nabla \times \mathbf{E} = -\dfrac{\partial \mathbf{B}}{\partial t}$	$\oint_C \mathbf{E} \cdot d\mathbf{l} = \iint_S \nabla \times \mathbf{E} \cdot d\mathbf{A} = -\iint_S \dfrac{\partial \mathbf{B}}{\partial t} \cdot d\mathbf{A}$ (with C and S not necessarily stationary)
Ampère's law (with Maxwell's extension):	$\nabla \times \mathbf{H} = \mathbf{J} + \dfrac{\partial \mathbf{D}}{\partial t}$	$\oint_C \mathbf{H} \cdot d\mathbf{l} = \iint_S \nabla \times \mathbf{H} \cdot d\mathbf{A}$ $= \iint_S \mathbf{J} \cdot d\mathbf{A} + \iint_S \dfrac{\partial \mathbf{D}}{\partial t} \cdot d\mathbf{A}$ (with C and S not necessarily stationary)

The above listed subset of Maxwell's equations are valid for electromagnetic fields expressed in SI units. In other systems of units, such as CGS or Gaussian units, the scaling factors for the terms differ. For example, in Gaussian units, Faraday's law of induction and Ampère's law take the forms

$$\nabla \times \mathbf{E} = -\frac{1}{c}\frac{\partial \mathbf{B}}{\partial t},$$

$$\nabla \times \mathbf{H} = \frac{1}{c}\frac{\partial \mathbf{D}}{\partial t} + \frac{4\pi}{c}\mathbf{J},$$

respectively, where c is the speed of light in vacuum.

Divergence Theorem

Likewise, the Ostrogradsky–Gauss theorem (also known as the divergence theorem or Gauss's theorem)

$$\int_{\text{Vol}} \nabla \cdot F \, d_{\text{Vol}} = \oint_{\partial \text{Vol}} F \cdot d\Sigma$$

is a special case if we identify a vector field with the $n - 1$ form obtained by contracting the vector field with the Euclidean volume form. An application of this is the case $\mathbf{F} = f\mathbf{c}$ where $\mathbf{c}$ is an arbitrary constant vector. Working out the divergence of the product gives

$$\mathbf{c} \cdot \int_{\text{Vol}} \nabla f \, d_{\text{Vol}} = \mathbf{c} \cdot \oint_{\partial \text{Vol}} f \, d\Sigma.$$

Since this holds for all $\vec{c}$, we find

$$\int_{\text{Vol}} \nabla f \, d_{\text{Vol}} = \oint_{\partial \text{Vol}} f \, d\Sigma.$$

Divergence Theorem

In vector calculus, the divergence theorem, also known as Gauss's theorem or Ostrogradsky's theorem, is a result that relates the flow (that is, flux) of a vector field through a surface to the behavior of the vector field inside the surface.

More precisely, the divergence theorem states that the outward flux of a vector field through a closed surface is equal to the volume integral of the divergence over the region inside the surface. Intuitively, it states that *the sum of all sources minus the sum of all sinks gives the net flow out of a region.*

The divergence theorem is an important result for the mathematics of engineering, in particular in electrostatics and fluid dynamics.

In physics and engineering, the divergence theorem is usually applied in three dimensions. However, it generalizes to any number of dimensions. In one dimension, it is equivalent to the fundamental theorem of calculus. In two dimensions, it is equivalent to Green's theorem.

The theorem is a special case of the more general Stokes' theorem.

Intuition

If a fluid is flowing in some area, then the rate at which fluid flows out of a certain region within that area can be calculated by adding up the sources inside the region and subtracting the sinks. The fluid flow is represented by a vector field, and the vector field's divergence at a given point describes the strength of the source or sink there. So, integrating the field's divergence over the interior of the region should equal the integral of the vector field over the region's boundary. The divergence theorem says that this is true.

The divergence theorem is employed in any conservation law which states that the volume total of all sinks and sources, that is the volume integral of the divergence, is equal to the net flow across the volume's boundary.

Mathematical Statement

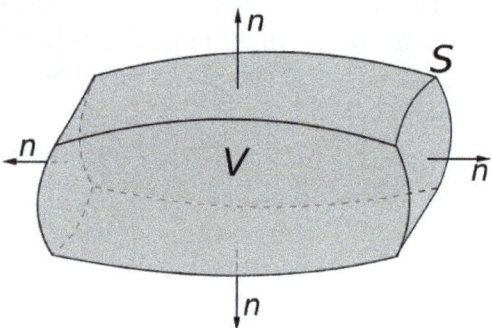

A region V bounded by the surface $S = \partial V$ with the surface normal n

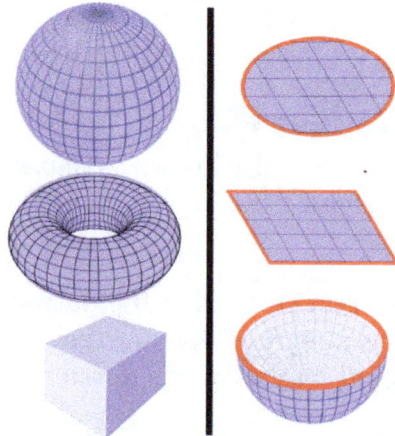

The divergence theorem can be used to calculate a flux through a closed surface that fully encloses a volume, like any of the surfaces on the left. It can *not* directly be used to calculate the flux through surfaces with boundaries, like those on the right. (Surfaces are blue, boundaries are red.)

Suppose V is a subset of $\mathbb{R}^n$ (in the case of $n = 3$, V represents a volume in 3D space) which is compact and has a piecewise smooth boundary S (also indicated with $\partial V = S$). If F is a continuously differentiable vector field defined on a neighborhood of V, then we have:

$$\iiint_V (\nabla \cdot \mathbf{F})dV = \oiint_S (\mathbf{F} \cdot \mathbf{n})dS.$$

The left side is a volume integral over the volume V, the right side is the surface integral over the boundary of the volume V. The closed manifold ∂V is quite generally the boundary of V oriented by outward-pointing normals, and n is the outward pointing unit normal field of the boundary ∂V. (dS may be used as a shorthand for ndS.) The symbol within the two integrals stresses once more that ∂V is a *closed* surface. In terms of the intuitive description above, the left-hand side of the equation represents the total of the sources in the volume V, and the right-hand side represents the total flow across the boundary S.

Corollaries

By replacing **F** in the divergence theorem with specific forms, other useful identities can be derived (cf. vector identities).

- With $\mathbf{F} \rightarrow \mathbf{F}g$ for a scalar function g and a vector field F,

$$\iiint_V \left[\mathbf{F} \cdot (\nabla g) + g(\nabla \cdot \mathbf{F})\right]dV = \oiint_S g\mathbf{F} \cdot \mathbf{n}dS.$$

  A special case of this is $F = \nabla f$, in which case the theorem is the basis for Green's identities.

- With $\mathbf{F} \rightarrow \mathbf{F} \times \mathbf{G}$ for two vector fields F and G,

$$\iiint \left[\mathbf{G} \cdot (\nabla \times \mathbf{F}) - \mathbf{F} \cdot (\nabla \times \mathbf{G})\right]dV = \oiint (\mathbf{F} \times \mathbf{G}) \cdot d\mathbf{S}$$

- With $\mathbf{F} \rightarrow f\mathbf{c}$ for a scalar function f and vector field c:

$$\iiint_V \mathbf{c} \cdot \nabla f \, dV = \oiint_S (\mathbf{c}f) \cdot d\mathbf{S} - \iiint_V f(\nabla \cdot \mathbf{c})dV.$$

  The last term on the right vanishes for constant **c** or any divergence free (solenoidal) vector field, e.g. Incompressible flows without sources or sinks such as phase change or chemical reactions etc.

- With $\mathbf{F} \rightarrow \mathbf{c} \times \mathbf{F}$ for vector field F and constant vector c:

$$\iiint_V \mathbf{c} \cdot (\nabla \times \mathbf{F})dV = \oiint_S (\mathbf{F} \times \mathbf{c}) \cdot d\mathbf{S}.$$

Example

Suppose we wish to evaluate

$$\oiint_S \mathbf{F} \cdot \mathbf{n}\, dS,$$

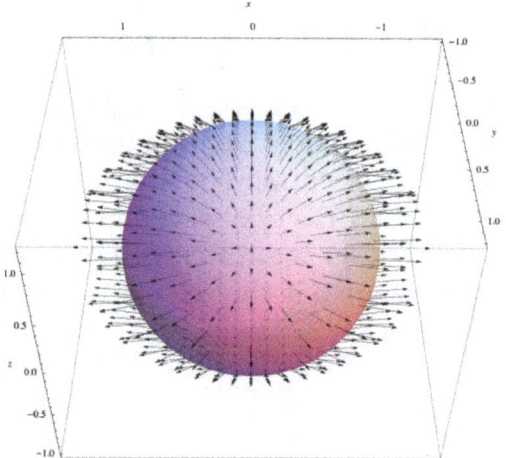

The vector field corresponding to the example shown. Note, vectors may point into or out of the sphere.

where S is the unit sphere defined by

$$S = \left\{ (x, y, z) \in \mathbb{R}^3 : x^2 + y^2 + z^2 = 1 \right\},$$

and F is the vector field

$$\mathbf{F} = 2x\mathbf{i} + y^2\mathbf{j} + z^2\mathbf{k}.$$

The direct computation of this integral is quite difficult, but we can simplify the derivation of the result using the divergence theorem, because the divergence theorem says that the integral is equal to:

$$\iiint_W (\nabla \cdot \mathbf{F})\, dV = 2\iiint_W (1 + y + z)\, dV = 2\iiint_W dV + 2\iiint_W y\, dV + 2\iiint_W z\, dV,$$

where W is the unit ball:

$$W = \left\{ (x, y, z) \in \mathbb{R}^3 : x^2 + y^2 + z^2 \leq 1 \right\}.$$

Since the function y is positive in one hemisphere of W and negative in the other, in an equal and opposite way, its total integral over W is zero. The same is true for z:

$$\iiint_W y\, dV = \iiint_W z\, dV = 0.$$

Therefore,

$$\oiint \mathbf{F} \cdot \mathbf{n}\, dS = 2\iiint_W dV = \frac{8\pi}{3},$$

because the unit ball W has volume $4\pi/3$.

Applications

Differential Form and Integral form of Physical Laws

As a result of the divergence theorem, a host of physical laws can be written in both a differential form (where one quantity is the divergence of another) and an integral form (where the flux of one quantity through a closed surface is equal to another quantity). Three examples are Gauss's law (in electrostatics), Gauss's law for magnetism, and Gauss's law for gravity.

Continuity Equations

Continuity equations offer more examples of laws with both differential and integral forms, related to each other by the divergence theorem. In fluid dynamics, electromagnetism, quantum mechanics, relativity theory, and a number of other fields, there are continuity equations that describe the conservation of mass, momentum, energy, probability, or other quantities. Generically, these equations state that the divergence of the flow of the conserved quantity is equal to the distribution of *sources* or *sinks* of that quantity. The divergence theorem states that any such continuity equation can be written in a differential form (in terms of a divergence) and an integral form (in terms of a flux).

Inverse-square Laws

Any *inverse-square law* can instead be written in a *Gauss' law*-type form (with a differential and integral form, as described above). Two examples are Gauss' law (in electrostatics), which follows from the inverse-square Coulomb's law, and Gauss' law for gravity, which follows from the inverse-square Newton's law of universal gravitation. The derivation of the Gauss' law-type equation from the inverse-square formulation (or vice versa) is exactly the same in both cases.

History

The theorem was first discovered by Lagrange in 1762, then later independently rediscovered by Gauss in 1813, by Ostrogradsky, who also gave the first proof of the general theorem, in 1826, by Green in 1828, etc. Subsequently, variations on the divergence theorem are correctly called Ostrogradsky's theorem, but also commonly Gauss's theorem, or Green's theorem.

Examples

To verify the planar variant of the divergence theorem for a region R:

$$R = \left\{ (x, y) \in \mathbb{R}^2 : x^2 + y^2 \leq 1 \right\},$$

and the vector field:

$$\mathbf{F}(x, y) = 2y\mathbf{i} + 5x\mathbf{j}.$$

The boundary of R is the unit circle, C, that can be represented parametrically by:

$$x = \cos(s), \quad y = \sin(s)$$

such that $0 \leq s \leq 2\pi$ where s units is the length arc from the point $s = 0$ to the point P on C. Then a vector equation of C is

$$C(s) = \cos(s)\mathbf{i} + \sin(s)\mathbf{j}.$$

At a point P on C:

$$P = (\cos(s), \sin(s)) \Rightarrow \mathbf{F} = 2\sin(s)\mathbf{i} + 5\cos(s)\mathbf{j}.$$

Therefore,

$$
\begin{aligned}
\oint_C \mathbf{F} \cdot \mathbf{n}\, ds &= \int_0^{2\pi} (2\sin(s)\mathbf{i} + 5\cos(s)\mathbf{j}) \cdot (\cos(s)\mathbf{i} + \sin(s)\mathbf{j})\, ds \\
&= \int_0^{2\pi} (2\sin(s)\cos(s) + 5\sin(s)\cos(s))\, ds \\
&= 7\int_0^{2\pi} \sin(s)\cos(s)\, ds \\
&= 0.
\end{aligned}
$$

Because $M = 2y$, $\partial M / \partial x = 0$, and because $N = 5x$, $\partial N / \partial y = 0$. Thus

$$\iint_R \nabla \cdot \mathbf{F}\, dA = \iint_R \left(\frac{\partial M}{\partial x} + \frac{\partial N}{\partial y} \right) dA = 0.$$

Generalizations

Multiple Dimensions

One can use the general Stokes' Theorem to equate the n-dimensional volume integral of the divergence of a vector field F over a region U to the $(n-1)$-dimensional surface integral of F over the boundary of U:

$$\underbrace{\int \cdots \int_U \nabla \cdot \mathbf{F}\, dV}_{n} = \underbrace{\oint \cdots \oint_{\partial U} \mathbf{F} \cdot \mathbf{n}\, dS}_{n-1}$$

This equation is also known as the Divergence theorem.

When $n = 2$, this is equivalent to Green's theorem.

When $n = 1$, it reduces to the Fundamental theorem of calculus.

Tensor Fields

Writing the theorem in Einstein notation:

$$\iiint_V \frac{\partial \mathbf{F}_i}{\partial x_i} dV = \oiint_S \mathbf{F}_i n_i \, dS$$

suggestively, replacing the vector field F with a rank-n tensor field T, this can be generalized to:

$$\iiint_V \frac{\partial T_{i_1 i_2 \cdots i_q \cdots i_n}}{\partial x_{i_q}} dV = \oiint_S T_{i_1 i_2 \cdots i_q \cdots i_n} n_{i_q} \, dS.$$

where on each side, tensor contraction occurs for at least one index. This form of the theorem is still in 3d, each index takes values 1, 2, and 3. It can be generalized further still to higher (or lower) dimensions (for example to 4d spacetime in general relativity).

Green's Theorem

In mathematics, Green's theorem gives the relationship between a line integral around a simple closed curve C and a double integral over the plane region D bounded by C. It is named after George Green and is the two-dimensional special case of the more general Kelvin–Stokes theorem.

Theorem

Let C be a positively oriented, piecewise smooth, simple closed curve in a plane, and let D be the region bounded by C. If L and M are functions of (x, y) defined on an open region containing D and have continuous partial derivatives there, then

$$\oint_C (L\,dx + M\,dy) = \iint_D \left(\frac{\partial M}{\partial x} - \frac{\partial L}{\partial y} \right) dx\,dy$$

where the path of integration along C is counterclockwise.

In physics, Green's theorem is mostly used to solve two-dimensional flow integrals, stating that the sum of fluid outflows from a volume is equal to the total outflow summed about an enclosing area. In plane geometry, and in particular, area surveying, Green's theorem can be used to determine the area and centroid of plane figures solely by integrating over the perimeter.

Proof When D is a Simple Region

The following is a proof of half of the theorem for the simplified area D, a type I region where C_1 and C_3 are curves connected by vertical lines (possibly of zero length). A similar proof exists for the other half of the theorem when D is a type II region where C_2 and C_4 are curves connected by horizontal lines (again, possibly of zero length). Putting these two parts together, the theorem is thus proven for regions of type III (defined as regions which are both type I and type II). The general case can then be deduced from this special case by decomposing D into a set of type III regions.

If it can be shown that if

$$\oint_C L\,dx = \iint_D \left(-\frac{\partial L}{\partial y}\right)dA \qquad (1)$$

and

$$\oint_C M\,dy = \iint_D \left(\frac{\partial M}{\partial x}\right)dA \qquad (2)$$

are true, then Green's theorem follows immediately for the region D. We can prove (1) easily for regions of type I, and (2) for regions of type II. Green's theorem then follows for regions of type III.

Assume region D is a type I region and can thus be characterized, as pictured on the right, by

$$D = \{(x, y) \mid a \le x \le b, g_1(x) \le y \le g_2(x)\}$$

where g_1 and g_2 are continuous functions on $[a, b]$. Compute the double integral in (1):

$$\iint_D \frac{\partial L}{\partial y}dA = \int_a^b \int_{g_1(x)}^{g_2(x)} \frac{\partial L}{\partial y}(x, y)\,dy\,dx$$

$$= \int_a^b \{L(x, g_2(x)) - L(x, g_1(x))\}\,dx. \qquad (3)$$

Now compute the line integral in (1). C can be rewritten as the union of four curves: C_1, C_2, C_3, C_4.

With C_1, use the parametric equations: $x = x$, $y = g_1(x)$, $a \le x \le b$. Then

$$\int_{C_1} L(x, y)\,dx = \int_a^b L(x, g_1(x))\,dx.$$

With C_3, use the parametric equations: $x = x$, $y = g_2(x)$, $a \le x \le b$. Then

$$\int_{C_3} L(x, y)\,dx = -\int_{-C_3} L(x, y)\,dx = -\int_a^b L(x, g_2(x))\,dx.$$

The integral over C_3 is negated because it goes in the negative direction from b to a, as C is oriented positively (counterclockwise). On C_2 and C_4, x remains constant, meaning

$$\int_{C_4} L(x, y)\,dx = \int_{C_2} L(x, y)\,dx = 0.$$

Therefore,

$$\int_C L\,dx = \int_{C_1} L(x, y)\,dx + \int_{C_2} L(x, y)\,dx + \int_{C_3} L(x, y)\,dx + \int_{C_4} L(x, y)\,dx$$

$$= -\int_a^b L(x, g_2(x))\,dx + \int_a^b L(x, g_1(x))\,dx. \qquad (4)$$

Combining (3) with (4), we get (1) for regions of type I. A similar treatment yields (2) for regions of type II. Putting the two together, we get the result for regions of type III.

Relationship to Stokes' Theorem

Green's theorem is a special case of the Kelvin–Stokes theorem, when applied to a region in the xy-plane:

We can augment the two-dimensional field into a three-dimensional field with a z component that is always 0. Write F for the vector-valued function $\mathbf{F} = (L, M, 0)$.. Start with the left side of Green's theorem:

$$\oint_C (L\,dx + M\,dy) = \oint_C (L, M, 0) \cdot (dx, dy, dz) = \oint_C \mathbf{F} \cdot d\mathbf{r}.$$

Kelvin–Stokes Theorem:

$$\oint_C \mathbf{F} \cdot d\mathbf{r} = \iint_S \nabla \times \mathbf{F} \cdot \hat{\mathbf{n}}\,dS.$$

The surface S is just the region in the plane D, with the unit normals $\hat{\mathbf{n}}$ pointing up (in the positive z direction) to match the "positive orientation" definitions for both theorems.

The expression inside the integral becomes

$$\nabla \times \mathbf{F} \cdot \hat{\mathbf{n}} = \left[\left(\frac{\partial 0}{\partial y} - \frac{\partial M}{\partial z} \right)\mathbf{i} + \left(\frac{\partial L}{\partial z} - \frac{\partial 0}{\partial x} \right)\mathbf{j} + \left(\frac{\partial M}{\partial x} - \frac{\partial L}{\partial y} \right)\mathbf{k} \right] \cdot \mathbf{k} = \left(\frac{\partial M}{\partial x} - \frac{\partial L}{\partial y} \right).$$

Thus we get the right side of Green's theorem

$$\iint_S \nabla \times \mathbf{F} \cdot \hat{\mathbf{n}}\,dS = \iint_D \left(\frac{\partial M}{\partial x} - \frac{\partial L}{\partial y} \right) dA.$$

Green's theorem is also a straightforward result of the general Stokes' theorem using differential forms and exterior derivatives:

$$\oint_C L\,dx + M\,dy = \oint_{\partial D} \omega = \int_D d\omega = \int_D \frac{\partial L}{\partial y}\,dy \wedge dx + \frac{\partial M}{\partial x}\,dx \wedge dy = \iint_D \left(\frac{\partial M}{\partial x} - \frac{\partial L}{\partial y} \right) dx\,dy.$$

Relationship to the Divergence Theorem

Considering only two-dimensional vector fields, Green's theorem is equivalent to the two-dimensional version of the divergence theorem:

$$\iint_D (\nabla \cdot \mathbf{F})\,dA = \oint_C \mathbf{F} \cdot \hat{\mathbf{n}}\,ds,$$

where $\nabla \cdot \mathbf{F}$ is the divergence on the two-dimensional vector field $\mathbf{F}$, and $\hat{\mathbf{n}}$ is the outward-pointing unit normal vector on the boundary.

To see this, consider the unit normal $\hat{\mathbf{n}}$ in the right side of the equation. Since in Green's theorem $d\mathbf{r} = (dx, dy)$ is a vector pointing tangential along the curve, and the curve C is the positively oriented (i.e. counterclockwise) curve along the boundary, an outward normal would be a vector which points 90° to the right of this; one choice would be $(dy, -dx)$. The length of this vector is

$$\sqrt{dx^2 + dy^2} = ds. \text{ So } (dy, -dx) = \hat{\mathbf{n}}ds.$$

Start with the left side of Green's theorem:

$$\oint_C (L\,dx + M\,dy) = \oint_C (M, -L) \cdot (dy, -dx) = \oint_C (M, -L) \cdot \hat{\mathbf{n}}\,ds.$$

Applying the two-dimensional divergence theorem with $\mathbf{F} = (M, -L)$, we get the right side of Green's theorem:

$$\oint_C (M, -L) \cdot \hat{\mathbf{n}}\,ds = \iint_D (\nabla \cdot (M, -L))\,dA = \iint_D \left(\frac{\partial M}{\partial x} - \frac{\partial L}{\partial y} \right) dA.$$

Area Calculation

Green's theorem can be used to compute area by line integral. The area of D is given by $A = \iint_D dA$. Then if we choose L and M such that $\dfrac{\partial M}{\partial x} - \dfrac{\partial L}{\partial y} = 1$, the area is given by $A = \oint_C (L\,dx + M\,dy)$. Possible formulas for the area of D include: $A = \oint_C x\,dy = -\oint_C y\,dx = \frac{1}{2}\oint_C (-y\,dx + x\,dy)$.

References

- Galbis, Antonio & Maestre, Manuel (2012). Vector Analysis Versus Vector Calculus. Springer. p. 12. ISBN 978-1-4614-2199-3. CS1 maint: Uses authors parameter (link)

- Richard Courant; Fritz John (14 December 1999). Introduction to Calculus and Analysis Volume II/2. Springer Science & Business Media. ISBN 978-3-540-66570-0.

- Stewart, James (2008), "Vector Calculus", Calculus: Early Transcendentals (6 ed.), Thomson Brooks/Cole, ISBN 978-0-495-01166-8

- Byron, Frederick; Fuller, Robert (1992), Mathematics of Classical and Quantum Physics, Dover Publications, p. 22, ISBN 978-0-486-67164-2

- M. R. Spiegel; S. Lipschutz; D. Spellman (2009). Vector Analysis. Schaum's Outlines (2nd ed.). USA: McGraw Hill. ISBN 978-0-07-161545-7.

- K.F. Riley; M.P. Hobson; S.J. Bence (2010). Mathematical methods for physics and engineering. Cambridge University Press. ISBN 978-0-521-86153-3.

- Mathematical methods for physics and engineering, K.F. Riley, M.P. Hobson, S.J. Bence, Cambridge University Press, 2010, ISBN 978-0-521-86153-3

- Vector Analysis (2nd Edition), M.R. Spiegel, S. Lipschutz, D. Spellman, Schaum's Outlines, McGraw Hill (USA), 2009, ISBN 978-0-07-161545-7

5

Understanding Fourier Series

This section discuses Fourier series in a critical manner providing key analysis to the subject matter. Fourier series characterize functions as the total of simple sine waves. Some of the aspects discussed herein are Fourier transform, convergence of Fourier series and pontryagin duality.

Fourier Series

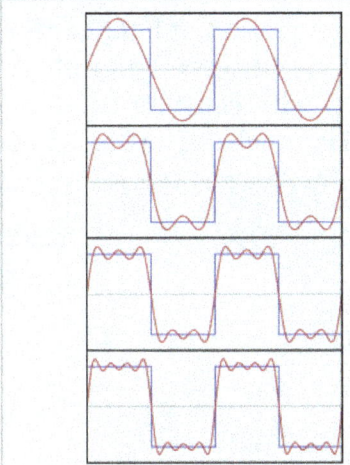

Each arrow starts at the vertical sum of all the arrows on its left. And the purple dot is the sum of all six. The arrows represent the amplitudes of sine functions with different peak-values and frequencies. They are the first six terms of a Fourier series derived from the square wave motion of the blue dot, which transitions between only two amplitudes every 2 seconds.

The first four partial sums of the Fourier series for a square wave

In mathematics, a Fourier series is a way to represent a (wave-like) function as the sum of simple sine waves. More formally, it decomposes any periodic function or periodic signal into the sum of a (possibly infinite) set of simple oscillating functions, namely sines and cosines (or, equivalently, complex exponentials). The discrete-time Fourier transform is a periodic function, often defined in terms of a Fourier series. The Z-transform, another example of application, reduces to a Fourier series for the important case $|z|=1$. Fourier series are also central to the original proof of the Nyquist–Shannon sampling theorem. The study of Fourier series is a branch of Fourier analysis.

History

The Fourier series is named in honour of Jean-Baptiste Joseph Fourier (1768–1830), who made important contributions to the study of trigonometric series, after preliminary investigations by Leonhard Euler, Jean le Rond d'Alembert, and Daniel Bernoulli. Fourier introduced the series for the purpose of solving the heat equation in a metal plate, publishing his initial results in his 1807 *Mémoire sur la propagation de la chaleur dans les corps solides* (*Treatise on the propagation of heat in solid bodies*), and publishing his *Théorie analytique de la chaleur* (*Analytical theory of heat*) in 1822. Early ideas of decomposing a periodic function into the sum of simple oscillating functions date back to the 3rd century BC, when ancient astronomers proposed an empiric model of planetary motions, based on deferents and epicycles.

The heat equation is a partial differential equation. Prior to Fourier's work, no solution to the heat equation was known in the general case, although particular solutions were known if the heat source behaved in a simple way, in particular, if the heat source was a sine or cosine wave. These simple solutions are now sometimes called eigensolutions. Fourier's idea was to model a complicated heat source as a superposition (or linear combination) of simple sine and cosine waves, and to write the solution as a superposition of the corresponding eigensolutions. This superposition or linear combination is called the Fourier series.

From a modern point of view, Fourier's results are somewhat informal, due to the lack of a precise notion of function and integral in the early nineteenth century. Later, Peter Gustav Lejeune Dirichlet and Bernhard Riemann expressed Fourier's results with greater precision and formality.

Although the original motivation was to solve the heat equation, it later became obvious that the same techniques could be applied to a wide array of mathematical and physical problems, and especially those involving linear differential equations with constant coefficients, for which the eigensolutions are sinusoids. The Fourier series has many such applications in electrical engineering, vibration analysis, acoustics, optics, signal processing, image processing, quantum mechanics, econometrics, thin-walled shell theory, etc.

Definition

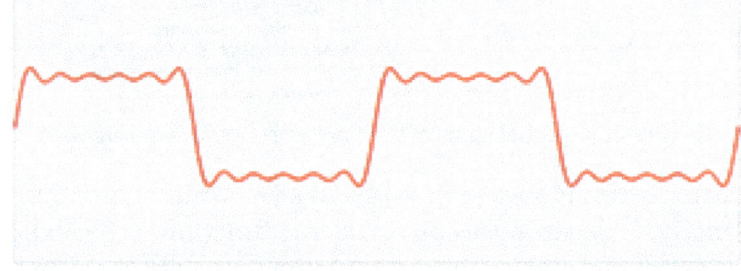

Function $s(x)$ (in red) is a sum of six sine functions of different amplitudes and harmonically related frequencies. Their summation is called a Fourier series. The Fourier transform, $S(f)$ (in blue), which depicts amplitude vs frequency, reveals the 6 frequencies and their amplitudes.

In this section, $s(x)$ denotes a function of the real variable x, and s is integrable on an interval $[x_0, x_0 + P]$, for real numbers x_0 and P. We will attempt to represent s in that interval as an infinite sum, or series, of harmonically related sinusoidal functions. Outside the interval, the series is periodic with period P (frequency $1/P$). It follows that if s also has that property, the approximation is valid on the entire real line. We can begin with a finite summation (or *partial sum*):

$$s_N(x) = \frac{A_0}{2} + \sum_{n=1}^{N} A_n \cdot \sin(\tfrac{2\pi nx}{P} + \phi_n), \quad \text{for integer } N \geq 1.$$

$s_N(x)$ is a periodic function with period P. Using the identities:

$$\sin(\tfrac{2\pi nx}{P} + \phi_n) \equiv \sin(\phi_n)\cos(\tfrac{2\pi nx}{P}) + \cos(\phi_n)\sin(\tfrac{2\pi nx}{P})$$

$$\sin(\tfrac{2\pi nx}{P} + \phi_n) \equiv \text{Re}\left\{\frac{1}{i} \cdot e^{i\left(\frac{2\pi nx}{P} + \phi_n\right)}\right\} = \frac{1}{2i} \cdot e^{i\left(\frac{2\pi nx}{P} + \phi_n\right)} + \left(\frac{1}{2i} \cdot e^{i\left(\frac{2\pi nx}{P} + \phi_n\right)}\right)^*,$$

we can also write the function in these equivalent forms:

$$s_N(x) = \overbrace{a_0}^{A_0}/2 + \sum_{n=1}^{N} \left(\overbrace{a_n}^{A_n\sin(\phi_n)}\cos(\tfrac{2\pi nx}{P}) + \overbrace{b_n}^{A_n\cos(\phi_n)}\sin(\tfrac{2\pi nx}{P})\right)$$

$$= \sum_{n=-N}^{N} c_n \cdot e^{i\frac{2\pi nx}{P}},$$

where:

$$c_n \overset{\text{def}}{=} \begin{cases} \dfrac{A_n}{2i}e^{i\phi_n} = \dfrac{1}{2}(a_n - ib_n) & \text{for } n > 0 \\[2mm] \dfrac{1}{2}A_0 = \dfrac{1}{2}a_0 & \text{for } n = 0 \\[2mm] c_{|n|}^* & \text{for } n < 0. \end{cases}$$

The inverse relationships between the coefficients are:

$$A_n = \sqrt{a_n^2 + b_n^2} \quad \phi_n = \text{atan}\,2\left(a_n, b_n\right).$$

When the coefficients (known as Fourier coefficients) are computed as follows:

$$a_n = \frac{2}{P}\int_{x_0}^{x_0+P} s(x) \cdot \cos(\tfrac{2\pi nx}{P})\, dx$$

$$c_n = \frac{1}{P}\int_{x_0}^{x_0+P} s(x) \cdot e^{-i\frac{2\pi nx}{P}}\, dx,$$

$$b_n = \frac{2}{P}\int_{x_0}^{x_0+P} s(x) \cdot \sin(\tfrac{2\pi nx}{P})\, dx$$

$s_N(x)$ approximates $s(x)$ on $[x_0, x_0 + P]$, and the approximation improves as $N \to \infty$. The infinite sum, $s_\infty(x)$, is called the Fourier series representation of s.

Complex-valued Functions

Both components of a complex-valued function are real-valued functions that can be represented by a Fourier series. The two sets of coefficients and the partial sum are given by:

$$C_{Rn} = \frac{1}{P}\int_{x_0}^{x_0+P} \text{Re}\{s(x)\} \cdot e^{-i\frac{2\pi nx}{P}}\, dx \quad \text{and} \quad C_{In} = \frac{1}{P}\int_{x_0}^{x_0+P} \text{Im}\{s(x)\} \cdot e^{-i\frac{2\pi nx}{P}}\, dx$$

$$s_N(x) = \sum_{n=-N}^{N} C_{Rn} \cdot e^{i\frac{2\pi nx}{P}} + i \cdot \sum_{n=-N}^{N} C_{In} \cdot e^{i\frac{2\pi nx}{P}} = \sum_{n=-N}^{N} \underbrace{\left(C_{Rn} + i \cdot C_{In}\right)}_{C_n} \cdot e^{i\frac{2\pi nx}{P}}.$$

This is the same formula as before except c_n and c_{-n} are no longer complex conjugates. The formula for c_n is also unchanged:

$$c_n = \frac{1}{P} \int_{x_0}^{x_0+P} \mathrm{Re}\{s(x)\} \cdot e^{-i\frac{2\pi nx}{P}} \, dx + i \cdot \frac{1}{P} \int_{x_0}^{x_0+P} \mathrm{Im}\{s(x)\} \cdot e^{-i\frac{2\pi nx}{P}} \, dx$$

$$= \frac{1}{P} \int_{x_0}^{x_0+P} \left(\mathrm{Re}\{s(x)\} + i \cdot \mathrm{Im}\{s(x)\}\right) \cdot e^{-i\frac{2\pi nx}{P}} \, dx = \frac{1}{P} \int_{x_0}^{x_0+P} s(x) \cdot e^{-i\frac{2\pi nx}{P}} \, dx.$$

Convergence

In engineering applications, the Fourier series is generally presumed to converge everywhere except at discontinuities, since the functions encountered in engineering are more well behaved than the ones that mathematicians can provide as counter-examples to this presumption. In particular, the Fourier series converges absolutely and uniformly to $s(x)$ whenever the derivative of $s(x)$ (which may not exist everywhere) is square integrable. If a function is square-integrable on the interval $[x_o, x_o+P]$, then the Fourier series converges to the function at *almost every* point. Convergence of Fourier series also depends on the finite number of maxima and minima in a function which is popularly known as one of the Dirichlet's condition for Fourier series. It is possible to define Fourier coeffi-cients for more general functions or distributions, in such cases convergence in norm or weak convergence is usually of interest.

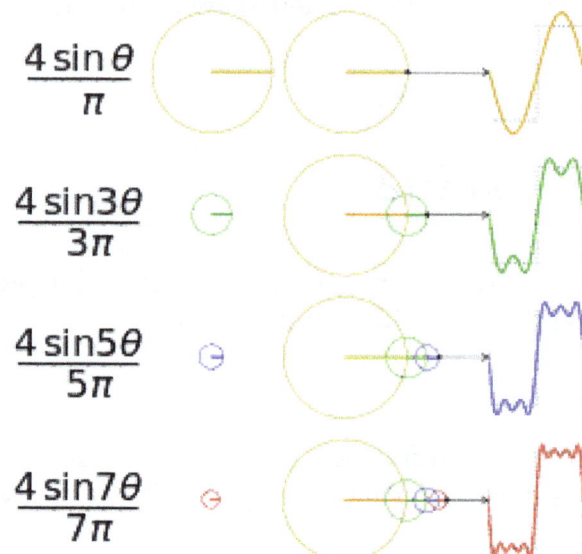

$$\frac{4\sin\theta}{\pi}$$

$$\frac{4\sin 3\theta}{3\pi}$$

$$\frac{4\sin 5\theta}{5\pi}$$

$$\frac{4\sin 7\theta}{7\pi}$$

Another visualisation of an approximation of a square wave by taking the first 1, 2, 3 and 4 terms of its Fourier series. (An interactive animation can be seen here)

$$\frac{2\sin\theta}{-\pi}$$

$$\frac{2\sin2\theta}{2\pi}$$

$$\frac{2\sin3\theta}{-3\pi}$$

$$\frac{2\sin4\theta}{4\pi}$$

A visualisation of an approximation of a sawtooth wave of the same amplitude and frequency for comparison

Example 1: a simple Fourier Series

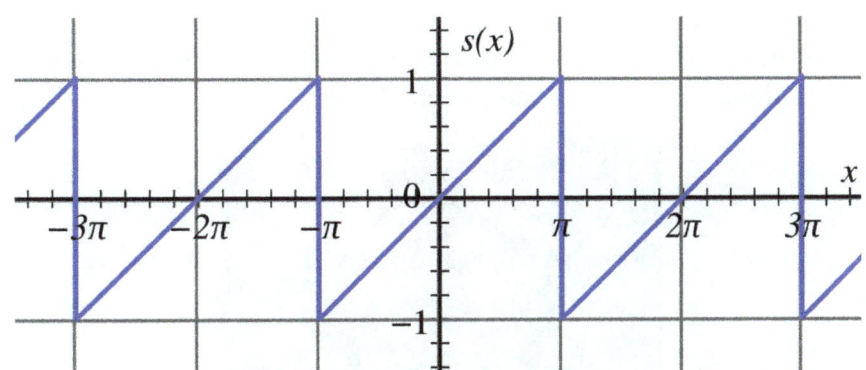

Plot of the sawtooth wave, a periodic continuation of the linear function on the interval

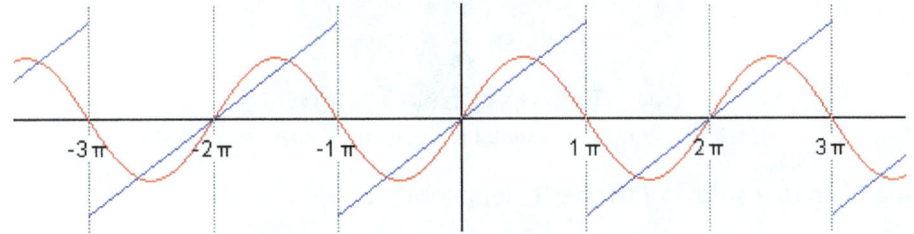

Animated plot of the first five successive partial Fourier series

We now use the formula above to give a Fourier series expansion of a very simple function. Consider a sawtooth wave

$$s(x) = \frac{x}{\pi}, \quad \text{for } -\pi < x < \pi,$$

$$s(x + 2\pi k) = s(x), \quad \text{for } -\infty < x < \infty \text{ and } k \in \mathbb{Z}.$$

In this case, the Fourier coefficients are given by

$$a_n = \frac{1}{\pi}\int_{-\pi}^{\pi} s(x)\cos(nx)dx = 0, \quad n \geq 0.$$

$$
\begin{aligned}
b_n &= \frac{1}{\pi}\int_{-\pi}^{\pi} s(x)\sin(nx)dx \\
&= -\frac{2}{\pi n}\cos(n\pi) + \frac{2}{\pi^2 n^2}\sin(n\pi) \\
&= \frac{2(-1)^{n+1}}{\pi n}, \quad n \geq 1.
\end{aligned}
$$

It can be proven that Fourier series converges to $s(x)$ at every point x where s is differentiable, and therefore:

$$
\begin{aligned}
s(x) &= \frac{a_0}{2} + \sum_{n=1}^{\infty}\left[a_n\cos\left(nx\right) + b_n\sin\left(nx\right)\right] \\
&= \frac{2}{\pi}\sum_{n=1}^{\infty}\frac{(-1)^{n+1}}{n}\sin(nx), \quad \text{for} \quad x - \pi \notin 2\pi\mathbf{Z}.
\end{aligned}
$$

(Eq.1)

When $x = \pi$, the Fourier series converges to 0, which is the half-sum of the left- and right-limit of s at $x = \pi$. This is a particular instance of the Dirichlet theorem for Fourier series.

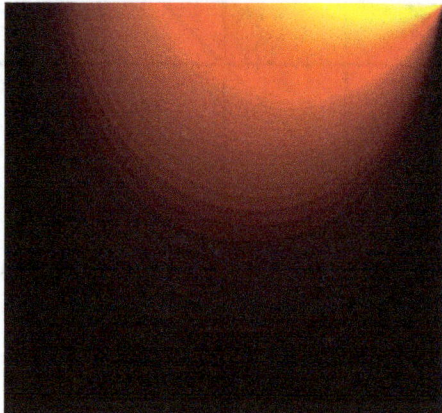

Heat distribution in a metal plate, using Fourier's method

This example leads us to a solution to the Basel problem.

Example 2: Fourier's Motivation

The Fourier series expansion of our function in Example 1 looks more complicated than the simple formula $s(x) = x/\pi$, so it is not immediately apparent why one would need the Fourier series. While there are many applications, Fourier's motivation was in solving the heat equation. For example, consider a metal plate in the shape of a square whose side measures π meters, with coordinates $(x, y) \in [0, \pi] \times [0, \pi]$. If there is no heat source within the plate, and if three of the four sides are held at 0 degrees Celsius, while the fourth side, given by $y = \pi$, is maintained at the temperature gradient $T(x, \pi) = x$ degrees Celsius, for x in $(0, \pi)$, then one can show that the stationary heat distribution (or the heat distribution after a long period of time has elapsed) is given by

$$T(x,y) = 2\sum_{n=1}^{\infty} \frac{(-1)^{n+1}}{n} \sin(nx) \frac{\sinh(ny)}{\sinh(n\pi)}.$$

Here, sinh is the hyperbolic sine function. This solution of the heat equation is obtained by multiplying each term of Eq.1 by $\sinh(ny)/\sinh(n\pi)$. While our example function $s(x)$ seems to have a needlessly complicated Fourier series, the heat distribution $T(x, y)$ is nontrivial. The function T cannot be written as a closed-form expression. This method of solving the heat problem was made possible by Fourier's work.

Other Applications

Another application of this Fourier series is to solve the Basel problem by using Parseval's theorem. The example generalizes and one may compute $\zeta(2n)$, for any positive integer n.

Other Common Notations

The notation c_n is inadequate for discussing the Fourier coefficients of several different functions. Therefore, it is customarily replaced by a modified form of the function (s, in this case), such as $\hat{s}$ or S, and functional notation often replaces subscripting:

$$\begin{aligned} s_{\infty}(x) &= \sum_{n=-\infty}^{\infty} \hat{s}(n)\cdot e^{i\frac{2\pi nx}{P}} \\ &= \sum_{n=-\infty}^{\infty} S[n]\cdot e^{j\frac{2\pi nx}{P}} \qquad \text{common engineering notation} \end{aligned}$$

In engineering, particularly when the variable x represents time, the coefficient sequence is called a frequency domain representation. Square brackets are often used to emphasize that the domain of this function is a discrete set of frequencies.

Another commonly used frequency domain representation uses the Fourier series coefficients to modulate a Dirac comb:

$$S(f) \stackrel{\text{def}}{=} \sum_{n=-\infty}^{\infty} S[n]\cdot \delta\left(f - \frac{n}{P}\right),$$

where f represents a continuous frequency domain. When variable x has units of seconds, f has units of hertz. The "teeth" of the comb are spaced at multiples (i.e. harmonics) of $1/P$, which is called the fundamental frequency. $s_{\infty}(x)$ can be recovered from this representation by an inverse Fourier transform:

$$\begin{aligned} \mathcal{F}^{-1}\{S(f)\} &= \int_{-\infty}^{\infty} \left(\sum_{n=-\infty}^{\infty} S[n]\cdot \delta\left(f - \frac{n}{P}\right)\right) e^{i2\pi fx}\, df, \\ &= \sum_{n=-\infty}^{\infty} S[n]\cdot \int_{-\infty}^{\infty} \delta\left(f - \frac{n}{P}\right) e^{i2\pi fx}\, df, \\ &= \sum_{n=-\infty}^{\infty} S[n]\cdot e^{i\frac{2\pi nx}{P}} \stackrel{\text{def}}{=} s_{\infty}(x). \end{aligned}$$

The constructed function $S(f)$ is therefore commonly referred to as a Fourier transform, even though the Fourier integral of a periodic function is not convergent at the harmonic frequencies.

Beginnings

"

$$\varphi(y) = a_0 \cos\frac{\pi y}{2} + a_1 \cos 3\frac{\pi y}{2} + a_2 \cos 5\frac{\pi y}{2} + \cdots.$$

Multiplying both sides by $\cos(2k+1)\dfrac{\pi y}{2}$, and then integrating from $y = -1$ to $y = +1$ yields:

$$a_k = \int_{-1}^{1} \varphi(y) \cos(2k+1)\frac{\pi y}{2} dy.$$

"

— Joseph Fourier, Mémoire sur la propagation de la chaleur dans les corps solides.
(1807)

This immediately gives any coefficient a_k of the trigonometrical series for $\phi(y)$ for any function which has such an expansion. It works because if ϕ has such an expansion, then (under suitable convergence assumptions) the integral

$$
\begin{aligned}
a_k &= \int_{-1}^{1} \varphi(y) \cos(2k+1)\frac{\pi y}{2} dy \\
&= \int_{-1}^{1} \left(a \cos\frac{\pi y}{2} \cos(2k+1)\frac{\pi y}{2} + a' \cos 3\frac{\pi y}{2} \cos(2k+1)\frac{\pi y}{2} + \cdots \right) dy
\end{aligned}
$$

can be carried out term-by-term. But all terms involving $\cos(2j+1)\dfrac{\pi y}{2}\cos(2k+1)\dfrac{\pi y}{2}$ for $j \neq k$ vanish when integrated from −1 to 1, leaving only the kth term.

In these few lines, which are close to the modern formalism used in Fourier series, Fourier revolutionized both mathematics and physics. Although similar trigonometric series were previously used by Euler, d'Alembert, Daniel Bernoulli and Gauss, Fourier believed that such trigonometric series could represent any arbitrary function. In what sense that is actually true is a somewhat subtle issue and the attempts over many years to clarify this idea have led to important discoveries in the theories of convergence, function spaces, and harmonic analysis.

When Fourier submitted a later competition essay in 1811, the committee (which included Lagrange, Laplace, Malus and Legendre, among others) concluded: *...the manner in which the author arrives at these equations is not exempt of difficulties and...his analysis to integrate them still leaves something to be desired on the score of generality and even rigour.*

Birth of Harmonic Analysis

Since Fourier's time, many different approaches to defining and understanding the concept of Fourier series have been discovered, all of which are consistent with one another, but each of which emphasizes different aspects of the topic. Some of the more powerful and elegant approaches are based on mathematical ideas and tools that were not available at the time Fourier completed his original work. Fourier originally defined the Fourier series for real-valued functions of real arguments, and using the sine and cosine functions as the basis set for the decomposition.

Many other Fourier-related transforms have since been defined, extending the initial idea to other applications. This general area of inquiry is now sometimes called harmonic analysis. A Fourier series, however, can be used only for periodic functions, or for functions on a bounded (compact) interval.

Extensions

Fourier Series on a Square

We can also define the Fourier series for functions of two variables x and y in the square $[-\pi, \pi] \times [-\pi, \pi]$:

$$f(x, y) = \sum_{j,k \in \mathbf{Z} \text{ (integers)}} c_{j,k} e^{ijx} e^{iky},$$

$$c_{j,k} = \frac{1}{4\pi^2} \int_{-\pi}^{\pi} \int_{-\pi}^{\pi} f(x, y) e^{-ijx} e^{-iky} \, dx dy.$$

Aside from being useful for solving partial differential equations such as the heat equation, one notable application of Fourier series on the square is in image compression. In particular, the jpeg image compression standard uses the two-dimensional discrete cosine transform, which is a Fourier transform using the cosine basis functions.

Fourier Series of Bravais-lattice-periodic-function

The Bravais lattice is defined as the set of vectors of the form:

$$\mathbf{R} = n_1 \mathbf{a}_1 + n_2 \mathbf{a}_2 + n_3 \mathbf{a}_3$$

where n_i are integers and a_i are three linearly independent vectors. Assuming we have some function, $f(\mathbf{r})$, such that it obeys the following condition for any Bravais lattice vector R: $f(\mathbf{r}) = f(\mathbf{r} + \mathbf{R})$, we could make a Fourier series of it. This kind of function can be, for example, the effective potential that one electron "feels" inside a periodic crystal. It is useful to make a Fourier series of the potential then when applying Bloch's theorem. First, we may write any arbitrary vector r in the coordinate-system of the lattice:

$$\mathbf{r} = x_1 \frac{\mathbf{a}_1}{a_1} + x_2 \frac{\mathbf{a}_2}{a_2} + x_3 \frac{\mathbf{a}_3}{a_3},$$

where $a_i = |\mathbf{a}_i|$.

Thus we can define a new function,

$$g(x_1, x_2, x_3) := f(\mathbf{r}) = f\left(x_1 \frac{\mathbf{a}_1}{a_1} + x_2 \frac{\mathbf{a}_2}{a_2} + x_3 \frac{\mathbf{a}_3}{a_3} \right).$$

This new function, $g(x_1, x_2, x_3)$, is now a function of three-variables, each of which has periodicity a_1, a_2, a_3 respectively: $g(x_1, x_2, x_3) = g(x_1 + a_1, x_2, x_3) = g(x_1, x_2 + a_2, x_3) = g(x_1, x_2, x_3 + a_3)$. If we write a series for g on the interval $[0, a_1]$ for x_1, we can define the following:

$$h^{\text{one}}(m_1, x_2, x_3) := \frac{1}{a_1} \int_0^{a_1} g(x_1, x_2, x_3) \cdot e^{-i2\pi \frac{m_1}{a_1} x_1} \, dx_1$$

And then we can write:

$$g(x_1, x_2, x_3) = \sum_{m_1=-\infty}^{\infty} h^{\text{one}}(m_1, x_2, x_3) \cdot e^{i2\pi \frac{m_1}{a_1} x_1}$$

Further defining:

$$h^{\text{two}}(m_1, m_2, x_3) := \frac{1}{a_2} \int_0^{a_2} h^{\text{one}}(m_1, x_2, x_3) \cdot e^{-i2\pi \frac{m_2}{a_2} x_2} \, dx_2$$

$$[12pt] \qquad = \frac{1}{a_2} \int_0^{a_2} dx_2 \frac{1}{a_1} \int_0^{a_1} dx_1 g(x_1, x_2, x_3) \cdot e^{-i2\pi \left(\frac{m_1}{a_1} x_1 + \frac{m_2}{a_2} x_2 \right)}$$

We can write g once again as:

$$g(x_1, x_2, x_3) = \sum_{m_1=-\infty}^{\infty} \sum_{m_2=-\infty}^{\infty} h^{\text{two}}(m_1, m_2, x_3) \cdot e^{i2\pi \frac{m_1}{a_1} x_1} \cdot e^{i2\pi \frac{m_2}{a_2} x_2}$$

Finally applying the same for the third coordinate, we define:

$$h^{\text{three}}(m_1, m_2, m_3) := \frac{1}{a_3} \int_0^{a_3} h^{\text{two}}(m_1, m_2, x_3) \cdot e^{-i2\pi \frac{m_3}{a_3} x_3} \, dx_3$$

$$= \frac{1}{a_3} \int_0^{a_3} dx_3 \frac{1}{a_2} \int_0^{a_2} dx_2 \frac{1}{a_1} \int_0^{a_1} dx_1 g(x_1, x_2, x_3) \cdot e^{-i2\pi \left(\frac{m_1}{a_1} x_1 + \frac{m_2}{a_2} x_2 + \frac{m_3}{a_3} x_3 \right)}$$

We write g as:

$$g(x_1, x_2, x_3) = \sum_{m_1=-\infty}^{\infty} \sum_{m_2=-\infty}^{\infty} \sum_{m_3=-\infty}^{\infty} h^{\text{three}}(m_1, m_2, m_3) \cdot e^{i2\pi \frac{m_1}{a_1} x_1} \cdot e^{i2\pi \frac{m_2}{a_2} x_2} \cdot e^{i2\pi \frac{m_3}{a_3} x_3}$$

Re-arranging:

$$g(x_1, x_2, x_3) = \sum_{m_1, m_2, m_3 \in \mathbb{Z}} h^{\text{three}}(m_1, m_2, m_3) \cdot e^{i2\pi \left(\frac{m_1}{a_1} x_1 + \frac{m_2}{a_2} x_2 + \frac{m_3}{a_3} x_3 \right)}.$$

Now, every *reciprocal* lattice vector can be written as $\mathbf{K} = l_1\mathbf{g}_1 + l_2\mathbf{g}_2 + l_3\mathbf{g}_3$, where l_i are integers and $\mathbf{g}_i$ are the reciprocal lattice vectors, we can use the fact that $\mathbf{g}_i \cdot \mathbf{a}_j = 2\pi\delta_{ij}$ to calculate that for any arbitrary reciprocal lattice vector K and arbitrary vector in space r, their scalar product is:

$$\mathbf{K} \cdot \mathbf{r} = \left(l_1\mathbf{g}_1 + l_2\mathbf{g}_2 + l_3\mathbf{g}_3 \right) \cdot \left(x_1 \frac{\mathbf{a}_1}{a_1} + x_2 \frac{\mathbf{a}_2}{a_2} + x_3 \frac{\mathbf{a}_3}{a_3} \right) = 2\pi \left(x_1 \frac{l_1}{a_1} + x_2 \frac{l_2}{a_2} + x_3 \frac{l_3}{a_3} \right).$$

And so it is clear that in our expansion, the sum is actually over reciprocal lattice vectors:

$$f(\mathbf{r}) = \sum_{\mathbf{K}} h(\mathbf{K}) \cdot e^{i\mathbf{K}\cdot\mathbf{r}},$$

where

$$h(\mathbf{K}) = \frac{1}{a_3}\int_0^{a_3} dx_3 \frac{1}{a_2}\int_0^{a_2} dx_2 \frac{1}{a_1}\int_0^{a_1} dx_1 f\left(x_1 \frac{\mathbf{a}_1}{a_1} + x_2 \frac{\mathbf{a}_2}{a_2} + x_3 \frac{\mathbf{a}_3}{a_3} \right) \cdot e^{-i\mathbf{K}\cdot\mathbf{r}}.$$

Assuming

$$\mathbf{r} = (x, y, z) = x_1 \frac{\mathbf{a}_1}{a_1} + x_2 \frac{\mathbf{a}_2}{a_2} + x_3 \frac{\mathbf{a}_3}{a_3},$$

we can solve this system of three linear equations for x, y, and z in terms of x_1, x_2 and x_3 in order to calculate the volume element in the original cartesian coordinate system. Once we have x, y, and z in terms of x_1, x_2 and x_3, we can calculate the Jacobian determinant:

$$\begin{bmatrix} \dfrac{\partial x_1}{\partial x} & \dfrac{\partial x_1}{\partial y} & \dfrac{\partial x_1}{\partial z} \\[2mm] \dfrac{\partial x_2}{\partial x} & \dfrac{\partial x_2}{\partial y} & \dfrac{\partial x_2}{\partial z} \\[2mm] \dfrac{\partial x_3}{\partial x} & \dfrac{\partial x_3}{\partial y} & \dfrac{\partial x_3}{\partial z} \end{bmatrix}$$

which after some calculation and applying some non-trivial cross-product identities can be shown to be equal to:

$$\frac{a_1 a_2 a_3}{\mathbf{a}_1 \cdot (\mathbf{a}_2 \times \mathbf{a}_3)}$$

(it may be advantageous for the sake of simplifying calculations, to work in such a cartesian coordinate system, in which it just so happens that a_1 is parallel to the x axis, a_2 lies in the *x-y* plane, and a_3 has components of all three axes). The denominator is exactly the volume of the primitive unit cell which is enclosed by the three primitive-vectors a_1, a_2 and a_3. In particular, we now know that

$$dx_1\,dx_2\,dx_3 = \frac{a_1 a_2 a_3}{\mathbf{a}_1 \cdot (\mathbf{a}_2 \times \mathbf{a}_3)} \cdot dx\,dy\,dz.$$

We can write now $h(K)$ as an integral with the traditional coordinate system over the volume of the primitive cell, instead of with the x_1, x_2 and x_3 variables:

$$h(\mathbf{K}) = \frac{1}{\mathbf{a}_1 \cdot (\mathbf{a}_2 \times \mathbf{a}_3)} \int_C d\mathbf{r}f(\mathbf{r}) \cdot e^{-i\mathbf{K}\cdot\mathbf{r}}$$

And C is the primitive unit cell, thus, $\mathbf{a}_1 \cdot (\mathbf{a}_2 \times \mathbf{a}_3)$ is the volume of the primitive unit cell.

Hilbert Space Interpretation

In the language of Hilbert spaces, the set of functions $\{e_n = e^{inx} ; n \in \mathbb{Z}\}$ is an orthonormal basis for the space $L^2([-\pi, \pi])$ of square-integrable functions of $[-\pi, \pi]$. This space is actually a Hilbert space with an inner product given for any two elements f and g by

$$\langle f, g \rangle \overset{\text{def}}{=} \frac{1}{2\pi} \int_{-\pi}^{\pi} f(x)\overline{g(x)}dx.$$

The basic Fourier series result for Hilbert spaces can be written as

$$f = \sum_{n=-\infty}^{\infty} \langle f, e_n \rangle e_n.$$

This corresponds exactly to the complex exponential formulation given above. The version with sines and cosines is also justified with the Hilbert space interpretation. Indeed, the sines and cosines form an orthogonal set:

$$\int_{-\pi}^{\pi} \cos(mx)\cos(nx)dx = \pi\delta_{mn}, \quad m, n \geq 1,$$

$$\int_{-\pi}^{\pi} \sin(mx)\sin(nx)dx = \pi\delta_{mn}, \quad m, n \geq 1$$

(where δ_{mn} is the Kronecker delta), and

$$\int_{-\pi}^{\pi} \cos(mx)\sin(nx)dx = 0;$$

furthermore, the sines and cosines are orthogonal to the constant function 1. An *orthonormal basis* for $L^2([-\pi,\pi])$ consisting of real functions is formed by the functions 1 and $\sqrt{2}\cos(nx)$, $\sqrt{2}\sin(nx)$ with $n = 1, 2,...$ The density of their span is a consequence of the Stone–Weierstrass theorem, but follows also from the properties of classical kernels like the Fejér kernel.

Properties

We say that f belongs to $C^k(\mathbb{T})$ if f is a 2π-periodic function on R which is k times differentiable, and its kth derivative is continuous.

- If f is a 2π-periodic odd function, then $a_n = 0$ for all n.

- If f is a 2π-periodic even function, then $b_n = 0$ for all n.

- If f is integrable, $\lim\limits_{|n|\to\infty} \hat{f}(n) = 0$, , $\lim\limits_{n\to+\infty} a_n = 0$ and $\lim\limits_{n\to+\infty} b_n = 0$. This result is known as the Riemann–Lebesgue lemma.

- A doubly infinite sequence $\{a_n\}$ in $c_0(\mathbf{Z})$ is the sequence of Fourier coefficients of a function in $L^1([0, 2\pi])$ if and only if it is a convolution of two sequences in $\ell^2(\mathbf{Z})$.

- If $f \in C^1(\mathbb{T})$, then the Fourier coefficients $\hat{f}'(n)$ of the derivative f' can be expressed in terms of the Fourier coefficients $\hat{f}(n)$ of the function f, via the formula $\hat{f}'(n) = in\hat{f}(n)$.

- If $f \in C^k(\mathbb{T})$,, then $\widehat{f^{(k)}}(n) = (in)^k \hat{f}(n)$. In particular, since $\widehat{f^{(k)}}(n)$ tends to zero, we have that $|n|^k \hat{f}(n)$ tends to zero, which means that the Fourier coefficients converge to zero faster than the kth power of n.

- Parseval's theorem. If f belongs to $L^2([-\pi, \pi])$, then $\sum\limits_{n=-\infty}^{\infty} |\hat{f}(n)|^2 = \dfrac{1}{2\pi}\int_{-\pi}^{\pi} |f(x)|^2\, dx$.

- Plancherel's theorem. If $c_0, c_{\pm 1}, c_{\pm 2}, \ldots$ are coefficients and $\sum\limits_{n=-\infty}^{\infty} |c_n|^2 < \infty$ then there is a unique function $f \in L^2([-\pi, \pi])$ such that $\hat{f}(n) = c_n$ for every n.

- The first convolution theorem states that if f and g are in $L^1([-\pi, \pi])$, the Fourier series coefficients of the 2π-periodic convolution of f and g are given by:

$$[\widehat{f *_{2\pi} g}](n) = 2\pi \cdot \hat{f}(n) \cdot \hat{g}(n),$$

where:

$$[f *_{2\pi} g](x) \stackrel{\text{def}}{=} \int_{-\pi}^{\pi} f(u) \cdot g[\text{pv}(x-u)]du, \qquad \left(\text{and } \underbrace{\text{pv}(x) \stackrel{\text{def}}{=} \text{Arg}\left(e^{ix}\right)}_{\text{principal value}}\right)$$

$$= \int_{-\pi}^{\pi} f(u) \cdot g(x-u)du, \qquad \text{when g(x) is } 2\pi\text{-periodic.}$$

$$= \int_{2\pi} f(u) \cdot g(x-u)du, \qquad \text{when both functions are } 2\pi\text{-periodic, and the integral is over any } 2\pi \text{ interval.}$$

- The second convolution theorem states that the Fourier series coefficients of the product of f and g are given by the discrete convolution of the $\hat{f}$ and $\hat{g}$ sequences:

$$[\widehat{f \cdot g}](n) = [\hat{f} * \hat{g}](n).$$

Compact Groups

One of the interesting properties of the Fourier transform which we have mentioned, is that it carries convolutions to pointwise products. If that is the property which we seek to preserve, one can produce Fourier series on any compact group. Typical examples include those classical groups that are compact. This generalizes the Fourier transform to all spaces of the form

$L^2(G)$, where G is a compact group, in such a way that the Fourier transform carries convolutions to pointwise products. The Fourier series exists and converges in similar ways to the $[-\pi,\pi]$ case.

An alternative extension to compact groups is the Peter–Weyl theorem, which proves results about representations of compact groups analogous to those about finite groups.

Riemannian Manifolds

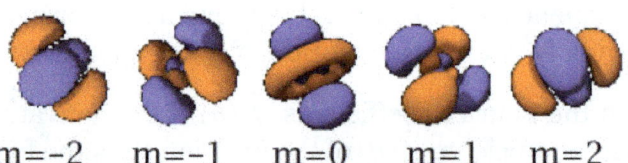

m=-2   m=-1   m=0   m=1   m=2

The atomic orbitals of chemistry are spherical harmonics and can be used to produce Fourier series on the sphere.

If the domain is not a group, then there is no intrinsically defined convolution. However, if X is a compact Riemannian manifold, it has a Laplace–Beltrami operator. The Laplace–Beltrami operator is the differential operator that corresponds to Laplace operator for the Riemannian manifold X. Then, by analogy, one can consider heat equations on X. Since Fourier arrived at his basis by attempting to solve the heat equation, the natural generalization is to use the eigensolutions of the Laplace–Beltrami operator as a basis. This generalizes Fourier series to spaces of the type $L^2(X)$, where X is a Riemannian manifold. The Fourier series converges in ways similar to the $[-\pi, \pi]$ case. A typical example is to take X to be the sphere with the usual metric, in which case the Fourier basis consists of spherical harmonics.

Locally Compact Abelian Groups

The generalization to compact groups discussed above does not generalize to noncompact, nonabelian groups. However, there is a straightfoward generalization to Locally Compact Abelian (LCA) groups.

This generalizes the Fourier transform to $L^1(G)$ or $L^2(G)$, where G is an LCA group. If G is compact, one also obtains a Fourier series, which converges similarly to the $[-\pi, \pi]$ case, but if G is noncompact, one obtains instead a Fourier integral. This generalization yields the usual Fourier transform when the underlying locally compact Abelian group is R.

Approximation and Convergence of Fourier Series

An important question for the theory as well as applications is that of convergence. In particular, it is often necessary in applications to replace the infinite series $\sum\limits_{-\infty}^{\infty}$ by a finite one,

$$f_N(x) = \sum_{n=-N}^{N} \hat{f}(n)e^{inx}.$$

This is called a *partial sum*. We would like to know, in which sense does $f_N(x)$ converge to $f(x)$ as $N \to \infty$.

Least Squares Property

We say that p is a trigonometric polynomial of degree N when it is of the form

$$p(x) = \sum_{n=-N}^{N} p_n e^{inx}.$$

Note that f_N is a trigonometric polynomial of degree N. Parseval's theorem implies that

Theorem. The trigonometric polynomial f_N is the unique best trigonometric polynomial of degree N approximating $f(x)$, in the sense that, for any trigonometric polynomial $p \neq f_N$ of degree N, we have

$$\| f_N - f \|_2 < \| p - f \|_2,$$

where the Hilbert space norm is defined as:

$$\| g \|_2 = \sqrt{\frac{1}{2\pi} \int_{-\pi}^{\pi} |g(x)|^2 \, dx}.$$

Convergence

Because of the least squares property, and because of the completeness of the Fourier basis, we obtain an elementary convergence result.

Theorem. If f belongs to $L^2([-\pi, \pi])$, then f_∞ converges to f in $L^2([-\pi, \pi])$, that is, $\| f_N - f \|_2$ converges to 0 as $N \to \infty$.

We have already mentioned that if f is continuously differentiable, then $(i \cdot n) \hat{f}(n)$ is the nth Fourier coefficient of the derivative f'. It follows, essentially from the Cauchy–Schwarz inequality, that f_∞ is absolutely summable. The sum of this series is a continuous function, equal to f, since the Fourier series converges in the mean to f:

Theorem. If $f \in C^1(\mathbb{T})$, then f_∞ converges to f uniformly (and hence also pointwise.)

This result can be proven easily if f is further assumed to be C^2, since in that case $n^2 \hat{f}(n)$ tends to zero as $n \to \infty$. More generally, the Fourier series is absolutely summable, thus converges uniformly to f, provided that f satisfies a Hölder condition of order $\alpha > \frac{1}{2}$. In the absolutely summable case, the inequality $\sup_x | f(x) - f_N(x) | \leq \sum_{|n|>N} |\hat{f}(n)|$? proves uniform convergence.

Many other results concerning the convergence of Fourier series are known, ranging from the moderately simple result that the series converges at x if f is differentiable at x, to Lennart Carleson's much more sophisticated result that the Fourier series of an L^2 function actually converges almost everywhere.

These theorems, and informal variations of them that don't specify the convergence conditions, are sometimes referred to generically as "Fourier's theorem" or "the Fourier theorem".

Divergence

Since Fourier series have such good convergence properties, many are often surprised by some of the negative results. For example, the Fourier series of a continuous T-periodic function need not converge pointwise. The uniform boundedness principle yields a simple non-constructive proof of this fact.

In 1922, Andrey Kolmogorov published an article titled "Une série de Fourier-Lebesgue divergente presque partout" in which he gave an example of a Lebesgue-integrable function whose Fourier series diverges almost everywhere. He later constructed an example of an integrable function whose Fourier series diverges everywhere (Katznelson 1976).

Fourier Transform

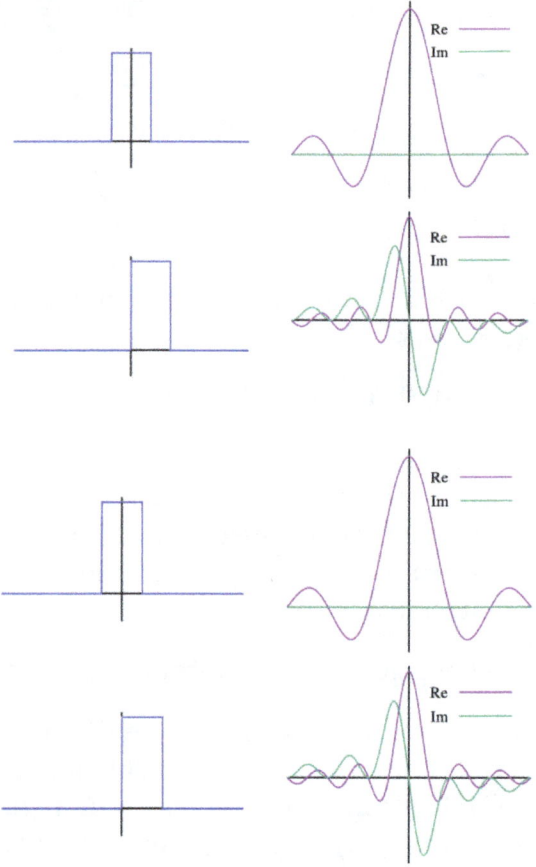

In the first row is the graph of the unit pulse function $f(t)$ and its Fourier transform $\hat{f}(\omega)$, a function of frequency ω. Translation (that is, delay) in the time domain goes over to complex phase shifts in the frequency domain. In the second row is shown $g(t)$, a delayed unit pulse, beside the real and imaginary parts of the Fourier transform. The Fourier transform decomposes a function into eigenfunctions for the group of translations.

The Fourier transform decomposes a function of time (a *signal*) into the frequencies that make it up, in a way similar to how a musical chord can be expressed as the amplitude (or loudness) of its constituent notes. The Fourier transform of a function of time itself is a complex-valued function of frequency, whose absolute value represents the amount of that frequency present in the original function, and whose complex argument is the phase offset of the basic sinusoid in that frequency. The Fourier transform is called the *frequency domain representation* of the original signal. The term *Fourier transform* refers to both the frequency domain representation and the mathematical operation that associates the frequency domain representation to a function of time. The Fourier transform is not limited to functions of time, but in order to have a unified language, the domain of the original function is commonly referred to as the *time domain*. For many functions of practical interest, one can define an operation that reverses this: the *inverse Fourier transformation*, also called *Fourier synthesis*, of a frequency domain representation combines the contributions of all the different frequencies to recover the original function of time.

Linear operations performed in one domain (time or frequency) have corresponding operations in the other domain, which are sometimes easier to perform. The operation of differentiation in the time domain corresponds to multiplication by the frequency, so some differential equations are easier to analyze in the frequency domain. Also, convolution in the time domain corresponds to ordinary multiplication in the frequency domain. Concretely, this means that any linear time-invariant system, such as a filter applied to a signal, can be expressed relatively simply as an operation on frequencies. After performing the desired operations, transformation of the result can be made back to the time domain. Harmonic analysis is the systematic study of the relationship between the frequency and time domains, including the kinds of functions or operations that are "simpler" in one or the other, and has deep connections to almost all areas of modern mathematics.

Functions that are localized in the time domain have Fourier transforms that are spread out across the frequency domain and vice versa, a phenomenon known as the uncertainty principle. The critical case for this principle is the Gaussian function, of substantial importance in probability theory and statistics as well as in the study of physical phenomena exhibiting normal distribution (e.g., diffusion). The Fourier transform of a Gaussian function is another Gaussian function. Joseph Fourier introduced the transform in his study of heat transfer, where Gaussian functions appear as solutions of the heat equation.

The Fourier transform can be formally defined as an improper Riemann integral, making it an integral transform, although this definition is not suitable for many applications requiring a more sophisticated integration theory. For example, many relatively simple applications use the Dirac delta function, which can be treated formally as if it were a function, but the justification requires a mathematically more sophisticated viewpoint. The Fourier transform can also be generalized to functions of several variables on Euclidean space, sending a function of 3-dimensional space to a function of 3-dimensional momentum (or a function of space and time to a function of 4-momentum). This idea makes the spatial Fourier transform very natural in the study of waves, as well as in quantum mechanics, where it is important to be able to represent wave solutions as functions of either space or momentum and sometimes both. In general, functions to which Fourier methods are applicable are complex-valued, and possibly vector-valued. Still further generalization is possible to functions on groups, which, besides the original Fourier transform on R or R^n (viewed as groups under addition), notably includes the discrete-time Fourier transform (DTFT, group = Z), the discrete Fourier transform (DFT, group = Z mod N) and the Fourier series or circular Fourier transform (group = S^1, the unit circle ≈ closed finite interval with endpoints identified). The latter is routinely employed to handle periodic functions. The fast Fourier transform (FFT) is an algorithm for computing the DFT.

Definition

The Fourier transform of the function f is traditionally denoted by adding a circumflex: $\hat{f}$. There are several common conventions for defining the Fourier transform of an integrable function $f: \rightarrow \mathbb{CR}$. This section will use the following definition:

$$\hat{f}(\xi) = \int_{-\infty}^{+\infty} f(x)\, e^{-2\pi i x \xi}\, dx, \text{ for any real number } \xi.$$

When the independent variable x represents *time* (with SI unit of seconds), the transform variable ξ represents frequency (in hertz). Under suitable conditions, f is determined by $\hat{f}$ via the inverse transform:

$$f(x) = \int_{-\infty}^{+\infty} \hat{f}(\xi)\, e^{2\pi i \xi x}\, d\xi, \text{ for any real number } x.$$

The statement that f can be reconstructed from $\hat{f}$ is known as the Fourier inversion theorem, and was first introduced in Fourier's *Analytical Theory of Heat*, although what would be considered a proof by modern standards was not given until much later. The functions f and $\hat{f}$ often are referred to as a *Fourier integral pair* or *Fourier transform pair*.

For other common conventions and notations, including using the angular frequency ω instead of the frequency ξ. The Fourier transform on Eu-clidean space is treated separately, in which the variable x often represents position and ξ momen-tum. The conventions chosen in this section are those of harmonic analysis, and are characterized as the unique conventions such that the Fourier transform is both unitary on L^2 and an algebra homomorphism from L^1 to L^∞, without normalizing the Lebesgue measure.

Many other characterizations of the Fourier transform exist. For example, one uses the Stone–von Neumann theorem: the Fourier transform is the unique unitary intertwiner for the symplectic and Euclidean Schrödinger representations of the Heisenberg group.

History

In 1822, Joseph Fourier showed that some functions could be written as an infinite sum of harmonics.

Introduction

One motivation for the Fourier transform comes from the study of Fourier series. In the study of Fourier series, complicated but periodic functions are written as the sum of simple waves mathematically represented by sines and cosines. The Fourier transform is an extension of the Fourier series that results when the period of the represented function is lengthened and allowed to approach infinity.

Due to the properties of sine and cosine, it is possible to recover the amplitude of each wave in a

Fourier series using an integral. In many cases it is desirable to use Euler's formula, which states that $e^{2\pi i \theta} = \cos(2\pi\theta) + i \sin(2\pi\theta)$, to write Fourier series in terms of the basic waves $e^{2\pi i \theta}$. This has the advantage of simplifying many of the formulas involved, and provides a formulation for Fourier series that more closely resembles the definition followed in this section. Re-writing sines and cosines as complex exponentials makes it necessary for the Fourier coefficients to be complex valued. The usual interpretation of this complex number is that it gives both the amplitude (or size) of the wave present in the function and the phase (or the initial angle) of the wave. These complex ex-ponentials sometimes contain negative "frequencies". If θ is measured in seconds, then the waves $e^{2\pi i \theta}$ and $e^{-2\pi i \theta}$ both complete one cycle per second, but they represent different frequencies in the Fourier transform. Hence, frequency no longer measures the number of cycles per unit time, but is still closely related.

There is a close connection between the definition of Fourier series and the Fourier transform for functions f that are zero outside an interval. For such a function, we can calculate its Fourier series on any interval that includes the points where f is not identically zero. The Fourier transform is also defined for such a function. As we increase the length of the interval on which we calculate the Fourier series, then the Fourier series coefficients begin to look like the Fourier transform and the sum of the Fourier series of f begins to look like the inverse Fourier transform. To explain this more precisely, suppose that T is large enough so that the interval $[-T/2, T/2]$ contains the interval on which f is not identically zero. Then the n-th series coefficient c_n is given by:

$$c_n = \frac{1}{T} \int_{-T/2}^{T/2} f(x)\, e^{-2\pi i(n/T)x}\, dx.$$

Comparing this to the definition of the Fourier transform, it follows that $c_n = (1/T)\hat{f}(n/T)$ since $f(x)$ is zero outside $[-T/2, T/2]$. Thus the Fourier coefficients are just the values of the Fourier transform sampled on a grid of width $1/T$, multiplied by the grid width $1/T$.

Under appropriate conditions, the Fourier series of f will equal the function f. In other words, f can be written:

$$f(x) = \sum_{n=-\infty}^{\infty} c_n\, e^{2\pi i(n/T)x} = \sum_{n=-\infty}^{\infty} \hat{f}(\xi_n)\, e^{2\pi i \xi_n x} \Delta\xi,$$

where the last sum is simply the first sum rewritten using the definitions $\xi_n = n/T$, and $\Delta\xi = (n + 1)/T - n/T = 1/T$.

This second sum is a Riemann sum, and so by letting $T \to \infty$ it will converge to the integral for the inverse Fourier transform given in the definition section. Under suitable conditions, this argument may be made precise.

In the study of Fourier series the numbers c_n could be thought of as the "amount" of the wave present in the Fourier series of f. Similarly, as seen above, the Fourier transform can be thought of as a function that measures how much of each individual frequency is present in our function f, and we can recombine these waves by using an integral (or "continuous sum") to reproduce the original function.

Example

The following figures provide a visual illustration of how the Fourier transform measures whether a frequency is present in a particular function. The function depicted $f(t) = \cos(6\pi t)\,e^{-\pi t2}$ oscillates at 3 Hz (if t measures seconds) and tends quickly to 0. (The second factor in this equation is an envelope function that shapes the continuous sinusoid into a short pulse. Its general form is a Gaussian function). This function was specially chosen to have a real Fourier transform that can easily be plotted. The first image contains its graph. In order to calculate $\hat{f}(3)$ we must integrate $e^{-2\pi i(3t)}f(t)$. The second image shows the plot of the real and imaginary parts of this function. The real part of the integrand is almost always positive, because when $f(t)$ is negative, the real part of $e^{-2\pi i(3t)}$ is negative as well. Because they oscillate at the same rate, when $f(t)$ is positive, so is the real part of $e^{-2\pi i(3t)}$. The result is that when you integrate the real part of the integrand you get a relatively large number (in this case 0.5). On the other hand, when you try to measure a frequency that is not present, as in the case when we look at $\hat{f}(5)$, you see that both real and imaginary component of this function vary rapidly between positive and negative values, as plotted in the third image. Therefore, in this case, the integrand oscillates fast enough so that the integral is very small and the value for the Fourier transform for that frequency is nearly zero. The general situation may be a bit more complicated than this, but this in spirit is how the Fourier transform measures how much of an individual frequency is present in a function $f(t)$.

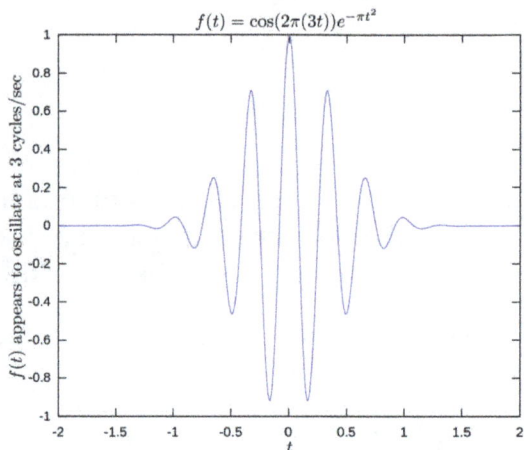

Original function showing oscillation 3 Hz.

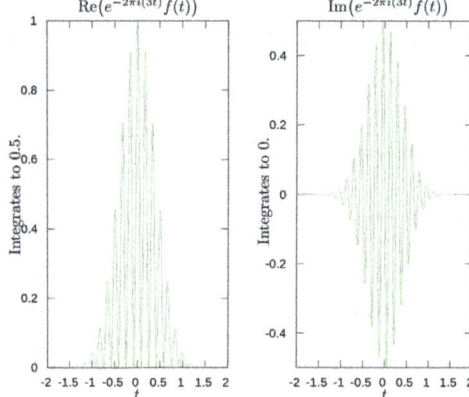

Real and imaginary parts of integrand for Fourier transform at 3 Hz

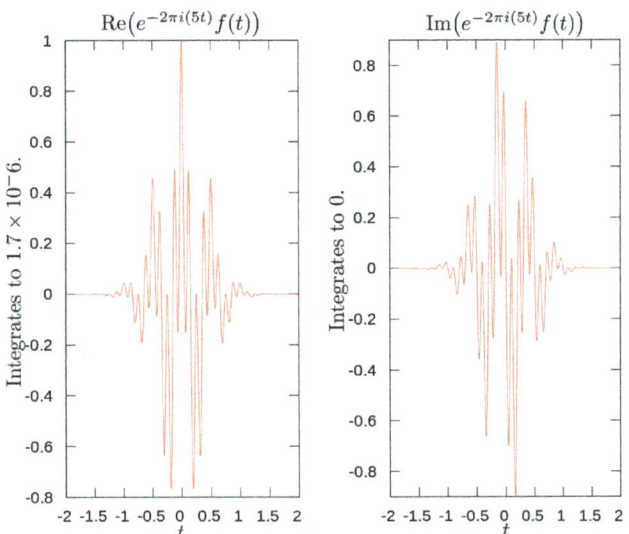

Real and imaginary parts of integrand for Fourier transform at 5 Hz

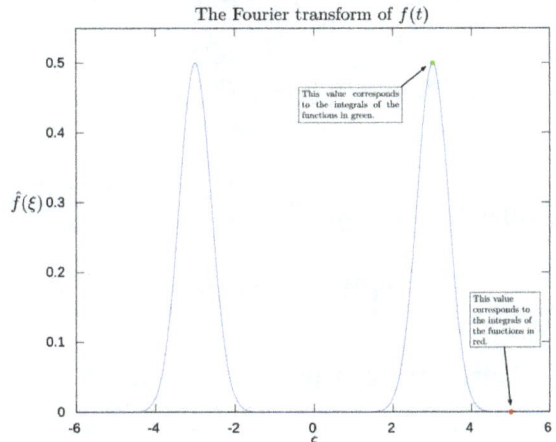

Fourier transform with 3 and 5 Hz labeled.

Properties of the Fourier Transform

Here we assume $f(x)$, $g(x)$ and $h(x)$ are *integrable functions*: Lebesgue-measurable on the real line satisfying:

$$\int_{-\infty}^{\infty} |f(x)| \, dx < \infty.$$

We denote the Fourier transforms of these functions by $\hat{f}(\xi)$, $\hat{g}(\xi)$ and $\hat{h}(\xi)$ respectively.

Basic Properties

The Fourier transform has the following basic properties:

Linearity

>   For any complex numbers a and b, if $h(x) = af(x) + bg(x)$, then $\hat{h}(\xi) = a \cdot \hat{f}(\xi) + b \cdot \hat{g}(\xi)$.

Translation/ Time-Shifting

For any real number x_0, if $h(x) = f(x - x_0)$, then $\hat{h}(\xi) = e^{-i2\pi x_0 \xi}\hat{f}(\xi)$.

Modulation/ Frequency shifting

For any real number ξ_0 if $h(x) = e^{i2\pi x \xi_0}f(x)$, then $\hat{h}(\xi) = \hat{f}(\xi - \xi_0)$.

Time Scaling

For a non-zero real number a, if $h(x) = f(ax)$, , then $\hat{h}(\xi) = \dfrac{1}{|a|}\hat{f}\left(\dfrac{\xi}{a}\right)$.

The case $a = -1$ leads to the *time-reversal* property, which states: if $h(x) = f(-x)$, then $\hat{h}(\xi) = \hat{f}(-\xi)$.

Conjugation

If $h(x) = \overline{f(x)}$, then $\hat{h}(\xi) = \overline{\hat{f}(-\xi)}$.

In particular, if f is real, then one has the *reality condition* $\hat{f}(-\xi) = \overline{\hat{f}(\xi)}$, that is, $\hat{f}$ is a Hermitian function.

And if f is purely imaginary, then $\hat{f}(-\xi) = -\overline{\hat{f}(\xi)}$.

Integration

Substituting $\xi = 0$ in the definition, we obtain

$$\hat{f}(0) = \int_{-\infty}^{\infty} f(x)dx.$$

That is, the evaluation of the Fourier transform in the origin ($\xi = 0$) equals the integral of f over all its domain.

Invertibility and Periodicity

Under suitable conditions on the function f, it can be recovered from its Fourier transform $\hat{f}$. Indeed, denoting the Fourier transform operator by $\mathcal{F}$, so $\mathcal{F}(f) := \hat{f}$, then for suitable functions, applying the Fourier transform twice simply flips the function: $\mathcal{F}^2(f)(x) = f(-x)$, , which can be interpreted as "reversing time". Since reversing time is two-periodic, applying this twice yields $\mathcal{F}^4(f) = f$, so the Fourier transform operator is four-periodic, and similarly the inverse Fourier transform can be obtained by applying the Fourier transform three times: $\mathcal{F}^3(\hat{f}) = f$. In particular the Fourier transform is invertible (under suitable conditions).

More precisely, defining the parity operator $\mathcal{P}$ that inverts time, $\mathcal{P}[f] : t \mapsto f(-t)$, :

$$\mathcal{F}^0 = \text{Id}, \qquad \mathcal{F}^1 = \mathcal{F}, \qquad \mathcal{F}^2 = \mathcal{P}, \qquad \mathcal{F}^4 = \text{Id}$$

$$\mathcal{F}^3 = \mathcal{F}^{-1} = \mathcal{P} \circ \mathcal{F} = \mathcal{F} \circ \mathcal{P}$$

These equalities of operators require careful definition of the space of functions in question, defining equality of functions (equality at every point? equality almost everywhere?) and defining equality of operators – that is, defining the topology on the function space and operator space in question. These are not true for all functions, but are true under various conditions, which are the content of the various forms of the Fourier inversion theorem.

This four-fold periodicity of the Fourier transform is similar to a rotation of the plane by 90°, particularly as the two-fold iteration yields a reversal, and in fact this analogy can be made precise. While the Fourier transform can simply be interpreted as switching the time domain and the frequency domain, with the inverse Fourier transform switching them back, more geometrically it can be interpreted as a rotation by 90° in the time–frequency domain (considering time as the x-axis and frequency as the y-axis), and the Fourier transform can be generalized to the fractional Fourier transform, which involves rotations by other angles. This can be further generalized to linear canonical transformations, which can be visualized as the action of the special linear group $SL_2(R)$ on the time–frequency plane, with the preserved symplectic form corresponding to the uncertainty principle, below. This approach is particularly studied in signal processing, under time–frequency analysis.

Units and Duality

In mathematics, one often does not think of any units as being attached to the two variables t and ξ. But in physical applications, ξ must have inverse units to the units of ξ. For example, if t is measured in seconds, ξ should be in cycles per second for the formulas here to be valid. If the scale of t is changed and t is measured in units of 2π seconds, then either ξ must be in the so-called "angular frequency", or one must insert some constant scale factor into some of the formulas. If t is measured in units of length, then ξ must be in inverse length, e.g., wavenumbers. That is to say, there are two copies of the real line: one measured in one set of units, where t ranges, and the other in inverse units to the units of t, and which is the range of ξ. So these are two distinct copies of the real line, and cannot be identified with each other. Therefore, the Fourier transform goes from one space of functions to a different space of functions: functions which have a different domain of definition.

In general, ξ must always be taken to be a linear form on the space of t s, which is to say that the second real line is the dual space of the first real line. This point of view becomes essential in generalisations of the Fourier transform to general symmetry groups, including the case of Fourier series.

That there is no one preferred way (often, one says "no canonical way") to compare the two copies of the real line which are involved in the Fourier transform—fixing the units on one line does not force the scale of the units on the other line—is the reason for the plethora of rival conventions on the definition of the Fourier transform. The various definitions resulting from different choices of units differ by various constants. If the units of t are in seconds but the units of ξ are in angular frequency, then the angular frequency variable is often denoted by one or another Greek letter, for example, $\omega = 2\pi\xi$ is quite common. Thus (writing $\hat{x}_1$ for the alternative definition and $\hat{x}$ for the definition adopted in this chapter)

$$\hat{x}_1(\omega) = \hat{x}\left(\frac{\omega}{2\pi}\right) = \int_{-\infty}^{\infty} x(t)e^{-i\omega t}\, dt$$

as before, but the corresponding alternative inversion formula would then have to be

$$x(t) = \frac{1}{2\pi} \int_{-\infty}^{\infty} \hat{x}_1(\omega) e^{it\omega} d\omega.$$

To have something involving angular frequency but with greater symmetry between the Fourier transform and the inversion formula, one very often sees still another alternative definition of the Fourier transform, with a factor of $\sqrt{2\pi}$,, thus

$$\hat{x}_2(\omega) = \frac{1}{\sqrt{2\pi}} \int_{-\infty}^{\infty} x(t) e^{-i\omega t} dt,$$

and the corresponding inversion formula then has to be

$$x(t) = \frac{1}{\sqrt{2\pi}} \int_{-\infty}^{\infty} \hat{x}_2(\omega) e^{it\omega} d\omega.$$

Furthermore, there is no way to fix which square root of negative one will be meant by the symbol i (it makes no sense to speak of "the positive square root" since only real numbers can be positive, similarly it makes no sense to say "rotation counter-clockwise", because until i is chosen, there is no fixed way to draw the complex plane), and hence one occasionally sees the Fourier transform written with i in the exponent instead of $-i$, , and vice versa for the inversion formula, a convention that is equally valid as the one chosen in this chapter, which is the more usual one.

For example, in probability theory, the characteristic function ϕ of the probability density function f of a random variable X of continuous type is defined without a negative sign in the exponential, and since the units of x are ignored, there is no 2π either:

$$\phi(\lambda) = \int_{-\infty}^{\infty} f(x) e^{i\lambda x} dx.$$

(In probability theory, and in mathematical statistics, the use of the Fourier—Stieltjes transform is preferred, because so many random variables are not of continuous type, and do not possess a density function, and one must treat discontinuous distribution functions, i.e., measures which possess "atoms".)

From the higher point of view of group characters, which is much more abstract, all these arbitrary choices disappear, as will be explained in the later section of this chapter, on the notion of the Fourier transform of a function on an Abelian locally compact group.

Uniform Continuity and the Riemann–lebesgue Lemma

The Fourier transform may be defined in some cases for non-integrable functions, but the Fourier transforms of integrable functions have several strong properties.

The Fourier transform, $\hat{f}$, of any integrable function f is uniformly continuous and $\| \hat{f} \|_{\infty} \leq \| f \|_1$ By the *Riemann–Lebesgue lemma*,

$$\hat{f}(\xi) \to 0 \text{ as } |\xi| \to \infty.$$

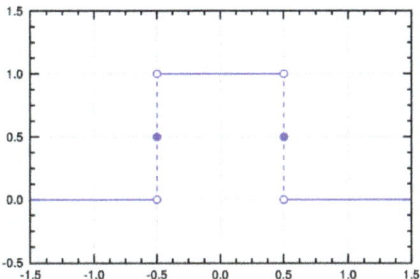

The rectangular function is Lebesgue integrable.

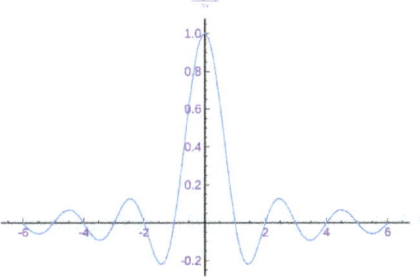

The sinc function, which is the Fourier transform of the rectangular function, is bounded and continuous, but not Lebesgue integrable.

However, $\hat{f}$ need not be integrable. For example, the Fourier transform of the rectangular function, which is integrable, is the sinc function, which is not Lebesgue integrable, because its improper integrals behave analogously to the alternating harmonic series, in converging to a sum without being absolutely convergent.

It is not generally possible to write the *inverse transform* as a Lebesgue integral. However, when both f and $\hat{f}$ are integrable, the inverse equality

$$f(x) = \int_{-\infty}^{\infty} \hat{f}(\xi) e^{2i\pi x\xi}\, d\xi$$

holds almost everywhere. That is, the Fourier transform is injective on $L^1(\mathbb{R})$. (But if f is continuous, then equality holds for every x.)

Plancherel Theorem and Parseval's Theorem

Let $f(x)$ and $g(x)$ be integrable, and let $\hat{f}(\xi)$ and $\hat{g}(\xi)$ be their Fourier transforms. If $f(x)$ and $g(x)$ are also square-integrable, then we have Parseval's Formula:

$$\int_{-\infty}^{\infty} f(x)\overline{g(x)}\mathrm{d}x = \int_{-\infty}^{\infty} \hat{f}(\xi)\overline{\hat{g}(\xi)}d\xi,$$

where the bar denotes complex conjugation.

The Plancherel theorem, which follows from the above, states that

$$\int_{-\infty}^{\infty} |f(x)|^2 \, dx = \int_{-\infty}^{\infty} |\hat{f}(\xi)|^2 \, d\xi.$$

Plancherel's theorem makes it possible to extend the Fourier transform, by a continuity argument, to a unitary operator on $L^2(R)$. On $L^1(R) \cap L^2(R)$, this extension agrees with original Fourier transform defined on $L^1(R)$, thus enlarging the domain of the Fourier transform to $L^1(R) + L^2(R)$ (and consequently to $L^p(R)$ for $1 \le p \le 2$). Plancherel's theorem has the interpretation in the sciences that the Fourier transform preserves the energy of the original quantity. The terminology of these formulas is not quite standardised. Parseval's theorem was proved only for Fourier series, and was first proved by Liapounoff. But Parseval's formula makes sense for the Fourier transform as well, and so even though in the context of the Fourier transform it was proved by Plancherel, it is still often referred to as Parseval's formula, or Parseval's relation, or even Parseval's theorem.

Poisson Summation Formula

The Poisson summation formula (PSF) is an equation that relates the Fourier series coefficients of the periodic summation of a function to values of the function's continuous Fourier transform. The Poisson summation formula says that for sufficiently regular functions f,

$$\sum_n \hat{f}(n) = \sum_n f(n).$$

It has a variety of useful forms that are derived from the basic one by application of the Fourier transform's scaling and time-shifting properties. The formula has applications in engineering, physics, and number theory. The frequency-domain dual of the standard Poisson summation formula is also called the discrete-time Fourier transform.

Poisson summation is generally associated with the physics of periodic media, such as heat conduction on a circle. The fundamental solution of the heat equation on a circle is called a theta function. It is used in number theory to prove the transformation properties of theta functions, which turn out to be a type of modular form, and it is connected more generally to the theory of automorphic forms where it appears on one side of the Selberg trace formula.

Differentiation

Suppose $f(x)$ is a differentiable function, and both f and its derivative f' are integrable. Then the Fourier transform of the derivative is given by

$$\widehat{f'}(\xi) = 2\pi i \xi \hat{f}(\xi).$$

More generally, the Fourier transformation of the n-th derivative $f^{(n)}$ is given by

$$\widehat{f^{(n)}}(\xi) = (2\pi i \xi)^n \hat{f}(\xi).$$

By applying the Fourier transform and using these formulas, some ordinary differential equations can be transformed into algebraic equations, which are much easier to solve. These formulas also give rise to the rule of thumb " $f(x)$ is smooth if and only if $\hat{f}(\xi)$ quickly falls down to 0 for $|\xi| \to \infty$." By using the analogous rules for the inverse Fourier transform, one can also say " $f(x)$ quickly falls down to 0 for $|x| \to \infty$ if and only if $\hat{f}(\xi)$ is smooth."

Convolution Theorem

The Fourier transform translates between convolution and multiplication of functions. If $f(x)$ and $g(x)$ are integrable functions with Fourier transforms $\hat{f}(\xi)$ and $\hat{g}(\xi)$ respectively, then the Fourier transform of the convolution is given by the product of the Fourier transforms $\hat{f}(\xi)$ and $\hat{g}(\xi)$ (under other conventions for the definition of the Fourier transform a constant factor may appear).

This means that if:

$$h(x) = (f * g)(x) = \int_{-\infty}^{\infty} f(y)g(x-y)dy,$$

where $*$ denotes the convolution operation, then:

$$\hat{h}(\xi) = \hat{f}(\xi) \cdot \hat{g}(\xi).$$

In linear time invariant (LTI) system theory, it is common to interpret $g(x)$ as the impulse response of an LTI system with input $f(x)$ and output $h(x)$, since substituting the unit impulse for $f(x)$ yields $h(x) = g(x)$. In this case, $\hat{g}(\xi)$ represents the frequency response of the system.

Conversely, if $f(x)$ can be decomposed as the product of two square integrable functions $p(x)$ and $q(x)$, then the Fourier transform of $f(x)$ is given by the convolution of the respective Fourier transforms $\hat{p}(\xi)$ and $\hat{q}(\xi)$.

Cross-correlation Theorem

In an analogous manner, it can be shown that if $h(x)$ is the cross-correlation of $f(x)$ and $g(x)$:

$$h(x) = (f \star g)(x) = \int_{-\infty}^{\infty} \overline{f(y)}g(x+y)dy$$

then the Fourier transform of $h(x)$ is:

$$\hat{h}(\xi) = \overline{\hat{f}(\xi)} \cdot \hat{g}(\xi).$$

As a special case, the autocorrelation of function $f(x)$ is:

$$h(x) = (f \star f)(x) = \int_{-\infty}^{\infty} \overline{f(y)}f(x+y)dy$$

for which

$$\hat{h}(\xi) = \overline{\hat{f}(\xi)}\hat{f}(\xi) = |\hat{f}(\xi)|^2 .$$

Eigenfunctions

One important choice of an orthonormal basis for $L^2(\mathbb{R})$ is given by the Hermite functions

$$\psi_n(x) = \frac{2^{1/4}}{\sqrt{n!}} e^{-\pi x^2} \mathrm{He}_n(2x\sqrt{\pi}),$$

where $\mathrm{He}_n(x)$ are the "probabilist's" Hermite polynomials, defined by

$$\mathrm{He}_n(x) = (-1)^n e^{\frac{x^2}{2}} \left(\frac{d}{dx}\right)^n e^{-\frac{x^2}{2}}$$

Under this convention for the Fourier transform, we have that

$$\hat{\psi}_n(\xi) = (-i)^n \psi_n(\xi).$$

In other words, the Hermite functions form a complete orthonormal system of eigenfunctions for the Fourier transform on $L^2(\mathbb{R})$. However, this choice of eigenfunctions is not unique. There are only four different eigenvalues of the Fourier transform (± 1 and $\pm i$) and any linear combination of eigenfunctions with the same eigenvalue gives another eigenfunction. As a consequence of this, it is possible to decompose $L^2(\mathbb{R})$ as a direct sum of four spaces H_0, H_1, H_2, and H_3 where the Fourier transform acts on He_k simply by multiplication by i^k.

Since the complete set of Hermite functions provides a resolution of the identity, the Fourier transform can be represented by such a sum of terms weighted by the above eigenvalues, and these sums can be explicitly summed. This approach to define the Fourier transform was first done by Norbert Wiener. Among other properties, Hermite functions decrease exponentially fast in both frequency and time domains, and they are thus used to define a generalization of the Fourier transform, namely the fractional Fourier transform used in time-frequency analysis. In physics, this transform was introduced by Edward Condon.

Connection with the Heisenberg Group

The Heisenberg group is a certain group of unitary operators on the Hilbert space $L^2(\mathbb{R})$ of square integrable complex valued functions f on the real line, generated by the translations $(T_y f)(x) = f(x + y)$ and multiplication by $e^{2\pi i x \xi}$, $(M_\xi f)(x) = e^{2\pi i x \xi} f(x)$. These operators do not commute, as their (group) commutator is

$$(M_\xi^{-1} T_y^{-1} M_\xi T_y f)(x) = e^{2\pi i y \xi} f(x)$$

which is multiplication by the constant (independent of x) $e^{2\pi i y \xi} \in U(1)$ (the circle group of unit modulus complex numbers). As an abstract group, the Heisenberg group is the three-dimensional Lie group of triples $(x, \xi, z) \in \mathbb{R}^2 \times U(1)$, with the group law

$$(x_1, \xi_1, t_1) \cdot (x_2, \xi_2, t_2) = (x_1 + x_2, \xi_1 + \xi_2, t_1 t_2 e^{2\pi i(x_1 \xi_1 + x_2 \xi_2 + x_1 \xi_2)}).$$

Denote the Heisenberg group by H_1. The above procedure describes not only the group structure, but also a standard unitary representation of H_1 on a Hilbert space, which we denote by $\rho : H_1 \to B(L^2(\mathbb{R}))$. Define the linear automorphism of $\mathbb{R}^2$ by

$$J\begin{pmatrix} x \\ \xi \end{pmatrix} = \begin{pmatrix} -\xi \\ x \end{pmatrix}$$

so that $J^2 = -I$. This J can be extended to a unique automorphism of H_1:

$$j(x,\xi,t) = (-\xi, x, t e^{-2\pi i x \xi}).$$

According to the Stone–von Neumann theorem, the unitary representations ρ and $\rho \circ j$ are unitarily equivalent, so there is a unique intertwiner $W \in U(L^2(\mathbb{R}))$ such that

$$\rho \circ j = W \rho W^*.$$

This operator W is the Fourier transform.

Many of the standard properties of the Fourier transform are immediate consequences of this more general framework. For example, the square of the Fourier transform, W^2, is an intertwiner associated to $J^2 = -I$, and so we have $(W^2 f)(x) = f(-x)$ is the reflection of the original function f.

Complex Domain

The integral for the Fourier transform

$$\hat{f}(\xi) = \int_{-\infty}^{\infty} e^{-2\pi i \xi t} f(t) dt$$

can be studied for complex values of its argument ξ. Depending on the properties of f, this might not converge off the real axis at all, or it might converge to a complex analytic function for all values of $\xi = \sigma + i\tau$, or something in between.

The Paley–Wiener theorem says that f is smooth (i.e., n-times differentiable for all positive integers n) and compactly supported if and only if $\hat{f}(\sigma + i\tau)$ is a holomorphic function for which there exists a constant $a > 0$ such that for any integer $n \geq 0$,

$$|\xi^n \hat{f}(\xi)| \leq C e^{a|\tau|}$$

for some constant C. (In this case, f is supported on $[-a, a]$.) This can be expressed by saying that $\hat{f}$ is an entire function which is rapidly decreasing in σ (for fixed τ) and of exponential growth in τ (uniformly in σ).

(If f is not smooth, but only L^2, the statement still holds provided $n = 0$.) The space of such functions of a complex variable is called the Paley–Wiener space. This theorem has been generalised to semi-simple Lie groups.

If f is supported on the half-line $t \geq 0$, then f is said to be "causal" because the impulse response function of a physically realisable filter must have this property, as no effect can precede its cause.

Paley and Wiener showed that then $\hat{f}$ extends to a holomorphic function on the complex lower half-plane $\tau < 0$ which tends to zero as τ goes to infinity. The converse is false and it is not known how to characterise the Fourier transform of a causal function.

Laplace Transform

The Fourier transform $\hat{f}(\xi)$ is intimately related with the Laplace transform $F(s)$, which is also used for the solution of differential equations and the analysis of filters. Chatfield, indeed, has said that "... the Laplace and the Fourier transforms [of a causal function] are the same, provided that the real part of s is zero."

It may happen that a function f for which the Fourier integral does not converge on the real axis at all, nevertheless has a complex Fourier transform defined in some region of the complex plane.

For example, if $f(t)$ is of exponential growth, i.e.,

$$| f(t) |< Ce^{a|t|}$$

for some constants $C, a \geq 0$, then

$$\hat{f}(i\tau) = \int_{-\infty}^{\infty} e^{2\pi\tau t} f(t)dt,$$

convergent for all $2\pi\tau < -a$, is the two-sided Laplace transform of f.

The more usual version ("one-sided") of the Laplace transform is

$$F(s) = \int_{0}^{\infty} f(t)e^{-st}\, dt.$$

If f is also causal, then

$$\hat{f}(i\tau) = F(-2\pi\tau).$$

Thus, extending the Fourier transform to the complex domain means it includes the Laplace transform as a special case—the case of causal functions—but with the change of variable $s = 2\pi i\xi$.

Inversion

If $\hat{f}$ has no poles for $a \leq \tau \leq b$, then

$$\int_{-\infty}^{\infty} \hat{f}(\sigma + ia)e^{2\pi i\xi t}\, d\sigma = \int_{-\infty}^{\infty} \hat{f}(\sigma + ib)e^{2\pi i\xi t}\, d\sigma$$

by Cauchy's integral theorem. Therefore, the Fourier inversion formula can use integration along different lines, parallel to the real axis.

Theorem: If $f(t) = 0$ for $t < 0$, and $| f(t) |< Ce^{a|t|}$ for some constants $C, a > 0$, then

$$f(t) = \int_{-\infty}^{\infty} \hat{f}(\sigma + i\tau)e^{2\pi i\xi t}\, d\sigma,$$

for any $\tau < -\dfrac{a}{2\pi}$.

This theorem implies the Mellin inversion formula for the Laplace transformation,

$$f(t) = \frac{1}{2\pi i} \int_{b-i\infty}^{b+i\infty} F(s)e^{st}\,ds$$

for any $b > a$, where $F(s)$ is the Laplace transform of $f(t)$.

The hypotheses can be weakened, as in the results of Carleman and Hunt, to $f(t)e^{-at}$ being L^1, provided that t is in the interior of a closed interval on which f is continuous and of bounded variation, and provided that the integrals are taken in the sense of Cauchy principal values.

L^2 versions of these inversion formulas are also available.

Fourier Transform on Euclidean Space

The Fourier transform can be defined in any arbitrary number of dimensions n. As with the one-dimensional case, there are many conventions. For an integrable function $f(x)$, this section takes the definition:

$$\hat{f}(\xi) = \mathcal{F}(f)(\xi) = \int_{\mathbb{R}^n} f(\mathbf{x})e^{-2\pi i \mathbf{x}\cdot\xi}\,d\mathbf{x}$$

where x and ξ are n-dimensional vectors, and x $\cdot$ ξ is the dot product of the vectors. The dot product is sometimes written as $\langle \mathbf{x}, \xi \rangle$.

All of the basic properties listed above hold for the n-dimensional Fourier transform, as do Plancherel's and Parseval's theorem. When the function is integrable, the Fourier transform is still uniformly continuous and the Riemann–Lebesgue lemma holds.

Uncertainty Principle

Generally speaking, the more concentrated $f(x)$ is, the more spread out its Fourier transform $\hat{f}(\xi)$ must be. In particular, the scaling property of the Fourier transform may be seen as saying: if we "squeeze" a function in x, its Fourier transform "stretches out" in ξ. It is not possible to arbitrarily concentrate both a function and its Fourier transform.

The trade-off between the compaction of a function and its Fourier transform can be formalized in the form of an uncertainty principle by viewing a function and its Fourier transform as conjugate variables with respect to the symplectic form on the time–frequency domain: from the point of view of the linear canonical transformation, the Fourier transform is rotation by 90° in the time–frequency domain, and preserves the symplectic form.

Suppose $f(x)$ is an integrable and square-integrable function. Without loss of generality, assume that $f(x)$ is normalized:

$$\int_{-\infty}^{\infty} |f(x)|^2\,dx = 1.$$

It follows from the Plancherel theorem that $\hat{f}(\xi)$ is also normalized.

The spread around $x = 0$ may be measured by the *dispersion about zero* defined by

$$D_0(f) = \int_{-\infty}^{\infty} x^2 \, | f(x) |^2 \, dx.$$

In probability terms, this is the second moment of $|f(x)|^2$ about zero.

The Uncertainty principle states that, if $f(x)$ is absolutely continuous and the functions $x \cdot f(x)$ and $f'(x)$ are square integrable, then

$$D_0(f)D_0(\hat{f}) \geq \frac{1}{16\pi^2}.$$

The equality is attained only in the case $f(x) = C_1 e^{-\pi x^2/\sigma^2}$ (hence $\hat{f}(\xi) = \sigma C_1 e^{-\pi\sigma^2\xi^2}$) where $\sigma > 0$ is arbitrary and $C_1 = \sqrt[4]{2}/\sqrt{\sigma}$ so that f is L^2–normalized. In other words, where f is a (normalized) Gaussian function with variance σ^2, centered at zero, and its Fourier transform is a Gaussian function with variance σ^{-2}.

In fact, this inequality implies that:

$$\left(\int_{-\infty}^{\infty} (x - x_0)^2 \, | f(x) |^2 \, dx \right)\left(\int_{-\infty}^{\infty} (\xi - \xi_0)^2 \, | \hat{f}(\xi) |^2 \, d\xi \right) \geq \frac{1}{16\pi^2}$$

for any $x_0, \xi_0 \in \mathbb{R}$.

In quantum mechanics, the momentum and position wave functions are Fourier transform pairs, to within a factor of Planck's constant. With this constant properly taken into account, the inequality above becomes the statement of the Heisenberg uncertainty principle.

A stronger uncertainty principle is the Hirschman uncertainty principle, which is expressed as:

$$H(| f |^2) + H(| \hat{f} |^2) \geq \log(e/2)$$

where $H(p)$ is the differential entropy of the probability density function $p(x)$:

$$H(p) = -\int_{-\infty}^{\infty} p(x) \log(p(x)) dx$$

where the logarithms may be in any base that is consistent. The equality is attained for a Gaussian, as in the previous case.

Sine and Cosine Transforms

Fourier's original formulation of the transform did not use complex numbers, but rather sines and cosines. Statisticians and others still use this form. An absolutely integrable function f for which Fourier inversion holds good can be expanded in terms of genuine frequencies (avoiding negative frequencies, which are sometimes considered hard to interpret physically) λ by

$$f(t) = \int_0^\infty \left[a(\lambda) \cos(2\pi\lambda t) + b(\lambda) \sin(2\pi\lambda t) \right] d\lambda.$$

This is called an expansion as a trigonometric integral, or a Fourier integral expansion. The coefficient functions a and b can be found by using variants of the Fourier cosine transform and the Fourier sine transform (the normalisations are, again, not standardised):

$$a(\lambda) = 2\int_{-\infty}^\infty f(t) \cos(2\pi\lambda t) dt$$

and

$$b(\lambda) = 2\int_{-\infty}^\infty f(t) \sin(2\pi\lambda t) dt.$$

Older literature refers to the two transform functions, the Fourier cosine transform, a, and the Fourier sine transform, b.

The function f can be recovered from the sine and cosine transform using

$$f(t) = 2\int_0^\infty \int_{-\infty}^\infty f(\tau) \cos(2\pi\lambda(\tau - t)) d\tau d\lambda.$$

together with trigonometric identities. This is referred to as Fourier's integral formula.

Spherical Harmonics

Let the set of homogeneous harmonic polynomials of degree k on $\mathbf{R}^n$ be denoted by A_k. The set A_k consists of the solid spherical harmonics of degree k. The solid spherical harmonics play a similar role in higher dimensions to the Hermite polynomials in dimension one. Specifically, if $f(x) = e^{-\pi|x|^2}P(x)$ for some $P(x)$ in A_k, then $\hat{f}(\xi) = i^{-k} f(\xi)$.. Let the set H_k be the closure in $L^2(\mathbf{R}^n)$ of linear combinations of functions of the form $f(|x|)P(x)$ where $P(x)$ is in A_k. The space $L^2(\mathbf{R}^n)$ is then a direct sum of the spaces H_k and the Fourier transform maps each space H_k to itself and is possible to characterize the action of the Fourier transform on each space H_k.

Let $f(x) = f_0(|x|)P(x)$ (with $P(x)$ in A_k), then $\hat{f}(\xi) = F_0(|\xi|)P(\xi)$ where

$$F_0(r) = 2\pi i^{-k} r^{-(n+2k-2)/2} \int_0^\infty f_0(s) J_{(n+2k-2)/2}(2\pi rs) s^{(n+2k)/2} ds.$$

Here $J_{(n+2k-2)/2}$ denotes the Bessel function of the first kind with order $(n + 2k - 2)/2$. When $k = 0$ this gives a useful formula for the Fourier transform of a radial function. Note that this is essentially the Hankel transform. Moreover, there is a simple recursion relating the cases $n + 2$ and n allowing to compute, e.g., the three-dimensional Fourier transform of a radial function from the one-dimensional one.

Restriction Problems

In higher dimensions it becomes interesting to study *restriction problems* for the Fourier transform. The Fourier transform of an integrable function is continuous and the restriction of this function to any set is defined. But for a square-integrable function the Fourier transform could be a general *class* of square integrable functions. As such, the restriction of the Fourier transform of an $L^2(\mathbf{R}^n)$ function cannot be defined on sets of measure 0. It is still an active area of study to understand restriction problems in L^p for $1 < p < 2$. Surprisingly, it is possible in some cases to define the restriction of a Fourier transform to a set S, provided S has non-zero curvature. The case when S is the unit sphere in $\mathbf{R}^n$ is of particular interest. In this case the Tomas–Stein restriction theorem states that the restriction of the Fourier transform to the unit sphere in $\mathbf{R}^n$ is a bounded operator on L^p provided $1 \le p \le (2n + 2)/(n + 3)$.

One notable difference between the Fourier transform in 1 dimension versus higher dimensions concerns the partial sum operator. Consider an increasing collection of measurable sets E_R indexed by $R \in (0,\infty)$: such as balls of radius R centered at the origin, or cubes of side $2R$. For a given integrable function f, consider the function f_R defined by:

$$f_R(x) = \int_{E_R} \hat{f}(\xi)e^{2\pi i x \cdot \xi}\, d\xi, \quad x \in \mathbf{R}^n.$$

Suppose in addition that $f \in L^p(\mathbf{R}^n)$. For $n = 1$ and $1 < p < \infty$, if one takes $E_R = (-R, R)$, then f_R converges to f in L^p as R tends to infinity, by the boundedness of the Hilbert transform. Naively one may hope the same holds true for $n > 1$. In the case that E_R is taken to be a cube with side length R, then convergence still holds. Another natural candidate is the Euclidean ball $E_R = \{\xi : |\xi| < R\}$. In order for this partial sum operator to converge, it is necessary that the multiplier for the unit ball be bounded in $L^p(\mathbf{R}^n)$. For $n \ge 2$ it is a celebrated theorem of Charles Fefferman that the multiplier for the unit ball is never bounded unless $p = 2$. In fact, when $p \ne 2$, this shows that not only may f_R fail to converge to f in L^p, but for some functions $f \in L^p(\mathbf{R}^n)$, f_R is not even an element of L^p.

Fourier Transform on Function Spaces

On L^p Spaces

On L^1

The definition of the Fourier transform by the integral formula

$$\hat{f}(\xi) = \int_{\mathbf{R}^n} f(x)e^{-2\pi i \xi \cdot x}\, dx$$

is valid for Lebesgue integrable functions f; that is, $f \in L^1(\mathbf{R}^n)$.

The Fourier transform $\mathcal{F} : L^1(\mathbf{R}^n) \to L^\infty(\mathbf{R}^n)$ is a bounded operator. This follows from the observation that

$$|\hat{f}(\xi)| \le \int_{\mathbf{R}^n} |f(x)|\, dx,$$

which shows that its operator norm is bounded by 1. Indeed, it equals 1, which can be seen, for example, from the transform of the rect function. The image of L^1 is a subset of the space $C_0(\mathbf{R}^n)$ of continuous functions that tend to zero at infinity (the Riemann–Lebesgue lemma), although it is not the entire space. Indeed, there is no simple characterization of the image.

On L^2

Since compactly supported smooth functions are integrable and dense in $L^2(\mathbb{R}^n)$, the Plancherel theorem allows us to extend the definition of the Fourier transform to general functions in $L^2(\mathbb{R}^n)$ by continuity arguments. The Fourier transform in $L^2(\mathbb{R}^n)$ is no longer given by an ordinary Lebesgue integral, although it can be computed by an improper integral, here meaning that for an L^2 function f,

$$\hat{f}(\xi) = \lim_{R\to\infty} \int_{|x|\le R} f(x) e^{-2\pi i x\cdot\xi}\, dx$$

where the limit is taken in the L^2 sense. Many of the properties of the Fourier transform in L^1 carry over to L^2, by a suitable limiting argument.

Furthermore, $\mathcal{F}: L^2(\mathbb{R}^n) \to L^2(\mathbb{R}^n)$ is a unitary operator. For an operator to be unitary it is sufficient to show that it is bijective and preserves the inner product, so in this case these follow from the Fourier inversion theorem combined with the fact that for any $f, g \in L^2(\mathbb{R}^n)$ we have

$$\int_{\mathbb{R}^n} f(x)\mathcal{F}g(x)dx = \int_{\mathbb{R}^n} \mathcal{F}f(x)g(x)dx.$$

In particular, the image of $L^2(\mathbb{R}^n)$ is itself under the Fourier transform.

On other L^p

The definition of the Fourier transform can be extended to functions in $L^p(\mathbb{R}^n)$ for $1 \le p \le 2$ by decomposing such functions into a fat tail part in L^2 plus a fat body part in L^1. In each of these spaces, the Fourier transform of a function in $L^p(\mathbb{R}^n)$ is in $L^q(\mathbb{R}^n)$, where $q = p/(p-1)$ is the Hölder conjugate of p (by the Hausdorff–Young inequality). However, except for $p = 2$, the image is not easily charac-terized. Further extensions become more technical. The Fourier transform of functions in L^p for the range $2 < p < \infty$ requires the study of distributions. In fact, it can be shown that there are functions in L^p with $p > 2$ so that the Fourier transform is not defined as a function.

Tempered Distributions

One might consider enlarging the domain of the Fourier transform from $L^1 + L^2$ by considering generalized functions, or distributions. A distribution on $\mathbb{R}^n$ is a continuous linear functional on the space $C_c(\mathbb{R}^n)$ of compactly supported smooth functions, equipped with a suitable topology. The strategy is then to consider the action of the Fourier transform on $C_c(\mathbb{R}^n)$ and pass to distributions by duality. The obstruction to do this is that the Fourier transform does not map $C_c(\mathbb{R}^n)$ to $C_c(\mathbb{R}^n)$. In fact the Fourier transform of an element in $C_c(\mathbb{R}^n)$ can not vanish on an open set. The right space here is the slightly larger space of Schwartz functions. The Fourier transform is an automorphism on the Schwartz space, as a topological vector space, and thus induces an automorphism on its dual, the space of tempered distributions. The tempered distributions include all the integrable functions mentioned above, as well as well-behaved functions of polynomial growth and distributions of compact support.

For the definition of the Fourier transform of a tempered distribution, let f and g be integrable functions, and let $\hat{f}$ and $\hat{g}$ be their Fourier transforms respectively. Then the Fourier transform obeys the following multiplication formula,

$$\int_{\mathbf{R}^n} \hat{f}(x)g(x)dx = \int_{\mathbf{R}^n} f(x)\hat{g}(x)dx.$$

Every integrable function f defines (induces) a distribution T_f by the relation

$$T_f(\varphi) = \int_{\mathbf{R}^n} f(x)\varphi(x)dx \quad \text{for all Schwartz functions } \phi.$$

So it makes sense to define Fourier transform $\hat{T}_f$ of T_f by

$$\hat{T}_f(\varphi) = T_f(\hat{\varphi})$$

for all Schwartz functions ϕ. Extending this to all tempered distributions T gives the general definition of the Fourier transform.

Distributions can be differentiated and the above-mentioned compatibility of the Fourier transform with differentiation and convolution remains true for tempered distributions.

Generalizations

Fourier–stieltjes Transform

The Fourier transform of a finite Borel measure μ on $\mathbf{R}^n$ is given by:

$$\hat{\mu}(\xi) = \int_{\mathbf{R}^n} e^{-2\pi i x \cdot \xi} \, d\mu.$$

This transform continues to enjoy many of the properties of the Fourier transform of integrable functions. One notable difference is that the Riemann–Lebesgue lemma fails for measures. In the case that $d\mu = f(x)\, dx$, then the formula above reduces to the usual definition for the Fourier transform of f. In the case that μ is the probability distribution associated to a random variable X, the Fourier–Stieltjes transform is closely related to the characteristic function, but the typical conventions in probability theory take $e^{ix \cdot \xi}$ instead of $e^{-2\pi ix \cdot \xi}$. In the case when the distribution has a probability density function this definition reduces to the Fourier transform applied to the probability density function, again with a different choice of constants.

The Fourier transform may be used to give a characterization of measures. Bochner's theorem characterizes which functions may arise as the Fourier–Stieltjes transform of a positive measure on the circle.

Furthermore, the Dirac delta function is not a function but it is a finite Borel measure. Its Fourier transform is a constant function (whose specific value depends upon the form of the Fourier transform used).

Locally Compact Abelian Groups

The Fourier transform may be generalized to any locally compact abelian group. A locally compact abelian group is an abelian group that is at the same time a locally compact Hausdorff topological space so that the group operation is continuous. If G is a locally compact abelian group, it has a translation invariant measure μ, called Haar measure. For a locally compact abelian group G, the set of irreducible, i.e. one-dimensional, unitary representations are called its characters. With its natural group structure and the topology of pointwise convergence, the set of characters $\hat{G}$ is itself a locally compact abelian group, called the *Pontryagin dual* of G. For a function f in $L^1(G)$, its Fourier transform is defined by

$$\hat{f}(\xi) = \int_G \xi(x) f(x) d\mu \qquad \text{for any } \xi \in \hat{G}.$$

The Riemann–Lebesgue lemma holds in this case; $\hat{f}(\xi)$ is a function vanishing at infinity on $\hat{G}$.

Gelfand Transform

The Fourier transform is also a special case of Gelfand transform. In this particular context, it is closely related to the Pontryagin duality map defined above.

Given an abelian locally compact Hausdorff topological group G, as before we consider space $L^1(G)$, defined using a Haar measure. With convolution as multiplication, $L^1(G)$ is an abelian Banach algebra. It also has an involution * given by

$$f^*(g) = \overline{f(g^{-1})}.$$

Taking the completion with respect to the largest possibly C*-norm gives its enveloping C*-algebra, called the group C*-algebra C*(G) of G. (Any C*-norm on $L^1(G)$ is bounded by the L^1 norm, therefore their supremum exists.)

Given any abelian C*-algebra A, the Gelfand transform gives an isomorphism between A and $C_0(A^\wedge)$, where $A^\wedge$ is the multiplicative linear functionals, i.e. one-dimensional representations, on A with the weak-* topology. The map is simply given by

$$a \mapsto (\varphi \mapsto \varphi(a))$$

It turns out that the multiplicative linear functionals of C*(G), after suitable identification, are exactly the characters of G, and the Gelfand transform, when restricted to the dense subset $L^1(G)$ is the Fourier-Pontryagin transform.

Compact Non-abelian Groups

The Fourier transform can also be defined for functions on a non-abelian group, provided that the group is compact. Removing the assumption that the underlying group is abelian, irreducible unitary representations need not always be one-dimensional. This means the Fourier transform on a non-abelian group takes values as Hilbert space operators. The Fourier transform on compact groups is a major tool in representation theory and non-commutative harmonic analysis.

Let G be a compact Hausdorff topological group. Let Σ denote the collection of all isomorphism classes of finite-dimensional irreducible unitary representations, along with a definite choice of representation $U^{(\sigma)}$ on the Hilbert space H_σ of finite dimension d_σ for each $\sigma \in \Sigma$. If μ is a finite Borel measure on G, then the Fourier–Stieltjes transform of μ is the operator on H_σ defined by

$$\langle \hat{\mu}\xi, \eta \rangle_{H_\sigma} = \int_G \langle \bar{U}_g^{(\sigma)}\xi, \eta \rangle d\mu(g)$$

where $\bar{U}^{(\sigma)}$ is the complex-conjugate representation of $U^{(\sigma)}$ acting on H_σ. If μ is absolutely continuous with respect to the left-invariant probability measure λ on G, represented as

$$d\mu = f\, d\lambda$$

for some $f \in L^1(\lambda)$, one identifies the Fourier transform of f with the Fourier–Stieltjes transform of μ.

The mapping $\mu \mapsto \hat{\mu}$ defines an isomorphism between the Banach space $M(G)$ of finite Borel measures and a closed subspace of the Banach space $C\,(\Sigma)$ consisting of all sequences $E = (E_\sigma)$ indexed by Σ of (bounded) linear operators $E : H_\sigma \to H_\sigma$ for which the norm

$$\| E \| = \sup_{\sigma \in \Sigma} \| E_\sigma \|$$

is finite. The "convolution theorem" asserts that, furthermore, this isomorphism of Banach spaces is in fact an isometric isomorphism of C* algebras into a subspace of $C_\infty(\Sigma)$. Multiplication on $M(G)$ is given by convolution of measures and the involution * defined by

$$f^*(g) = \overline{f(g^{-1})},$$

and $C_\infty(\Sigma)$ has a natural C*-algebra structure as Hilbert space operators.

The Peter–Weyl theorem holds, and a version of the Fourier inversion formula (Plancherel's theorem) follows: if $f \in L^2(G)$, then

$$f(g) = \sum_{\sigma \in \Sigma} d_\sigma \, \mathrm{tr}(\hat{f}(\sigma)U_g^{(\sigma)})$$

where the summation is understood as convergent in the L^2 sense.

The generalization of the Fourier transform to the noncommutative situation has also in part contributed to the development of noncommutative geometry. In this context, a categorical generalization of the Fourier transform to noncommutative groups is Tannaka–Krein duality, which replaces the group of characters with the category of representations. However, this loses the connection with harmonic functions.

Alternatives

In signal processing terms, a function (of time) is a representation of a signal with perfect *time resolution*, but no frequency information, while the Fourier transform has perfect *frequency res-olution*, but no time information: the magnitude of the Fourier transform at a point is how much frequency content there is, but location is only given by phase (argument of the Fourier transform at a point), and standing waves are not localized in time – a sine wave continues out to infinity, without decaying. This limits the usefulness of the Fourier transform for analyzing signals that are localized in time, notably transients, or any signal of finite extent.

As alternatives to the Fourier transform, in time-frequency analysis, one uses time-frequency transforms or time-frequency distributions to represent signals in a form that has some time in-formation and some frequency information – by the uncertainty principle, there is a trade-off be-tween these. These can be generalizations of the Fourier transform, such as the short-time Fourier transform or fractional Fourier transform, or other functions to represent signals, as in wavelet transforms and chirplet transforms, with the wavelet analog of the (continuous) Fourier transform being the continuous wavelet transform.

Applications

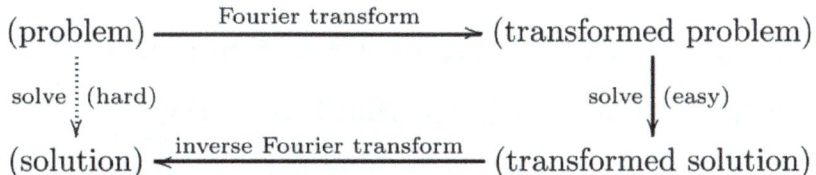

Some problems, such as certain differential equations, become easier to solve when the Fourier transform is applied.
In that case the solution to the original problem is recovered using the inverse Fourier transform.

Analysis of Differential Equations

Perhaps the most important use of the Fourier transformation is to solve partial differential equa-tions. Many of the equations of the mathematical physics of the nineteenth century can be treated this way. Fourier studied the heat equation, which in one dimension and in dimensionless units is

$$\frac{\partial^2 y(x,t)}{\partial^2 x} = \frac{\partial y(x,t)}{\partial t}.$$

The example we will give, a slightly more difficult one, is the wave equation in one dimension,

$$\frac{\partial^2 y(x,t)}{\partial^2 x} = \frac{\partial^2 y(x,t)}{\partial^2 t}.$$

As usual, the problem is not to find a solution: there are infinitely many. The problem is that of the so-called "boundary problem": find a solution which satisfies the "boundary conditions"

$$y(x,0) = f(x), \frac{\partial y(x,0)}{\partial t} = g(x).$$

Here, f and g are given functions. For the heat equation, only one boundary condition can be required (usually the first one). But for the wave equation, there are still infinitely many solutions

y which satisfy the first boundary condition. But when one imposes both conditions, there is only one possible solution.

It is easier to find the Fourier transform $\hat{y}$ of the solution than to find the solution directly. This is because the Fourier transformation takes differentiation into multiplication by the variable, and so a partial differential equation applied to the original function is transformed into multiplication by polynomial functions of the dual variables applied to the transformed function. After $\hat{y}$ is determined, we can apply the inverse Fourier transformation to find y.

Fourier's method is as follows. First, note that any function of the forms

$$\cos\big(2\pi\xi(x\pm t)\big) \ \text{ or } \ \sin\big(2\pi\xi(x\pm t)\big)$$

satisfies the wave equation. These are called "the elementary solutions."

Second, note that therefore any integral

$$y(x,t) = \int_0^\infty a_+(\xi)\cos\big(2\pi\xi(x+t)\big) + a_-(\xi)\cos\big(2\pi\xi(x-t)\big)$$
$$+ b_+(\xi)\sin\big(2\pi\xi(x+t)\big) + b_-(\xi)\sin\big(2\pi\xi(x-t)\big)d\xi$$

(for arbitrary a_+, a_-, b_+, and b_-) satisfies the wave equation. (This integral is just a kind of continuous linear combination, and the equation is linear.)

Now this resembles the formula for the Fourier synthesis of a function. In fact, this is the real inverse Fourier transform of $a_\pm$ and $b_\pm$ in the variable x.

The third step is to examine how to find the specific unknown coefficient functions $a_\pm$ and $b_\pm$ that will lead to y's satisfying the boundary conditions. We are interested in the values of these solutions at $t = 0..$ So we will set $t = 0$. Assuming that the conditions needed for Fourier inversion are satisfied, we can then find the Fourier sine and cosine transforms (in the variable x) of both sides and obtain

$$2\int_{-\infty}^{\infty} y(x,0)\cos\big(2\pi\xi x\big)dx = a_+ + a_-$$

and

$$2\int_{-\infty}^{\infty} y(x,0)\sin\big(2\pi\xi x\big)dx = b_+ + b_-.$$

Similarly, taking the derivative of y with respect to t and then applying the Fourier sine and cosine transformations yields

$$2\int_{-\infty}^{\infty} \frac{\partial y(u,0)}{\partial t}\sin(2\pi\xi x)dx = (2\pi\xi)(-a_+ + a_-)$$

and

$$2\int_{-\infty}^{\infty} \frac{\partial y(u,0)}{\partial t} \cos(2\pi\xi x)dx = (2\pi\xi)(b_+ - b_-).$$

These are four linear equations for the four unknowns $a_\pm$ and $b_\pm$, in terms of the Fourier sine and cosine transforms of the boundary conditions, which are easily solved by elementary algebra, provided that these transforms can be found.

In summary, we chose a set of elementary solutions, parametrised by ξ, of which the general solution would be a (continuous) linear combination in the form of an integral over the parameter ξ. But this integral was in the form of a Fourier integral. The next step was to express the boundary conditions in terms of these integrals, and set them equal to the given functions f and g. But these expressions also took the form of a Fourier integral because of the properties of the Fourier transform of a derivative. The last step was to exploit Fourier inversion by applying the Fourier transformation to both sides, thus obtaining expressions for the coefficient functions $a_\pm$ and $b_\pm$ in terms of the given boundary conditions f and g.

From a higher point of view, Fourier's procedure can be reformulated more conceptually. Since there are two variables, we will use the Fourier transformation in both x and t rather than operate as Fourier did, who only transformed in the spatial variables. Note that $\hat{y}$ must be considered in the sense of a distribution since $y(x,t)$ is not going to be L^1: as a wave, it will persist through time and thus is not a transient phenomenon. But it will be bounded and so its Fourier transform can be defined as a distribution. The operational properties of the Fourier transformation that are relevant to this equation are that it takes differentiation in x to multiplication by $2\pi i\xi$ and differentiation with respect to t to multiplication by $2\pi i f$ where f is the frequency. Then the wave equation becomes an algebraic equation in $\hat{y}$:

$$\xi^2 \hat{y}(\xi,f) = f^2 \hat{y}(\xi,f).$$

This is equivalent to requiring $\hat{y}(\xi,f) = 0$ unless $\xi = \pm f$. Right away, this explains why the choice of elementary solutions we made earlier worked so well: obviously $\hat{f} = \delta(\xi \pm f)$ will be solutions. Applying Fourier inversion to these delta functions, we obtain the elementary solutions we picked earlier. But from the higher point of view, one does not pick elementary solutions, but rather considers the space of all distributions which are supported on the (degenerate) conic $\xi^2 - f^2 = 0.$.

We may as well consider the distributions supported on the conic that are given by distributions of one variable on the line $\xi = f$ plus distributions on the line $\xi = -f$ as follows: if ϕ is any test function,

$$\iint \hat{y}\phi(\xi,f)d\xi df = \int s_+\phi(\xi,\xi)d\xi + \int s_-\phi(\xi,-\xi)d\xi,$$

where s_+, and s_-, are distributions of one variable.

Then Fourier inversion gives, for the boundary conditions, something very similar to what we had more concretely above (put $\phi(\xi,f) = e^{2\pi i(x\xi+tf)}$, which is clearly of polynomial growth):

$$y(x,0) = \int \{s_+(\xi) + s_-(\xi)\}e^{2\pi i\xi x+0}\, d\xi$$

and

$$\frac{\partial y(x,0)}{\partial t} = \int \{s_+(\xi) - s_-(\xi)\} 2\pi i \xi e^{2\pi i \xi x + 0} \, d\xi.$$

Now, as before, applying the one-variable Fourier transformation in the variable x to these functions of x yields two equations in the two unknown distributions $s_\pm$ (which can be taken to be ordinary functions if the boundary conditions are L^1 or L^2).

From a calculational point of view, the drawback of course is that one must first calculate the Fourier transforms of the boundary conditions, then assemble the solution from these, and then calculate an inverse Fourier transform. Closed form formulas are rare, except when there is some geometric symmetry that can be exploited, and the numerical calculations are difficult because of the oscillatory nature of the integrals, which makes convergence slow and hard to estimate. For practical calculations, other methods are often used.

The twentieth century has seen the extension of these methods to all linear partial differential equations with polynomial coefficients, and by extending the notion of Fourier transformation to include Fourier integral operators, some non-linear equations as well.

Fourier Transform Spectroscopy

The Fourier transform is also used in nuclear magnetic resonance (NMR) and in other kinds of spectroscopy, e.g. infrared (FTIR). In NMR an exponentially shaped free induction decay (FID) signal is acquired in the time domain and Fourier-transformed to a Lorentzian line-shape in the frequency domain. The Fourier transform is also used in magnetic resonance imaging (MRI) and mass spectrometry.

Quantum Mechanics

The Fourier transform is useful in Quantum Mechanics in two different ways. To begin with, the basic conceptual structure of Quantum Mechanics postulates the existence of pairs of complementary variables, connected by the Heisenberg uncertainty principle. For example, in one dimension, the spatial variable q of, say, a particle, can only be measured by the quantum mechanical "position operator" at the cost of losing information about the momentum p of the particle. Therefore, the physical state of the particle can either be described by a function, called "the wave function", of q or by a function of p but not by a function of both variables. The variable p is called the conjugate variable to q. In Classical Mechanics, the physical state of a particle (existing in one dimension, for simplicity of exposition) would be given by assigning definite values to both p and q simultaneously. Thus, the set of all possible physical states is the two-dimensional real vector space with a p-axis and a q-axis called the phase space.

In contrast, quantum mechanics chooses a polarisation of this space in the sense that it picks a subspace of one-half the dimension, for example, the q-axis alone, but instead of considering only points, takes the set of all complex-valued "wave functions" on this axis. Nevertheless, choosing the p-axis is an equally valid polarisation, yielding a different representation of the set of possible physical states of the particle which is related to the first representation by the Fourier transformation

$$\phi(p) = \int \psi(q) e^{\frac{2\pi i p q}{h}} \, dq.$$

Physically realisable states are L^2, and so by the Plancherel theorem, their Fourier transforms are also L^2. (Note that since q is in units of distance and p is in units of momentum, the presence of Planck's constant in the exponent makes the exponent dimensionless, as it should be.)

Therefore, the Fourier transform can be used to pass from one way of representing the state of the particle, by a wave function of position, to another way of representing the state of the particle: by a wave function of momentum. Infinitely many different polarisations are possible, and all are equally valid. Being able to transform states from one representation to another is sometimes convenient.

The other use of the Fourier transform in both Quantum Mechanics and Quantum Field Theory is to solve the applicable wave equation. In non-relativistic Quantum Mechanics, Schroedinger's equation for a time-varying wave function in one-dimension, not subject to external forces, is

$$\frac{\partial^2}{\partial x^2} \psi(x,t) = i \frac{h}{2\pi} \frac{\partial}{\partial t} \psi(x,t).$$

This is the same as the heat equation except for the presence of the imaginary unit i. Fourier methods can be used to solve this equation.

In the presence of a potential, given by the potential energy function $V(x)$, the equation becomes

$$\frac{\partial^2}{\partial x^2} \psi(x,t) + V(x)\psi(x,t) = i \frac{h}{2\pi} \frac{\partial}{\partial t} \psi(x,t).$$

The "elementary solutions", as we referred to them above, are the so-called "stationary states" of the particle, and Fourier's algorithm, as described above, can still be used to solve the boundary value problem of the future evolution of ψ given its values for $t = 0$. Neither of these approaches is of much practical use in Quantum Mechanics. Boundary value problems and the time-evolution of the wave function is not of much practical interest: it is the stationary states that are most important.

In relativistic Quantum Mechanics, Schroedinger's equation becomes a wave equation as was usual in classical physics, except that complex-valued waves are considered. A simple example, in the absence of interactions with other particles or fields, is the free one-dimensional Klein—Gordon—Schroedinger—Fock equation, this time in dimensionless units,

$$\left(\frac{\partial^2}{\partial x^2} + 1 \right) \psi(x,t) = \frac{\partial^2}{\partial t^2} \psi(x,t).$$

This is, from the mathematical point of view, the same as the wave equation of classical physics solved above (but with a complex-valued wave, which makes no difference in the methods). This is of great use in Quantum Field Theory: each separate Fourier component of a wave can be treated as a separate harmonic oscillator and then quantised, a procedure known as "second quantisation". Fourier methods have been adapted to also deal with non-trivial interactions.

Signal Processing

The Fourier transform is used for the spectral analysis of time-series. The subject of statistical signal processing does not, however, usually apply the Fourier transformation to the signal itself. Even if a real signal is indeed transient, it has been found in practice advisable to model a signal by a function (or, alternatively, a stochastic process) which is stationary in the sense that its characteristic properties are constant over all time. The Fourier transform of such a function does not exist in the usual sense, and it has been found more useful for the analysis of signals to instead take the Fourier transform of its auto-correlation function.

The auto-correlation function R of a function f is defined by

$$\mathbb{R}_f(\tau) = \lim_{T \to \infty} \frac{1}{2T} \int_{-T}^{T} f(t) f(t + \tau) dt.$$

This function is a function of the time-lag τ elapsing between the values of f to be correlated.

For most functions f that occur in practice, R is a bounded even function of the time-lag τ and for typical noisy signals it turns out to be uniformly continuous with a maximum at $\tau = $ zero.

The auto-correlation function, more properly called the auto-covariance function unless it is normalised in some appropriate fashion, measures the strength of the correlation between the values of f separated by a time-lag. This is a way of searching for the correlation of f with its own past. It is useful even for other statistical tasks besides the analysis of signals. For example, if $f(t)$ represents the temperature at time t, one expects a strong correlation with the temperature at a time-lag of 24 hours.

It possesses a Fourier transform,

$$P_f(\xi) = \int_{-\infty}^{\infty} R_f(\tau) e^{-2\pi i \xi \tau} \, d\tau.$$

This Fourier transform is called the power spectral density function of f. (Unless all periodic components are first filtered out from f, this integral will diverge, but it is easy to filter out such periodicities.)

The power spectrum, as indicated by this density function P, measures the amount of variance contributed to the data by the frequency ξ. In electrical signals, the variance is proportional to the average power (energy per unit time), and so the power spectrum describes how much the different frequencies contribute to the average power of the signal. This process is called the spectral analysis of time-series and is analogous to the usual analysis of variance of data that is not a time-series (ANOVA).

Knowledge of which frequencies are "important" in this sense is crucial for the proper design of filters and for the proper evaluation of measuring apparatuses. It can also be useful for the scientific analysis of the phenomena responsible for producing the data.

The power spectrum of a signal can also be approximately measured directly by measuring the average power that remains in a signal after all the frequencies outside a narrow band have been filtered out.

Spectral analysis is carried out for visual signals as well. The power spectrum ignores all phase relations, which is good enough for many purposes, but for video signals other types of spectral analysis must also be employed, still using the Fourier transform as a tool.

Other Notations

Other common notations for $\hat{f}(\xi)$ include:

$$\tilde{f}(\xi), \tilde{f}(\omega), F(\xi), \mathcal{F}(f)(\xi), (\mathcal{F}f)(\xi), \mathcal{F}(f), \mathcal{F}(\omega), F(\omega), \mathcal{F}(j\omega), \mathcal{F}\{f\}, \mathcal{F}(f(t)), \mathcal{F}\{f(t)\}.$$

Denoting the Fourier transform by a capital letter corresponding to the letter of function being transformed (such as $f(x)$ and $F(\xi)$) is especially common in the sciences and engineering. In electronics, the omega (ω) is often used instead of ξ due to its interpretation as angular frequency, sometimes it is written as $F(j\omega)$, where j is the imaginary unit, to indicate its relationship with the Laplace transform, and sometimes it is written informally as $F(2\pi f)$ in order to use ordinary frequency.

The interpretation of the complex function $\hat{f}(\xi)$ may be aided by expressing it in polar coordinate form

$$\hat{f}(\xi) = A(\xi)e^{i\varphi(\xi)}$$

in terms of the two real functions $A(\xi)$ and $\varphi(\xi)$ where:

$$A(\xi) = |\hat{f}(\xi)|,$$

is the amplitude and

$$\varphi(\xi) = \arg\big(\hat{f}(\xi)\big),$$

is the phase.

Then the inverse transform can be written:

$$f(x) = \int_{-\infty}^{\infty} A(\xi)\, e^{i(2\pi\xi x + \varphi(\xi))}\, d\xi,$$

which is a recombination of all the frequency components of $f(x)$. Each component is a complex sinusoid of the form $e^{2\pi i x\xi}$ whose amplitude is $A(\xi)$ and whose initial phase angle (at $x = 0$) is $\varphi(\xi)$.

The Fourier transform may be thought of as a mapping on function spaces. This mapping is here denoted $\mathcal{F}$ and $\mathcal{F}(f)$ is used to denote the Fourier transform of the function f. This mapping is linear, which means that $\mathcal{F}$ can also be seen as a linear transformation on the function space and implies that the standard notation in linear algebra of applying a linear transformation to a vector (here the function f) can be used to write $\mathcal{F}f$ instead of $\mathcal{F}(f)$. Since the result of applying the Fourier transform is again a function, we can be interested in the value of this function evaluated at the value ξ for its variable, and this is denoted either as $\mathcal{F}f(\xi)$ or as $\mathcal{F}f(\xi)$. Notice that in the former case, it is implicitly understood that $\mathcal{F}$ is applied first to f and then the resulting function is evaluated at ξ, not the other way around.

In mathematics and various applied sciences, it is often necessary to distinguish between a function f and the value of f when its variable equals x, denoted $f(x)$. This means that a notation like $\mathcal{F}(f(x))$ formally can be interpreted as the Fourier transform of the values of f at x. Despite this flaw, the previous notation appears frequently, often when a particular function or a function of a particular variable is to be transformed.

For example, $\mathcal{F}(\text{rect}(x)) = \text{sinc}(\xi)$ is sometimes used to express that the Fourier transform of a rectangular function is a sinc function,

or $\mathcal{F}(f(x+x_0)) = \mathcal{F}(f(x))e^{2\pi i \xi x_0}$ is used to express the shift property of the Fourier transform.

Notice, that the last example is only correct under the assumption that the transformed function is a function of x, not of x_0.

Other Conventions

The Fourier transform can also be written in terms of angular frequency:

$$\omega = 2\pi\xi,$$

whose units are radians per second.

The substitution $\xi = \omega/(2\pi)$ into the formulas above produces this convention:

$$\hat{f}(\omega) = \int_{\mathbf{R}^n} f(x)e^{-i\omega \cdot x}\,dx.$$

Under this convention, the inverse transform becomes:

$$f(x) = \frac{1}{(2\pi)^n}\int_{\mathbf{R}^n} \hat{f}(\omega)e^{i\omega \cdot x}\,d\omega.$$

Unlike the convention followed in this section, when the Fourier transform is defined this way, it is no longer a unitary transformation on $L^2(\mathbf{R}^n)$. There is also less symmetry between the formulas for the Fourier transform and its inverse.

Another convention is to split the factor of $(2\pi)^n$ evenly between the Fourier transform and its inverse, which leads to definitions:

$$\hat{f}(\omega) = \frac{1}{(2\pi)^{n/2}}\int_{\mathbf{R}^n} f(x)e^{-i\omega \cdot x}\,dx,$$

$$f(x) = \frac{1}{(2\pi)^{n/2}} \int_{\mathbf{R}^n} \hat{f}(\omega) e^{i\omega \cdot x} \, d\omega.$$

Under this convention, the Fourier transform is again a unitary transformation on $L^2(\mathbf{R}^n)$. It also restores the symmetry between the Fourier transform and its inverse.

Variations of all three conventions can be created by conjugating the complex-exponential kernel of both the forward and the reverse transform. The signs must be opposites. Other than that, the choice is (again) a matter of convention.

Summary of popular forms of the Fourier transform

ordinary frequency ξ (Hz)	unitary	$\hat{f_1}(\xi) \stackrel{\text{def}}{=} \int_{\mathbf{R}^n} f(x) e^{-2\pi i x \cdot \xi} \, dx = \hat{f_2}(2\pi\xi) = (2\pi)^{n/2} \hat{f_3}(2\pi\xi)$ $f(x) = \int_{\mathbf{R}^n} \hat{f_1}(\xi) e^{2\pi i x \cdot \xi} \, d\xi$
angular frequency ω (rad/s)	unitary	$\hat{f_3}(\omega) \stackrel{\text{def}}{=} \frac{1}{(2\pi)^{n/2}} \int_{\mathbf{R}^n} f(x) e^{-i\omega \cdot x} \, dx = \frac{1}{(2\pi)^{n/2}} \hat{f_1}\left(\frac{\omega}{2\pi}\right) = \frac{1}{(2\pi)^{n/2}} \hat{f_2}(\omega)$ $f(x) = \frac{1}{(2\pi)^{n/2}} \int_{\mathbf{R}^n} \hat{f_3}(\omega) e^{i\omega \cdot x} \, d\omega$
	non-unitary	$\hat{f_2}(\omega) \stackrel{\text{def}}{=} \int_{\mathbf{R}^n} f(x) e^{-i\omega \cdot x} \, dx = \hat{f_1}\left(\frac{\omega}{2\pi}\right) = (2\pi)^{n/2} \hat{f_3}(\omega)$ $f(x) = \frac{1}{(2\pi)^n} \int_{\mathbf{R}^n} \hat{f_2}(\omega) e^{i\omega \cdot x} \, d\omega$

As discussed above, the characteristic function of a random variable is the same as the Fourier–Stieltjes transform of its distribution measure, but in this context it is typical to take a different convention for the constants. Typically characteristic function is defined $E(e^{it \cdot X}) = \int e^{it \cdot x} \, d\mu_X(x)$.

As in the case of the "non-unitary angular frequency" convention above, there is no factor of 2π appearing in either of the integral, or in the exponential. Unlike any of the conventions appearing above, this convention takes the opposite sign in the exponential.

Computation Methods

The appropriate computation method largely depends how the original mathematical function is represented and the desired form of the output function.

Since the fundamental definition of a Fourier transform is an integral, functions that can be expressed as closed-form expressions are commonly computed by working the integral analytically to yield a closed-form expression in the Fourier transform conjugate variable as the result. This is the method used to generate tables of Fourier transforms, including those found in the table below (Fourier transform#Tables of important Fourier transforms).

Many computer algebra systems such as Matlab and Mathematica that are capable of symbolic integration are capable of computing Fourier transforms analytically. For example, to compute the Fourier transform of $f(t) = \cos(6\pi t)\, e^{-\pi t2}$ one might enter the command "integrate cos(6*pi*t) exp(−pi*t^2) exp(-i*2*pi*f*t) from -inf to inf" into Wolfram Alpha.

Numerical Integration of Closed-form Functions

If the input function is in closed-form and the desired output function is a series of ordered pairs (for example a table of values from which a graph can be generated) over a specified domain, then the Fourier transform can be generated by numerical integration at each value of the Fourier conjugate variable (frequency, for example) for which a value of the output variable is desired. Note that this method requires computing a separate numerical integration for each value of frequency for which a value of the Fourier transform is desired. The numerical integration approach works on a much broader class of functions than the analytic approach, because it yields results for functions that do not have closed form Fourier transform integrals.

Numerical Integration of a Series of Ordered Pairs

If the input function is a series of ordered pairs (for example, a time series from measuring an output variable repeatedly over a time interval) then the output function must also be a series of ordered pairs (for example, a complex number vs. frequency over a specified domain of frequencies), unless certain assumptions and approximations are made allowing the output function to be approximated by a closed-form expression. In the general case where the available input series of ordered pairs are assumed be samples representing a continuous function over an interval (amplitude vs. time, for example), the series of ordered pairs representing the desired output function can be obtained by numerical integration of the input data over the available interval at each value of the Fourier conjugate variable (frequency, for example) for which the value of the Fourier transform is desired.

Explicit numerical integration over the ordered pairs can yield the Fourier transform output value for any desired value of the conjugate Fourier transform variable (frequency, for example), so that a spectrum can be produced at any desired step size and over any desired variable range for accurate determination of amplitudes, frequencies, and phases corresponding to isolated peaks. Unlike limitations in DFT and FFT methods, explicit numerical integration can have any desired step size and compute the Fourier transform over any desired range of the congugate Fourier transform variable (for example, frequency).

Discrete Fourier Transforms and Fast Fourier Transforms

If the ordered pairs representing the original input function are equally spaced in their input variable (for example, equal time steps), then the Fourier transform is known as a discrete Fourier transform (DFT), which can be computed either by explicit numerical integration, by explicit evaluation of the DFT definition, or by fast Fourier transform (FFT) methods. In contrast to explicit integration of input data, use of the DFT and FFT methods produces Fourier transforms described by ordered pairs of step size equal to the reciprocal of the original sampling interval. For example, if the input data is sampled for 10 seconds, the output of DFT and FFT methods will have a 0.1 Hz frequency spacing.

Convergence of Fourier Series

In mathematics, the question of whether the Fourier series of a periodic function converges to the given function is researched by a field known as classical harmonic analysis, a branch of pure mathematics. Convergence is not necessarily given in the general case, and certain criteria must be met for convergence to occur.

Determination of convergence requires the comprehension of pointwise convergence, uniform convergence, absolute convergence, L^p spaces, summability methods and the Cesàro mean.

Preliminaries

Consider f an integrable function on the interval $[0,2\pi]$. For such an f the Fourier coefficients $\hat{f}(n)$ are defined by the formula

$$\hat{f}(n) = \frac{1}{2\pi} \int_0^{2\pi} f(t)e^{-int}\, dt, \quad n \in \mathbf{Z}.$$

It is common to describe the connection between f and its Fourier series by

$$f \sim \sum_n \hat{f}(n)e^{int}.$$

The notation ~ here means that the sum represents the function in some sense. To investigate this more carefully, the partial sums must be defined:

$$S_N(f;t) = \sum_{n=-N}^{N} \hat{f}(n)e^{int}.$$

The question here is: Do the functions $S_N(f)$ (which are functions of the variable t we omitted in the notation) converge to f and in which sense? Are there conditions on f ensuring this or that type of convergence? This is the main problem discussed in this chapter.

Before continuing, the Dirichlet kernel must be introduced. Taking the formula for $\hat{f}(n)$, inserting it into the formula for S_N and doing some algebra gives that

$$S_N(f) = f * D_N$$

where $*$ stands for the periodic convolution and D_N is the Dirichlet kernel, which has an explicit formula,

$$D_n(t) = \frac{\sin((n+\frac{1}{2})t)}{\sin(t/2)}.$$

The Dirichlet kernel is *not* a positive kernel, and in fact, its norm diverges, namely

$$\int |D_n(t)|\, dt \to \infty$$

a fact that plays a crucial role in the discussion. The norm of D_n in $L^1(T)$ coincides with the norm of the convolution operator with D_n, acting on the space $C(T)$ of periodic continuous functions, or with the norm of the linear functional $f \to (S_n f)(0)$ on $C(T)$. Hence, this family of linear functionals on $C(T)$ is unbounded, when $n \to \infty$.

Magnitude of Fourier Coefficients

In applications, it is often useful to know the size of the Fourier coefficient.

If f is an absolutely continuous function,

$$\left| \hat{f}(n) \right| \le \frac{K}{|n|}$$

for K a constant that only depends on f.

If f is a bounded variation function,

$$\left| \hat{f}(n) \right| \le \frac{\mathrm{var}(f)}{2\pi |n|}.$$

If $f \in C^p$

$$\left| \hat{f}(n) \right| \le \frac{\| f^{(p)} \|_{L_1}}{|n|^p}.$$

If $f \in C^p$ and $f^{(p)}$ has modulus of continuity ω_p,

$$\left| \hat{f}(n) \right| \le \frac{\omega(2\pi/n)}{|n|^p}$$

and therefore, if f is in the α-Hölder class

$$\left| \hat{f}(n) \right| \le \frac{K}{|n|^\alpha}.$$

Pointwise convergence

There are many known sufficient conditions for the Fourier series of a function to converge at a given point x, for example if the function is differentiable at x. Even a jump discontinuity does not pose a problem: if the function has left and right derivatives at x, then the Fourier series converges to the average of the left and right limits.

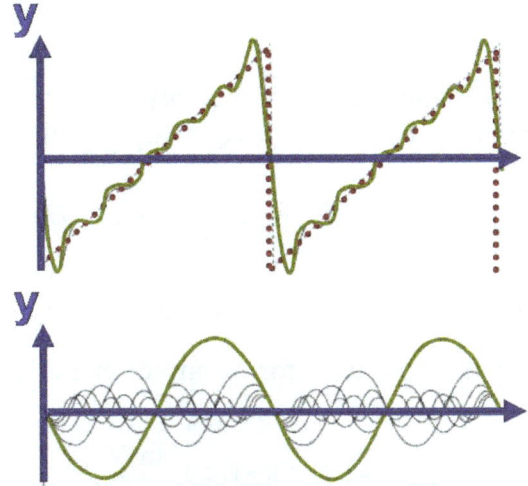

Superposition of sinusoidal wave basis functions (bottom) to form a sawtooth wave (top); the basis functions have wavelengths λ/k (k=integer) shorter than the wavelength λ of the sawtooth itself (except for k=1). All basis functions have nodes at the nodes of the sawtooth, but all but the fundamental have additional nodes. The oscillation about the sawtooth is called the Gibbs phenomenon

The Dirichlet–Dini Criterion states that: if f is 2π–periodic, locally integrable and satisfies

$$\int_0^\pi \left| \frac{f(x_0 + t) + f(x_0 - t)}{2} - \ell \right| \frac{dt}{t} < \infty,$$

then $(S_n f)(x_0)$ converges to ℓ. This implies that for any function f of any Hölder class $\alpha > 0$, the Fourier series converges everywhere to $f(x)$.

It is also known that for any periodic function of bounded variation, the Fourier series converges everywhere. In general, the most common criteria for pointwise convergence of a periodic function f are as follows:

- If f satisfies a Holder condition, then its Fourier series converges uniformly.

- If f is of bounded variation, then its Fourier series converges everywhere.

- If f is continuous and its Fourier coefficients are absolutely summable, then the Fourier series converges uniformly.

There exist continuous functions whose Fourier series converges pointwise but not uniformly.

However, the Fourier series of a continuous function need not converge pointwise. Perhaps the easiest proof uses the non-boundedness of Dirichlet's kernel in $L^1(T)$ and the Banach–Steinhaus uniform boundedness principle. As typical for existence arguments invoking the Baire category theorem, this proof is nonconstructive. It shows that the family of continuous functions whose Fourier series converges at a given x is of first Baire category, in the Banach space of continuous functions on the circle. So in some sense pointwise convergence is *atypical*, and for most continuous functions the Fourier series does not converge at a given point.

However Carleson's theorem shows that for a given continuous function the Fourier series converges almost everywhere.

Uniform Convergence

Suppose $f \in C^p$, and $f^{(p)}$ has modulus of continuity ω (we assume here that ω is also non decreasing), then the partial sum of the Fourier series converges to the function with speed

$$| f(x) - (S_N f)(x) | \leq K \frac{\ln N}{N^p} \omega(2\pi / N)$$

for a constant K that does not depend upon f, nor p, nor N.

This theorem, first proved by D Jackson, tells, for example, that if f satisfies the α-Hölder condition, then

$$| f(x) - (S_N f)(x) | \leq K \frac{\ln N}{N^\alpha}.$$

If f is 2π periodic and absolutely continuous on $[0, 2\pi]$, then the Fourier series of f converges uniformly, but not necessarily absolutely, to f.

Absolute Convergence

A function f has an absolutely converging Fourier series if

$$\| f \|_A := \sum_{n=-\infty}^{\infty} |\hat{f}(n)| < \infty.$$

Obviously, if this condition holds then $(S_N f)(t)$ converges absolutely for every t and on the other hand, it is enough that $(S_N f)(t)$ converges absolutely for even one t, then this condition holds. In other words, for absolute convergence there is no issue of *where* the sum converges absolutely — if it converges absolutely at one point then it does so everywhere.

The family of all functions with absolutely converging Fourier series is a Banach algebra (the operation of multiplication in the algebra is a simple multiplication of functions). It is called the Wiener algebra, after Norbert Wiener, who proved that if f has absolutely converging Fourier series and is never zero, then $1/f$ has absolutely converging Fourier series. The original proof of Wiener's theorem was difficult; a simplification using the theory of Banach algebras was given by Israel Gelfand. Finally, a short elementary proof was given by Donald J. Newman in 1975.

If f belongs to a α-Hölder class for α > 1/2 then

$$\| f \|_A \leq c_\alpha \| f \|_{\text{Lip}_\alpha}, \qquad \| f \|_K := \sum_{n=-\infty}^{+\infty} |n| |\hat{f}(n)|^2 \leq c_\alpha \| f \|_{\text{Lip}_\alpha}^2$$

for $\| f \|_{\text{Lip}_\alpha}$ the constant in the Hölder condition, c_α a constant only dependent on α; $\| f \|_K$ is the norm of the Krein algebra. Notice that the 1/2 here is essential—there are 1/2-Hölder functions, which do not belong to the Wiener algebra. Besides, this theorem cannot improve the best known bound on the size of the Fourier coefficient of a α-Hölder function—that is only $O(1/n^\alpha)$ and then not summable.

If f is of bounded variation *and* belongs to a α-Hölder class for some α > 0, it belongs to the Wiener algebra.

Norm Convergence

The simplest case is that of L^2, which is a direct transcription of general Hilbert space results. According to the Riesz–Fischer theorem, if f is square-integrable then

$$\lim_{N\to\infty}\int_0^{2\pi}\left|f(x)-S_N(f)\right|^2 dx = 0$$

i.e., $S_N f$ converges to f in the norm of L^2. It is easy to see that the converse is also true: if the limit above is zero, f must be in L^2. So this is an if and only if condition.

If 2 in the exponents above is replaced with some p, the question becomes much harder. It turns out that the convergence still holds if $1 < p < \infty$. In other words, for f in L^p, $S_N(f)$ converges to f in the L^p norm. The original proof uses properties of holomorphic functions and Hardy spaces, and another proof, due to Salomon Bochner relies upon the Riesz–Thorin interpolation theorem. For $p = 1$ and infinity, the result is not true. The construction of an example of divergence in L^1 was first done by Andrey Kolmogorov. For infinity, the result is a corollary of the uniform boundedness principle.

If the partial summation operator S_N is replaced by a suitable summability kernel (for example the *Fejér sum* obtained by convolution with the Fejér kernel), basic functional analytic techniques can be applied to show that norm convergence holds for $1 \le p < \infty$.

Convergence Almost Everywhere

The problem whether the Fourier series of any continuous function converges almost everywhere was posed by Nikolai Lusin in the 1920s and remained open until finally resolved positively in 1966 by Lennart Carleson. Indeed, Carleson showed that the Fourier expansion of any function in L^2 converges almost everywhere. This result is now known as Carleson's theorem. Later on Richard Hunt generalized this to L^p for any $p > 1$. Despite a number of attempts at simplifying the proof, it is still one of the most difficult results in analysis.

Contrariwise, Andrey Kolmogorov, as a student at the age of 19, in his very first scientific work, constructed an example of a function in L^1 whose Fourier series diverges almost everywhere (later improved to diverge everywhere).

It might be interesting to note that Jean-Pierre Kahane and Yitzhak Katznelson proved that for any given set E of measure zero, there exists a continuous function f such that the Fourier series of f fails to converge on any point of E.

Summability

Does the sequence 0,1,0,1,0,1,... (the partial sums of Grandi's series) converge to ½? This does not seem like a very unreasonable generalization of the notion of convergence. Hence we say that any sequence a_n is Cesàro summable to some a if

$$\lim_{n\to\infty}\frac{1}{n}\sum_{k=1}^{n}a_k = a.$$

It is not difficult to see that if a sequence converges to some a then it is also Cesàro summable to it.

To discuss summability of Fourier series, we must replace S_N with an appropriate notion. Hence we define

$$K_N(f;t)=\frac{1}{N}\sum_{n=0}^{N-1}S_n(f;t), \quad N\geq 1,$$

and ask: does $K_N(f)$ converge to f? K_N is no longer associated with Dirichlet's kernel, but with Fejér's kernel, namely

$$K_N(f)=f*F_N$$

where F_N is Fejér's kernel,

$$F_N=\frac{1}{N}\sum_{n=0}^{N-1}D_n.$$

The main difference is that Fejér's kernel is a positive kernel. Fejér's theorem states that the above sequence of partial sums converge uniformly to f. This implies much better convergence properties

- If f is continuous at t then the Fourier series of f is summable at t to $f(t)$. If f is continuous, its Fourier series is uniformly summable (i.e. $K_N f$ converges uniformly to f).

- For any integrable f, $K_N f$ converges to f in the L^1 norm.

- There is no Gibbs phenomenon.

Results about summability can also imply results about regular convergence. For example, we learn that if f is continuous at t, then the Fourier series of f cannot converge to a value different from $f(t)$. It may either converge to $f(t)$ or diverge. This is because, if $S_N(f;t)$ converges to some value x, it is also summable to it, so from the first summability property above, $x=f(t)$.

Order of Growth

The order of growth of Dirichlet's kernel is logarithmic, i.e.

$$\int |D_N(t)|\,dt = \frac{4}{\pi^2}\log N + O(1).$$

Big O notation is the notation $O(1)$. It should be noted that the actual value $4/\pi^2$ is both difficult to calculate and of almost no use. The fact that for *some* constant c we have

$$\int |D_N(t)|\,dt > c\log N + O(1)$$

is quite clear when one examines the graph of Dirichlet's kernel. The integral over the n-th peak is bigger than c/n and therefore the estimate for the harmonic sum gives the logarithmic estimate.

This estimate entails quantitative versions of some of the previous results. For any continuous function f and any t one has

$$\lim_{N \to \infty} \frac{S_N(f;t)}{\log N} = 0.$$

However, for any order of growth $\omega(n)$ smaller than log, this no longer holds and it is possible to find a continuous function f such that for some t,

$$\varlimsup_{N \to \infty} \frac{S_N(f;t)}{\omega(N)} = \infty.$$

The equivalent problem for divergence everywhere is open. Sergei Konyagin managed to construct an integrable function such that for *every* t one has

$$\varlimsup_{N \to \infty} \frac{S_N(f;t)}{\sqrt{\log N}} = \infty.$$

It is not known whether this example is best possible. The only bound from the other direction known is $\log n$.

Multiple Dimensions

Upon examining the equivalent problem in more than one dimension, it is necessary to specify the precise order of summation one uses. For example, in two dimensions, one may define

$$S_N(f;t_1,t_2) = \sum_{|n_1| \le N, |n_2| \le N} \hat{f}(n_1,n_2) e^{i(n_1 t_1 + n_2 t_2)}$$

which are known as "square partial sums". Replacing the sum above with

$$\sum_{n_1^2 + n_2^2 \le N^2}$$

lead to "circular partial sums". The difference between these two definitions is quite notable. For example, the norm of the corresponding Dirichlet kernel for square partial sums is of the order of $\log^2 N$ while for circular partial sums it is of the order of $\sqrt{N}$.

Many of the results true for one dimension are wrong or unknown in multiple dimensions. In particular, the equivalent of Carleson's theorem is still open for circular partial sums. Almost everywhere convergence of "square partial sums" (as well as more general polygonal partial sums) in multiple dimensions was established around 1970 by Charles Fefferman.

Pontryagin Duality

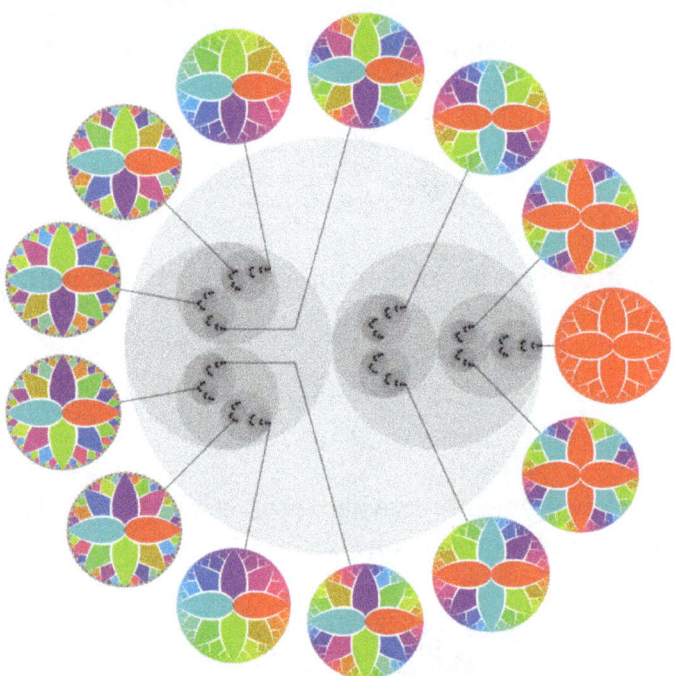

The 2-adic integers, with selected corresponding characters on their Pontryagin dual group

In mathematics, specifically in harmonic analysis and the theory of topological groups, Pontryagin duality explains the general properties of the Fourier transform on locally compact groups, such as $\mathbb{R}$, , the circle, or finite cyclic groups. The Pontryagin duality theorem itself states that locally compact abelian groups identify naturally with their bidual.

The subject is named after Lev Semenovich Pontryagin who laid down the foundations for the theory of locally compact abelian groups and their duality during his early mathematical works in 1934. Pontryagin's treatment relied on the group being second-countable and either compact or discrete. This was improved to cover the general locally compact abelian groups by Egbert van Kampen in 1935 and André Weil in 1940.

Introduction

Pontryagin duality places in a unified context a number of observations about functions on the real line or on finite abelian groups:

- Suitably regular complex-valued periodic functions on the real line have Fourier series and these functions can be recovered from their Fourier series;

- Suitably regular complex-valued functions on the real line have Fourier transforms that are also functions on the real line and, just as for periodic functions, these functions can be recovered from their Fourier transforms; and

- Complex-valued functions on a finite abelian group have discrete Fourier transforms which are functions on the dual group, which is a (non-canonically) isomorphic group. Moreover, any function on a finite group can be recovered from its discrete Fourier transform.

The theory, introduced by Lev Pontryagin and combined with Haar measure introduced by John von Neumann, André Weil and others depends on the theory of the dual group of a locally compact abelian group.

It is analogous to the dual vector space of a vector space: a finite-dimensional vector space V and its dual vector space V^* are not naturally isomorphic, but their endomorphism algebras (matrix algebras) are: $\text{End}(V) \cong \text{End}(V^*)$, via the transpose. Similarly, a group G and its dual group are $\hat{G}$ not in general isomorphic, but their group algebras are: $C(G) \cong C(\hat{G})$ via the Fourier transform, though one must carefully define these algebras analytically. More categorically, this is not just an isomorphism of endomorphism algebras, but an isomorphism of categories.

Locally Compact Abelian Groups

A topological group is called *locally compact* if the underlying topological space is locally compact and Hausdorff. It's called *abelian* if the underlying group is abelian.

Examples

Examples of locally compact abelian groups are:

- $\mathbb{R}^n$ for n a positive integer, with vector addition as group operation.

- The positive real numbers $\mathbb{R}^+$ with multiplication as operation. This group is isomorphic to $(\mathbb{R}, +)$ by the exponential map.

- Any finite abelian group, with the discrete topology. By the structure theorem for finite abelian groups, all such groups are products of cyclic groups.

- The integers $\mathbb{Z}$ under addition, again with the discrete topology.

- The circle group, denoted $\mathbb{T}$ for torus. This is the group of complex numbers of modulus 1. $\mathbb{T}$ is isomorphic as a topological group to the quotient group $\mathbb{R}/\mathbb{Z}$.

- The field $\mathbb{Q}_p$ of p-adic numbers under addition, with the usual p-adic topology.

The Dual Group

If G is a locally compact *abelian* group, a character of G is a continuous group homomorphism from G with values in the circle group $\mathbb{T}$. The set of all characters on G can be made into a locally compact abelian group, called the *dual group* of G and denoted $\hat{G}$. The group operation on the dual group is given by pointwise multiplication of characters, the inverse of a character is its complex conjugate and the topology on the space of characters is that of uniform convergence on compact sets (i.e., the compact-open topology, viewing $\hat{G}$ as a subset of the space of all continuous functions from G to $\mathbb{T}$.). This topology in general is not metrizable. However, if the group G is a separable locally compact abelian group, then the dual group is metrizable.

This is analogous to the dual space in linear algebra: just as for a vector space V over a field K, the dual space is $\mathrm{Hom}(V, K)$, so too is the dual group $\mathrm{Hom}(G, \mathbb{T})$. More abstractly, these are both examples of representable functors, being represented respectively by K and $\mathbb{T}$.

A group that is isomorphic (as topological groups) to its dual group is called *self-dual*. While the reals and finite cyclic groups are self-dual, the group and the dual group are not *naturally* isomorphic, and should be thought of as two different groups.

Examples of Dual Groups

The dual of $\mathbb{Z}$ is isomorphic to the circle group $\mathbb{T}$. A character on the infinite cyclic group of integers $\mathbb{Z}$ under addition is determined by its value at the generator 1. Thus for any character χ on $\mathbb{Z}$, $\chi(n) = \chi(1)^n$. Moreover, this formula defines a character for any choice of $\chi(1)$ in $\mathbb{T}$. The topology of uniform convergence on compact sets is in this case the topology of pointwise convergence. This is the topology of the circle group inherited from the complex numbers.

The dual of $\mathbb{T}$ is canonically isomorphic with $\mathbb{Z}$. Indeed, a character on $\mathbb{T}$ is of the form $z \mapsto z^n$ for n an integer. Since $\mathbb{T}$ is compact, the topology on the dual group is that of uniform convergence, which turns out to be the discrete topology.

The group of real numbers $\mathbb{R}$, is isomorphic to its own dual; the characters on $\mathbb{T}$ are of the form $r \mapsto e^{i\theta r}$ for θ a real number. With these dualities, the version of the Fourier transform to be introduced next coincides with the classical Fourier transform on $\mathbb{R}$.

Analogously, the group of p-adic numbers $\mathbb{Q}_p$ is isomorphic to its dual. (In fact, any finite extension of $\mathbb{Q}_p$ is also self-dual.) It follows that the adeles are self-dual.

The Pontryagin Duality Theorem

> Theorem. There is a canonical isomorphism $G \cong \hat{\hat{G}}$ between any locally compact abelian group G and its double dual.

Canonical means that there is a naturally defined map $ev_G : G \to \hat{\hat{G}}$; more importantly, the map should be functorial in G. The canonical isomorphism is defined on $x \in G$ as follows:

$$ ev_G(x) = \{\chi \mapsto \chi(x)\} \text{ i.e. } ev_G(x)(\chi) := \chi(x) \in \mathbb{T}. $$

In other words, each group element x is identified to the evaluation character on the dual. This is strongly analogous to the canonical isomorphism between a finite-dimensional vector space and its double dual, $V \cong V^{**}$. However, there is also a difference: V is isomorphic to its dual space V^*, although not canonically so, while many groups are not isomorphic to their dual groups (for instance $\hat{\mathbb{T}} = \mathbb{Z}$ but $\mathbb{T} \ncong \mathbb{Z}$ as topological groups). If G is a finite abelian group, then $G \cong \hat{G}$ but this isomorphism is not canonical. Making this statement precise (in general) requires thinking about dualizing not only on groups, but also on maps between the groups, in order to treat dualization as a functor and prove the identity functor and the dualization functor are not naturally equivalent. Also it should be noted that the duality theorem implies that for any group (not necessarily finite) the dualization functor is an exact functor.

Pontryagin Duality and the Fourier Transform

Haar Measure

One of the most remarkable facts about a locally compact group G is that it carries an essentially unique natural measure, the Haar measure, which allows one to consistently measure the "size" of sufficiently regular subsets of G. "Sufficiently regular subset" here means a Borel set; that is, an element of the σ-algebra generated by the compact sets. More precisely, a right Haar measure on a locally compact group G is a countably additive measure μ defined on the Borel sets of G which is *right invariant* in the sense that μ(Ax) = μ(A) for x an element of G and A a Borel subset of G and also satisfies some regularity conditions. Except for positive scaling factors, a Haar measure on G is unique.

The Haar measure on G allows us to define the notion of integral for (complex-valued) Borel functions defined on the group. In particular, one may consider various L^p spaces associated to the Haar measure μ. Specifically,

$$L^p_\mu(G) = \left\{ f : G \to \mathbb{C} : \int_G |f(x)|^p \, d\mu(x) < \infty \right\}.$$

Note that, since any two Haar measures on G are equal up to a scaling factor, this L^p-space is independent of the choice of Haar measure and thus perhaps could be written as $L^p(G)$. However, the L^p-norm on this space depends on the choice of Haar measure, so if one wants to talk about isometries it is important to keep track of the Haar measure being used.

Fourier Transform and Fourier inversion formula for L^1-functions

The dual group of a locally compact abelian group is used as the underlying space for an abstract version of the Fourier transform. If $f \in L^1(G)$, , then the Fourier transform is the function $\hat{f}$ on $\hat{G}$ defined by

$$\hat{f}(\chi) = \int_G f(x) \overline{\chi(x)} \, d\mu(x),$$

where the integral is relative to Haar measure μ on G. This is also denoted $(\mathcal{F}f)(\chi)$. Note the Fourier transform depends on the choice of Haar measure. It is not too difficult to show that the Fourier transform of an L^1 function on G is a bounded continuous function on $\hat{G}$ which vanishes at infinity.

Fourier Inversion Formula for L^1-Functions. For each Haar measure μ on G there is a unique Haar measure v on $\hat{G}$ such that whenever $f \in L^1(G)$ and $\hat{f} \in L^1(\hat{G})$,, we have

$$f(x) = \int_{\hat{G}} \hat{f}(\chi) \chi(x) \, dv(\chi) \qquad \mu\text{-almost everywhere}$$

If f is continuous then this identity holds for all x.

The *inverse Fourier transform* of an integrable function on $\hat{G}$ is given by

$$\overset{\text{i}}{g}(x) = \int_{\hat{G}} g(\chi)\chi(x)\, dv(\chi),$$

where the integral is relative to the Haar measure v on the dual group $\hat{G}$. The measure v on $\hat{G}$ that appears in the Fourier inversion formula is called the dual measure to μ and may be denoted $\hat{\mu}$.

The various Fourier transforms can be classified in terms of their domain and transform domain (the group and dual group) as follows:

Transform	Original domain	Transform domain
Fourier transform	$\mathbb{R}$	$\mathbb{R}$
Fourier series	$\mathbb{T}$	$\mathbb{Z}$
Discrete-time Fourier transform (DTFT)	$\mathbb{Z}$	$\mathbb{T}$
Discrete Fourier transform (DFT)	$\mathbb{Z}/(n)$	$\mathbb{Z}/(n)$

As an example, suppose $G = \mathbb{R}^n$, , so we can think about $\hat{G}$ as $\mathbb{R}^n$ by the pairing $(\mathbf{v}, \mathbf{w}) \mapsto e^{i\mathbf{v}\cdot\mathbf{w}}$. If μ is the Lebesgue measure on Euclidean space, we obtain the ordinary Fourier transform on $\mathbb{R}^n$ and the dual measure needed for the Fourier inversion formula is $\hat{\mu} = (2\pi)^{-n}\mu$. . If we want to get a Fourier inversion formula with the same measure on both sides (that is, since we can think about $\mathbb{R}^n$ as its own dual space we can ask for $\hat{\mu}$ to equal μ) then we need to use

$$\mu = (2\pi)^{-\frac{n}{2}} \times \text{Lebesgue measure}$$

$$\hat{\mu} = (2\pi)^{-\frac{n}{2}} \times \text{Lebesgue measure}$$

However, if we change the way we identify $\mathbb{R}^n$ with its dual group, by using the pairing

$$(\mathbf{v}, \mathbf{w}) \mapsto e^{2\pi i \mathbf{v}\cdot\mathbf{w}},$$

then Lebesgue measure on $\mathbb{R}^n$ is equal to its own dual measure. This convention minimizes the number of factors of 2π that show up in various places when computing Fourier transforms or inverse Fourier transforms on Euclidean space. (In effect it limits the 2π only to the exponent rather than as some messy factor outside the integral sign.) Note that the choice of how to identify $\mathbb{R}^n$ with its dual group affects the meaning of the term "self-dual function", which is a function on $\mathbb{R}^n$ equal to its own Fourier transform: using the classical pairing $(\mathbf{v}, \mathbf{w}) \mapsto e^{i\mathbf{v}\cdot\mathbf{w}}$ the function $e^{-\frac{x^2}{2}}$ is self-dual, but using the (cleaner) pairing $(\mathbf{v}, \mathbf{w}) \mapsto e^{2\pi i \mathbf{v}\cdot\mathbf{w}}$ makes $e^{-\pi x^2}$ self-dual instead.

The Group Algebra

The space of integrable functions on a locally compact abelian group G is an algebra, where multiplication is convolution: the convolution of two integrable functions f and g is defined as

$$(f * g)(x) = \int_G f(x - y)g(y)\, d\mu(y).$$

Theorem. The Banach space $L^1(G)$ is an associative and commutative algebra under convolution.

This algebra is referred to as the *Group Algebra* of G. By the Fubini–Tonelli theorem, the convolution is submultiplicative with respect to the L^1 norm, making $L^1(G)$ a Banach algebra. The Banach algebra $L^1(G)$ has a multiplicative identity element if and only if G is a discrete group, namely the function that is 1 at the identity and zero elsewhere. In general, however, it has an approximate identity which is a net (or generalized sequence) $\{e_i\}_{i \in I}$ indexed on a directed set I such that $f * e_i \to f$.

The Fourier transform takes convolution to multiplication, i.e. it is a homomorphism of abelian Banach algebras $L^1(G) \to C_0(\hat{G})$ (of norm ≤ 1):

$$\mathcal{F}(f * g)(\chi) = \mathcal{F}(f)(\chi) \cdot \mathcal{F}(g)(\chi).$$

In particular, to every group character on G corresponds a unique *multiplicative linear functional* on the group algebra defined by

$$f \mapsto \hat{f}(\chi).$$

It is an important property of the group algebra that these exhaust the set of non-trivial (that is, not identically zero) multiplicative linear functionals on the group algebra. This means the Fourier transform is a special case of the Gelfand transform.

Plancherel and L^2 Fourier Inversion Theorems

As we have stated, the dual group of a locally compact abelian group is a locally compact abelian group in its own right and thus has a Haar measure, or more precisely a whole family of scale-related Haar measures.

Theorem. Choose a Haar measure μ on G and let ν be the dual measure on $\hat{G}$ as defined above. If $f : G \to \mathbb{C}$ is continuous with compact support then $\hat{f} \in L^2(\hat{G})$ and

$$\int_G |f(x)|^2\, d\mu(x) = \int_{\hat{G}} \left| \hat{f}(\chi) \right|^2\, d\nu(\chi).$$

In particular, the Fourier transform is an L^2 isometry from the complex-valued continuous functions of compact support on G to the L^2--functions on $\hat{G}$ (using the L^2-norm with respect to μ for functions on G and the L^2-norm with respect to ν for functions on $\hat{G}$).

Since the complex-valued continuous functions of compact support on G are L^2-dense, there is a unique extension of the Fourier transform from that space to a unitary operator

$$\mathcal{F} : L^2_\mu(G) \to L^2_\nu(\hat{G}).$$

and we have the formula

$$\forall f \in L^2(G): \qquad \int_G |f(x)|^2 \, d\mu(x) = \int_{\hat{G}} |\hat{f}(\chi)|^2 \, d\nu(\chi).$$

Note that for non-compact locally compact groups G the space $L^1(G)$ does not contain $L^2(G)$, so the Fourier transform of general L^2-functions on G is "not" given by any kind of integration formula (or really any explicit formula). To define the L^2 Fourier transform one has to resort to some technical trick such as starting on a dense subspace like the continuous functions with compact support and then extending the isometry by continuity to the whole space. This unitary extension of the Fourier transform is what we mean by the Fourier transform on the space of square integrable functions.

The dual group also has an inverse Fourier transform in its own right; it can be characterized as the inverse (or adjoint, since it is unitary) of the L^2 Fourier transform. This is the content of the L^2 Fourier inversion formula which follows.

> Theorem. The adjoint of the Fourier transform restricted to continuous functions of compact support is the inverse Fourier transform
>
> $$L^2_\nu(\hat{G}) \to L^2_\mu(G)$$
>
> where ν is the dual measure to μ.

In the case $G = \mathbb{T}$ the dual group $\hat{G}$ is naturally isomorphic to the group of integers $\mathbb{Z}$ and the Fourier transform specializes to the computation of coefficients of Fourier series of periodic functions.

If G is a finite group, we recover the discrete Fourier transform. Note that this case is very easy to prove directly.

Bohr Compactification and Almost-periodicity

One important application of Pontryagin duality is the following characterization of compact abelian topological groups:

> Theorem. A locally compact *abelian* group G is compact if and only if the dual group $\hat{G}$ is discrete. Conversely, G is discrete if and only if $\hat{G}$ is compact.

That G being compact implies $\hat{G}$ is discrete or that G being discrete implies that $\hat{G}$ is compact is an elementary consequence of the definition of the compact-open topology on $\hat{G}$ and does not need Pontryagin duality. One uses Pontryagin duality to prove the converses.

The Bohr compactification is defined for any topological group G, regardless of whether G is locally compact or abelian. One use made of Pontryagin duality between compact abelian groups and discrete abelian groups is to characterize the Bohr compactification of an arbitrary abelian *locally compact* topological group. The *Bohr compactification* $B(G)$ of G is $\hat{H}$, where H has the group structure $\hat{G}$, but given the discrete topology. Since the inclusion map

$$\iota : H \to \hat{G}$$

is continuous and a homomorphism, the dual morphism

$$G \sim \hat{\hat{G}} \to \hat{H}$$

is a morphism into a compact group which is easily shown to satisfy the requisite universal property.

Categorical Considerations

It is useful to regard the dual group functorially. In what follows, LCA is the category of locally compact abelian groups and continuous group homomorphisms. The dual group construction of $\hat{G}$ is a contravariant functor LCA $\to$ LCA, represented (in the sense of representable functors) by the circle group $\mathbb{T}$ as $\hat{G} = \mathrm{Hom}(G, \mathbb{T})$. In particular, the double dual functor $G \to \hat{G}$ is *covariant*.

Theorem. The dual group functor is an equivalence of categories from LCA to LCA^{op}.

Theorem. The double dual functor is naturally isomorphic to the identity functor on LCA.

This isomorphism is analogous to the double dual of finite-dimensional vector spaces (a special case, for real and complex vector spaces).

The duality interchanges the subcategories of discrete groups and compact groups. If R is a ring and G is a left R-module, the dual group $\hat{G}$ will become a right R-module; in this way we can also see that discrete left R-modules will be Pontryagin dual to compact right R-modules. The ring $\mathrm{End}(G)$ of endomorphisms in LCA is changed by duality into its opposite ring (change the multiplication to the other order). For example, if G is an infinite cyclic discrete group, $\hat{G}$ is a circle group: the former has $\mathrm{End}(G) = \mathbb{Z}$ so this is true also of the latter.

Generalizations

Non-commutative Theory

Such a theory cannot exist in the same form for non-commutative groups G, since in that case the appropriate dual object $\hat{G}$ of isomorphism classes of representations cannot only contain one-dimensional representations, and will fail to be a group. The generalisation that has been found useful in category theory is called Tannaka–Krein duality; but this diverges from the connection with harmonic analysis, which needs to tackle the question of the *Plancherel measure* on $\hat{G}$.

There are analogues of duality theory for noncommutative groups, some of which are formulated in the language of C*-algebras.

Others

When G is a Hausdorff abelian topological group, the group $\hat{G}$ with the compact-open topology is a Hausdorff abelian topological group and the natural mapping from G to its double-dual $G\hat{\ }\hat{\ }$ makes sense. If this mapping is an isomorphism, we say that G satisfies Pontryagin duality. This has been extended in a number directions beyond the case that G is locally compact.

- S.Kaplan, in "Extensions of the Pontryagin duality" ("part I: infinite products", Duke Math. J. 15 (1948) 649–658, and "part II: direct and inverse limits", same journal, 17 (1950), 419–435) showed that arbitrary products and countable inverse limits of locally compact (Hausdorff) abelian groups satisfy Pontryagin duality. Note that an infinite product of locally compact non-compact spaces is not locally compact.

- Later, in 1975, R.Venkataraman ("Extensions of Pontryagin Duality", Math. Z. 143, 105-112) showed, among other facts, that every open subgroup of an abelian topological group which satisfies Pontryagin duality itself satisfies Pontryagin duality.

- More recently, S. Ardanza-Trevijano and M.J. Chasco have extended the results of Kaplan mentioned above. They showed, in "The Pontryagin duality of sequential limits of topological Abelian groups", Journal of Pure and Applied Algebra 202 (2005), 11–21, that direct and inverse limits of sequences of abelian groups satisfying Pontryagin duality also satisfy Pontryagin duality if the groups are metrizable or k_ω-spaces but not necessarily locally compact, provided some extra conditions are satisfied by the sequences.

However, there is a fundamental aspect that changes if we want to consider Pontryagin duality beyond the locally compact case. In E. Martin-Peinador, *A reflexible admissible topological group must be locally compact*, Proc. Amer. Math. Soc. 123 (1995), 3563–3566, it is proved that if G is a Hausdorff abelian topological group that satisfies Pontryagin duality and the natural evaluation pairing:

$$\begin{cases} G \times \hat{G} \to \mathbb{T} \\ (x, \chi) \mapsto \chi(x) \end{cases}$$

is continuous, then G is locally compact. Thus any non-locally compact example of Pontryagin duality is a group where the natural evaluation pairing of G and $\hat{G}$ is not continuous.

References

- Nerlove, Marc; Grether, David M.; Carvalho, Jose L. (1995). Analysis of Economic Time Series. Economic Theory, Econometrics, and Mathematical Economics. Elsevier. ISBN 0-12-515751-7.

- Dorf, Richard C.; Tallarida, Ronald J. (1993-07-15). Pocket Book of Electrical Engineering Formulas (1 ed.). Boca Raton,FL: CRC Press. pp. 171–174. ISBN 0849344735.

- L. Marton; Claire Marton (1990). Advances in Electronics and Electron Physics. Academic Press. p. 369. ISBN 978-0-12-014650-5.

- Karl H. Pribram; Kunio Yasue; Mari Jibu (1991). Brain and perception. Lawrence Erlbaum Associates. p. 26. ISBN 978-0-89859-995-4.

- Boashash, B., ed. (2003), Time-Frequency Signal Analysis and Processing: A Comprehensive Reference, Oxford: Elsevier Science, ISBN 0-08-044335-4 .

- Knapp, Anthony W. (2001), Representation Theory of Semisimple Groups: An Overview Based on Examples, Princeton University Press, ISBN 978-0-691-09089-4 .

- Müller, Meinard (2015), The Fourier Transform in a Nutshell. (PDF), In Fundamentals of Music Processing, Section 2.1, pages 40-56: Springer, doi:10.1007/978-3-319-21945-5, ISBN 978-3-319-21944-8 .

- Stein, Elias; Weiss, Guido (1971), Introduction to Fourier Analysis on Euclidean Spaces, Princeton, N.J.: Princeton University Press, ISBN 978-0-691-08078-9 .

- Taneja, H.C. (2008), "Chapter 18: Fourier integrals and Fourier transforms", Advanced Engineering Mathematics, Vol. 2, New Delhi, India: I. K. International Pvt Ltd, ISBN 8189866567 .

- Titchmarsh, E. (1986) [1948], Introduction to the theory of Fourier integrals (2nd ed.), Oxford University: Clarendon Press, ISBN 978-0-8284-0324-5 .

- Vretblad, Anders (2000), Fourier Analysis and its Applications, Graduate Texts in Mathematics, 223, New York: Springer, ISBN 0-387-00836-5 .

- Wilson, R. G. (1995), Fourier Series and Optical Transform Techniques in Contemporary Optics, New York: Wiley, ISBN 0-471-30357-7 .

6

Mathematical Model: An Essential Aspect

Systems that use mathematical concepts and mathematical languages are known as mathematical models. It is used in engineering, social sciences and in natural sciences. The topics discussed in the section are of great importance to broaden the existing knowledge on mathematical model.

Mathematical Model

A mathematical model is a description of a system using mathematical concepts and language. The process of developing a mathematical model is termed mathematical modeling. Mathematical models are used in the natural sciences (such as physics, biology, earth science, meteorology) and engineering disciplines (such as computer science, artificial intelligence), as well as in the social sciences (such as economics, psychology, sociology, political science). Physicists, engineers, statisticians, operations research analysts, and economists use mathematical models most extensively. A model may help to explain a system and to study the effects of different components, and to make predictions about behaviour.

Elements of a Mathematical Model

Mathematical models can take many forms, including dynamical systems, statistical models, differential equations, or game theoretic models. These and other types of models can overlap, with a given model involving a variety of abstract structures. In general, mathematical models may include logical models. In many cases, the quality of a scientific field depends on how well the mathematical models developed on the theoretical side agree with results of repeatable experiments. Lack of agreement between theoretical mathematical models and experimental measurements often leads to important advances as better theories are developed.

The traditional mathematical model contains four major elements. These are

1. Governing equations

2. Constitutive equations

3. Constraints

4. Kinematic equations

Classifications

Mathematical models are usually composed of relationships and *variables*. Relationships can be described by *operators*, such as algebraic operators, functions, differential operators, etc. Variables are abstractions of system parameters of interest, that can be quantified. Several classification criteria can be used for mathematical models according to their structure:

- Linear vs. nonlinear: If all the operators in a mathematical model exhibit linearity, the resulting mathematical model is defined as linear. A model is considered to be nonlinear otherwise. The definition of linearity and nonlinearity is dependent on context, and linear models may have nonlinear expressions in them. For example, in a statistical linear model, it is assumed that a relationship is linear in the parameters, but it may be nonlinear in the predictor variables. Similarly, a differential equation is said to be linear if it can be written with linear differential operators, but it can still have nonlinear expressions in it. In a mathematical programming model, if the objective functions and constraints are represented entirely by linear equations, then the model is regarded as a linear model. If one or more of the objective functions or constraints are represented with a nonlinear equation, then the model is known as a nonlinear model. Nonlinearity, even in fairly simple systems, is often associated with phenomena such as chaos and irreversibility. Although there are exceptions, nonlinear systems and models tend to be more difficult to study than linear ones. A common approach to nonlinear problems is linearization, but this can be problematic if one is trying to study aspects such as irreversibility, which are strongly tied to nonlinearity.

- Static vs. dynamic: A *dynamic* model accounts for time-dependent changes in the state of the system, while a *static* (or steady-state) model calculates the system in equilibrium, and thus is time-invariant. Dynamic models typically are represented by differential equations.

- Explicit vs. implicit: If all of the input parameters of the overall model are known, and the output parameters can be calculated by a finite series of computations (known as linear programming), the model is said to be *explicit*. But sometimes it is the *output* parameters which are known, and the corresponding inputs must be solved for by an iterative procedure, such as Newton's method (if the model is linear) or Broyden's method (if non-linear). For example, a jet engine's physical properties such as turbine and nozzle throat areas can be explicitly calculated given a design thermodynamic cycle (air and fuel flow rates, pressures, and temperatures) at a specific flight condition and power setting, but the engine's operating cycles at other flight conditions and power settings cannot be explicitly calculated from the constant physical properties.

- Discrete vs. continuous: A discrete model treats objects as discrete, such as the particles in a molecular model or the states in a statistical model; while a continuous model represents the objects in a continuous manner, such as the velocity field of fluid in pipe flows, temperatures and stresses in a solid, and electric field that applies continuously over the entire model due to a point charge.

- Deterministic vs. probabilistic (stochastic): A deterministic model is one in which every set of variable states is uniquely determined by parameters in the model and by sets of previous states of these variables; therefore, a deterministic model always performs the same way for a given set of initial conditions. Conversely, in a stochastic model—usually called a "statistical model"—randomness is present, and variable states are not described by unique values, but rather by probability distributions.

- Deductive, inductive, or floating: A deductive model is a logical structure based on a theory. An inductive model arises from empirical findings and generalization from them. The floating model rests on neither theory nor observation, but is merely the invocation of expected structure. Application of mathematics in social sciences outside of economics has been criticized for unfounded models. Application of catastrophe theory in science has been characterized as a floating model.

Significance in the Natural Sciences

Mathematical models are of great importance in the natural sciences, particularly in physics. Physical theories are almost invariably expressed using mathematical models.

Throughout history, more and more accurate mathematical models have been developed. Newton's laws accurately describe many everyday phenomena, but at certain limits relativity theory and quantum mechanics must be used; even these do not apply to all situations and need further refinement. It is possible to obtain the less accurate models in appropriate limits, for example relativistic mechanics reduces to Newtonian mechanics at speeds much less than the speed of light. Quantum mechanics reduces to classical physics when the quantum numbers are high. For example, the de Broglie wavelength of a tennis ball is insignificantly small, so classical physics is a good approximation to use in this case.

It is common to use idealized models in physics to simplify things. Massless ropes, point particles, ideal gases and the particle in a box are among the many simplified models used in physics. The laws of physics are represented with simple equations such as Newton's laws, Maxwell's equations and the Schrödinger equation. These laws are such as a basis for making mathematical models of real situations. Many real situations are very complex and thus modeled approximate on a computer, a model that is computationally feasible to compute is made from the basic laws or from approximate models made from the basic laws. For example, molecules can be modeled by molecular orbital models that are approximate solutions to the Schrödinger equation. In engineering, physics models are often made by mathematical methods such as finite element analysis.

Different mathematical models use different geometries that are not necessarily accurate descriptions of the geometry of the universe. Euclidean geometry is much used in classical physics, while special relativity and general relativity are examples of theories that use geometries which are not Euclidean.

Some Applications

Since prehistorical times simple models such as maps and diagrams have been used.

Often when engineers analyze a system to be controlled or optimized, they use a mathematical model. In analysis, engineers can build a descriptive model of the system as a hypothesis of how the system could work, or try to estimate how an unforeseeable event could affect the system. Similarly, in control of a system, engineers can try out different control approaches in simulations.

A mathematical model usually describes a system by a set of variables and a set of equations that establish relationships between the variables. Variables may be of many types; real or integer numbers, boolean values or strings, for example. The variables represent some properties of the system, for example, measured system outputs often in the form of signals, timing data, counters, and event occurrence (yes/no). The actual model is the set of functions that describe the relations between the different variables.

Building Blocks

In business and engineering, mathematical models may be used to maximize a certain output. The system under consideration will require certain inputs. The system relating inputs to outputs depends on other variables too: decision variables, state variables, exogenous variables, and random variables.

Decision variables are sometimes known as independent variables. Exogenous variables are sometimes known as parameters or constants. The variables are not independent of each other as the state variables are dependent on the decision, input, random, and exogenous variables. Furthermore, the output variables are dependent on the state of the system (represented by the state variables).

Objectives and constraints of the system and its users can be represented as functions of the output variables or state variables. The objective functions will depend on the perspective of the model's user. Depending on the context, an objective function is also known as an *index of performance*, as it is some measure of interest to the user. Although there is no limit to the number of objective functions and constraints a model can have, using or optimizing the model becomes more involved (computationally) as the number increases.

For example, in economics students often apply linear algebra when using input-output models. Complicated mathematical models that have many variables may be consolidated by use of vectors where one symbol represents several variables.

A Priori Information

To analyse something with a typical "black box approach", only the behavior of the stimulus/response will be accounted for, to infer the (unknown) *box*. The usual representation of this *black box system* is a data flow diagram centered in the box.

Mathematical modeling problems are often classified into black box or white box models, according to how much a priori information on the system is available. A black-box model is a system of which there is no a priori information available. A white-box model (also called glass box or clear box) is a system where all necessary information is available. Practically all systems are somewhere between the black-box and white-box models, so this concept is useful only as an intuitive guide for deciding which approach to take.

Usually it is preferable to use as much a priori information as possible to make the model more accurate. Therefore, the white-box models are usually considered easier, because if you have used the information correctly, then the model will behave correctly. Often the a priori information comes in forms of knowing the type of functions relating different variables. For example, if we make a model of how a medicine works in a human system, we know that usually the amount of medicine in the blood is an exponentially decaying function. But we are still left with several unknown parameters; how rapidly does the medicine amount decay, and what is the initial amount of medicine in blood? This example is therefore not a completely white-box model. These parameters have to be estimated through some means before one can use the model.

In black-box models one tries to estimate both the functional form of relations between variables and the numerical parameters in those functions. Using a priori information we could end up, for example, with a set of functions that probably could describe the system adequately. If there is no a priori information we would try to use functions as general as possible to cover all different models. An often used approach for black-box models are neural networks which usually do not make assumptions about incoming data. Alternatively the NARMAX (Nonlinear AutoRegressive Moving Average model with eXogenous inputs) algorithms which were developed as part of nonlinear system identificationcan be used to select the model terms, determine the model structure, and estimate the unknown parameters in the presence of correlated and nonlinear noise. The advantage of NARMAX models compared to neural networks is that NARMAX produces models that can be written down and related to the underlying process, whereas neural networks produce an approximation that is opaque.

Subjective Information

Sometimes it is useful to incorporate subjective information into a mathematical model. This can be done based on intuition, experience, or expert opinion, or based on convenience of mathematical form. Bayesian statistics provides a theoretical framework for incorporating such subjectivity into a rigorous analysis: we specify a prior probability distribution (which can be subjective), and then update this distribution based on empirical data.

An example of when such approach would be necessary is a situation in which an experimenter bends a coin slightly and tosses it once, recording whether it comes up heads, and is then given the task of predicting the probability that the next flip comes up heads. After bending the coin, the true probability that the coin will come up heads is unknown; so the experimenter would need to make a decision (perhaps by looking at the shape of the coin) about what prior distribution to use. Incorporation of such subjective information might be important to get an accurate estimate of the probability.

Complexity

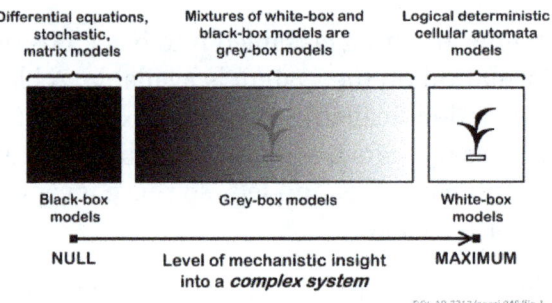

DOI: 10.7717/peerj.948/fig-1

This is a schematic representation of three types of mathematical models of complex systems with the level of their mechanistic understanding.

In general, model complexity involves a trade-off between simplicity and accuracy of the model. Occam's razor is a principle particularly relevant to modeling; the essential idea being that among models with roughly equal predictive power, the simplest one is the most desirable. While added complexity usually improves the realism of a model, it can make the model difficult to understand and analyze, and can also pose computational problems, including numerical instability. Thomas Kuhn argues that as science progresses, explanations tend to become more complex before a paradigm shift offers radical simplification.

For example, when modeling the flight of an aircraft, we could embed each mechanical part of the aircraft into our model and would thus acquire an almost white-box model of the system. However, the computational cost of adding such a huge amount of detail would effectively inhibit the usage of such a model. Additionally, the uncertainty would increase due to an overly complex system, because each separate part induces some amount of variance into the model. It is therefore usually appropriate to make some approximations to reduce the model to a sensible size. Engineers often can accept some approximations in order to get a more robust and simple model. For example, Newton's classical mechanics is an approximated model of the real world. Still, Newton's model is quite sufficient for most ordinary-life situations, that is, as long as particle speeds are well below the speed of light, and we study macro-particles only.

Training

Any model which is not pure white-box contains some parameters that can be used to fit the model to the system it is intended to describe. If the modeling is done by a neural network, the optimization of parameters is called *training*. In more conventional modeling through explicitly given mathematical functions, parameters are determined by curve fitting.

Model Evaluation

A crucial part of the modeling process is the evaluation of whether or not a given mathematical model describes a system accurately. This question can be difficult to answer as it involves several different types of evaluation.

Fit to Empirical Data

Usually the easiest part of model evaluation is checking whether a model fits experimental measurements or other empirical data. In models with parameters, a common approach to test this fit is to split the data into two disjoint subsets: training data and verification data. The training data are used to estimate the model parameters. An accurate model will closely match the verification data even though these data were not used to set the model's parameters. This practice is referred to as cross-validation in statistics.

Defining a metric to measure distances between observed and predicted data is a useful tool of assessing model fit. In statistics, decision theory, and some economic models, a loss function plays a similar role.

While it is rather straightforward to test the appropriateness of parameters, it can be more difficult to test the validity of the general mathematical form of a model. In general, more mathematical tools have been developed to test the fit of statistical models than models involving differential equations. Tools from non-parametric statistics can sometimes be used to evaluate how well the data fit a known distribution or to come up with a general model that makes only minimal assumptions about the model's mathematical form.

Scope of the Model

Assessing the scope of a model, that is, determining what situations the model is applicable to, can be less straightforward. If the model was constructed based on a set of data, one must determine for which systems or situations the known data is a "typical" set of data.

The question of whether the model describes well the properties of the system between data points is called interpolation, and the same question for events or data points outside the observed data is called extrapolation.

As an example of the typical limitations of the scope of a model, in evaluating Newtonian classical mechanics, we can note that Newton made his measurements without advanced equipment, so he could not measure properties of particles travelling at speeds close to the speed of light. Likewise, he did not measure the movements of molecules and other small particles, but macro particles only. It is then not surprising that his model does not extrapolate well into these domains, even though his model is quite sufficient for ordinary life physics.

Philosophical Considerations

Many types of modeling implicitly involve claims about causality. This is usually (but not always) true of models involving differential equations. As the purpose of modeling is to increase our understanding of the world, the validity of a model rests not only on its fit to empirical observations, but also on its ability to extrapolate to situations or data beyond those originally described in the model. One can think of this as the differentiation between qualitative and quantitative predictions. One can also argue that a model is worthless unless it provides some insight which goes beyond what is already known from direct investigation of the phenomenon being studied.

An example of such criticism is the argument that the mathematical models of Optimal foraging theory do not offer insight that goes beyond the common-sense conclusions of evolution and other basic principles of ecology.

Examples

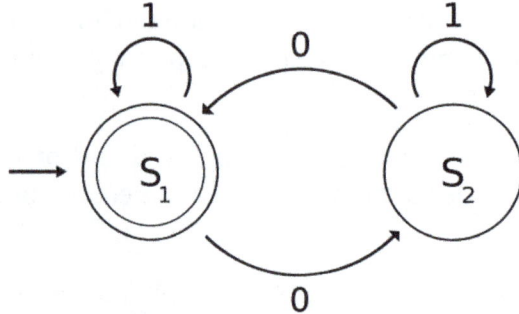

The state diagram for M

- One of the popular examples in computer science is the mathematical models of various machines, an example is the deterministic finite automaton which is defined as an abstract mathematical concept, but due to the deterministic nature of a DFA, it is implementable in hardware and software for solving various specific problems. For example, the following is a DFA M with a binary alphabet, which requires that the input contains an even number of 0s.

$M = (Q, \Sigma, \delta, q_0, F)$ where

- $Q = \{S_1, S_2\}$,

- $\Sigma = \{0, 1\}$,

- $q_0 = S_1$,

- $F = \{S_1\}$, and

- δ is defined by the following state transition table:

	0	1
S_1	S_2	S_1
S_2	S_1	S_2

The state S_1 represents that there has been an even number of 0s in the input so far, while S_2 signifies an odd number. A 1 in the input does not change the state of the automaton. When the input ends, the state will show whether the input contained an even number of 0s or not. If the input did contain an even number of 0s, M will finish in state S_1, an accepting state, so the input string will be accepted.

The language recognized by M is the regular language given by the regular expression 1*(0 (1*) 0 (1*))*, where "*" is the Kleene star, e.g., 1* denotes any non-negative number (possibly zero) of symbols "1".

- Many everyday activities carried out without a thought are uses of mathematical models. A geographical map projection of a region of the earth onto a small, plane surface is a model which can be used for many purposes such as planning travel.

- Another simple activity is predicting the position of a vehicle from its initial position, direction and speed of travel, using the equation that distance traveled is the product of time and speed. This is known as dead reckoning when used more formally. Mathematical modeling in this way does not necessarily require formal mathematics; animals have been shown to use dead reckoning.

- *Population Growth*. A simple (though approximate) model of population growth is the Malthusian growth model. A slightly more realistic and largely used population growth model is the logistic function, and its extensions.

- *Individual-based cellular automata models of population growth*

- *Model of a particle in a potential-field*. In this model we consider a particle as being a point of mass which describes a trajectory in space which is modeled by a function giving its coordinates in space as a function of time. The potential field is given by a function $V: \mathbb{R}^3 \to \mathbb{R}$ and the trajectory, that is a function $\mathbf{r}: \mathbb{R} \to \mathbb{R}^3$, is the solution of the differential equation:

$$-\frac{d^2\mathbf{r}(t)}{dt^2} m = \frac{\partial V[\mathbf{r}(t)]}{\partial x} \hat{\mathbf{x}} + \frac{\partial V[\mathbf{r}(t)]}{\partial y} \hat{\mathbf{y}} + \frac{\partial V[\mathbf{r}(t)]}{\partial z} \hat{\mathbf{z}},$$

that can be written also as:

$$m \frac{d^2\mathbf{r}(t)}{dt^2} = -\nabla V[\mathbf{r}(t)].$$

Note this model assumes the particle is a point mass, which is certainly known to be false in many cases in which we use this model; for example, as a model of planetary motion.

- *Model of rational behavior for a consumer.* In this model we assume a consumer faces a choice of n commodities labeled $1,2,...,n$ each with a market price $p_1, p_2,..., p_n$. The consumer is assumed to have a *cardinal* utility function U (cardinal in the sense that it assigns numerical values to utilities), depending on the amounts of commodities $x_1, x_2,..., x_n$ consumed. The model further assumes that the consumer has a budget M which is used to purchase a vector $x_1, x_2,..., x_n$ in such a way as to maximize $U(x_1, x_2,..., x_n)$. The problem of rational behavior in this model then becomes an optimization problem, that is:

$$\max U(x_1, x_2, \ldots, x_n)$$

subject to:

$$\sum_{i=1}^{n} p_i x_i \leq M.$$

$$x_i \geq 0 \quad \forall i \in \{1, 2, \ldots, n\}$$

This model has been used in general equilibrium theory, particularly to show existence and Pareto efficiency of economic equilibria. However, the fact that this particular formulation assigns *numerical values* to levels of satisfaction is the source of criticism (and even ridicule). However, it is not an essential ingredient of the theory and again this is an idealization.

- *Neighbour-sensing model* explains the mushroom formation from the initially chaotic fungal network.

- *Computer science*: models in Computer Networks, data models, surface model,...

- *Mechanics*: movement of rocket model,...

Modeling requires selecting and identifying relevant aspects of a situation in the real world.

Governing Equation

Mathematical models can take many forms, including dynamical systems, statistical models, differential equations, game theoretic models, recurrence relations, an algorithm for calculation of a sequence of related states (e.g. equilibrium states) and possibly even more forms. The governing equations of a mathematical model describes how the unknown variables (i.e. the dependent variables) will change. The change of variables w.r.t. time may be explicit (i.e. a governing equation includes derivative with respect to time) or implicit (e.g. a governing equation has velocity or flux as unknown variable or an algorithm). The classic governing equations in continuum mechanics are

- Balance of mass

- Balance of (linear) momentum

- Balance of angular momentum

- Balance of energy

- Balance of entropy

For isolated systems the upper four equations are the familiar conservation equations in physics. A governing equation may also take the form of a flux equation like the diffusion equation or the heat conduction equation. In these cases the flux itself is a variable describing change of the unknown variable or property (e.g. mole concentration or internal energy or temperature). A governing equation may also be an approximation and adaption of the above basic equations to the situation or model in question. A governing equation may also be derived directly from experimental results and therefore be an empiric equation. A governing equation may also be an equation describing the state of the system, and thus actually be a constitutive equation that has "stepped up the ranks" because the model in question was not meant to include a time-dependent term in the equation. This is the case for a model of a petroleum processing plant. Results from one thermodynamic equilibrium calculation are input data to the next equilibrium calculation together with some new state parameters and so on. In this case the algorithm and sequence of input data form a chain of actions, or calculations, that describes change of states from the first state (based solely on input data) to the last state that finally comes out of the calculation sequence.

Some examples using differential equations are

- Lotka-Volterra equations are predator-prey equations

- Hele-Shaw flow

- Plate theory

  o Kirchhoff–Love plate theory or Bending of Kirchhoff-Love plates

  o Mindlin–Reissner plate theory or Bending of thick Mindlin plates or Bending of Reissner-Stein cantilever plates

- Vortex shedding

- Annular fin

- Astronautics

- Finite volume method for unsteady flow

- Acoustic theory

- Precipitation hardening

- Kelvin's circulation theorem

- Kernel function for solving integral equation of surface radiation exchanges

- Nonlinear acoustics

- Large eddy simulation

- Föppl–von Kármán equations

- Timoshenko beam theory

Constitutive Equation

In physics and engineering, a constitutive equation or constitutive relation is a relation between two physical quantities (especially kinetic quantities as related to kinematic quantities) that is specific to a material or substance, and approximates the response of that material to external stimuli, usually as applied fields or forces. They are combined with other equations governing physical laws to solve physical problems; for example in fluid mechanics the flow of a fluid in a pipe, in solid state physics the response of a crystal to an electric field, or in structural analysis, the connection between applied stresses or forces to strains or deformations.

Some constitutive equations are simply phenomenological; others are derived from first principles. A common approximate constitutive equation frequently is expressed as a simple proportionality using a parameter taken to be a property of the material, such as electrical conductivity or a spring constant. However, it is often necessary to account for the directional dependence of the material, and the scalar parameter is generalized to a tensor. Constitutive relations are also modified to account for the rate of response of materials and their non-linear behavior.

Mechanical Properties of Matter

The first constitutive equation (constitutive law) was developed by Robert Hooke and is known as Hooke's law. It deals with the case of linear elastic materials. Following this discovery, this type of equation, often called a "stress-strain relation" in this example, but also called a "constitutive assumption" or an "equation of state" was commonly used. Walter Noll advanced the use of constitutive equations, clarifying their classification and the role of invariance requirements, constraints, and definitions of terms like "material", "isotropic", "aeolotropic", etc. The class of "constitutive relations" of the form *stress rate = f (velocity gradient, stress, density)* was the subject of Walter Noll's dissertation in 1954 under Clifford Truesdell.

In modern condensed matter physics, the constitutive equation plays a major role.

Definitions

Quantity (common name/s)	(Common) symbol/s	Defining equation	SI units	Dimension
General stress, Pressure	P, σ	$\sigma = F / A$ F may be any perpendicular force applied to area A	Pa = N m^{-2}	[M] [T]$^{-2}$[L]$^{-1}$
General strain	ε	$\varepsilon = \Delta D / D$ • D = dimension (length, area, volume) • ΔK = change in material	dimensionless	dimensionless
General elastic modulus	E_{mod}	$E_{mod} = \sigma / \varepsilon$	Pa = N m^{-2}	[M] [T]$^{-2}$ [L]$^{-1}$
Young's modulus	E, Y	$Y = \sigma / (\Delta L / L)$	Pa = N m^{-2}	[M] [T] $^{-2}$[L]$^{-1}$
Shear modulus	G	$G = \Delta x / L$	Pa = N m^{-2}	[M] [T]$^{-2}$[L]$^{-1}$
Bulk modulus	K, B	$B = P / (\Delta V / V)$	Pa = N m^{-2}	[M] [T]$^{-2}$[L]$^{-1}$
Compressibility	C	$C = 1 / B$	Pa^{-1} = m^2 N^{-1}	[L] [T]^2[M]$^{-1}$

Deformation of Solids

Friction

Friction is a complicated phenomenon. Macroscopically the friction force F between the interface of two materials can be modelled as proportional to the reaction force R at a point of contact between two interfaces, through a dimensionless coefficient of friction μ_f which depends on the pair of materials:

$$F = \mu_f R.$$

This can be applied to static friction (friction preventing two stationary objects from slipping on their own), kinetic friction (friction between two objects scraping/sliding past each other), or rolling (frictional force which prevents slipping but causes a torque to exert on a round object). Surprisingly, the friction force does not depend on the surface area of common contact.

Stress and Strain

The stress-strain constitutive relation for linear materials is commonly known as Hooke's law. In its simplest form, the law defines the spring constant (or elasticity constant) k in a scalar equation, stating the tensile/compressive force is proportional to the extended (or contracted) displacement x:

$$F_i = -kx_i$$

meaning the material responds linearly. Equivalently, in terms of the stress σ, Young's modulus E, and strain ε (dimensionless):

$$\sigma = E\varepsilon$$

In general, forces which deform solids can be normal to a surface of the material (normal forces), or tangential (shear forces), this can be described mathematically using the stress tensor:

$$\sigma_{ij} = C_{ijkl}\,\varepsilon_{kl} \rightleftharpoons \varepsilon_{ij} = S_{ijkl}\,\sigma_{kl}$$

where C is the elasticity tensor and S is the compliance tensor

Solid-state Deformations

Several classes of deformations in elastic materials are the following:

- *Elastic*: The material recovers its initial shape after deformation.

- *Anelastic*: if the material is close to elastic, but the applied force induces additional time-dependent resistive forces (i.e. depend on rate of change of extension/compression, in addition to the extension/compression). Metals and ceramics have this characteristic, but it is usually negligible, although not so much when heating due to friction occurs (such as vibrations or shear stresses in machines).

- *Viscoelastic*: If the time-dependent resistive contributions are large, and cannot be neglected. Rubbers and plastics have this property, and certainly do not satisfy Hooke's law. In fact, elastic hysteresis occurs.

- *Plastic*: The applied force induces non-recoverable deformations in the material when the stress (or elastic strain) reaches a critical magnitude, called the yield point.

- *Hyperelastic*: The applied force induces displacements in the material following a strain energy density function.

Collisions

The relative speed of separation $v_{separation}$ of an object A after a collision with another object B is related to the relative speed of approach $v_{approach}$ by the coefficient of restitution, defined by Newton's experimental impact law:

$$e = \frac{|\mathbf{v}|_{separation}}{|\mathbf{v}|_{approach}}$$

which depends the materials A and B are made from, since the collision involves interactions at the surfaces of A and B. Usually $0 \le e \le 1$, in which $e = 1$ for completely elastic collisions, and $e = 0$ for completely inelastic collisions. It's possible for $e \ge 1$ to occur – for superelastic (or explosive) collisions.

Deformation of Fluids

The drag equation gives the drag force D on an object of cross-section area A moving through a fluid of density ρ at velocity v (relative to the fluid)

$$D = \frac{1}{2} c_d \rho A v^2$$

where the drag coefficient (dimensionless) c_d depends on the geometry of the object and the drag forces at the interface between the fluid and object.

For a Newtonian fluid of viscosity μ, the shear stress τ is linearly related to the strain rate (transverse flow velocity gradient) $\partial u/\partial y$ (units s^{-1}). In a uniform shear flow:

$$\tau = \mu \frac{\partial u}{\partial y},$$

with $u(y)$ the variation of the flow velocity u in the cross-flow (transverse) direction y. In general, for a Newtonian fluid, the relationship between the elements τ_{ij} of the shear stress tensor and the deformation of the fluid is given by

$$\tau_{ij} = 2\mu \left(e_{ij} - \frac{1}{3}\Delta \delta_{ij} \right) \quad \text{with} \quad e_{ij} = \frac{1}{2}\left(\frac{\partial v_i}{\partial x_j} + \frac{\partial v_j}{\partial x_i} \right) \quad \text{and} \quad \Delta = \sum_k e_{kk} = \operatorname{div} \mathbf{v},$$

where v_i are the components of the flow velocity vector in the corresponding x_i coordinate directions, e_{ij} are the components of the strain rate tensor, Δ is the volumetric strain rate (or dilatation rate) and δ_{ij} is the Kronecker delta.

The *ideal gas law* is a constitutive relation in the sense the pressure p and volume V are related to the temperature T, via the number of moles n of gas:

$$pV = nRT$$

where R is the gas constant (J K^{-1} mol^{-1}).

Electromagnetism

Constitutive Equations in Electromagnetism and Related Areas

In both classical and quantum physics, the precise dynamics of a system form a set of coupled differential equations, which are almost always too complicated to be solved exactly, even at the level of statistical mechanics. In the context of electromagnetism, this remark applies to not only the dynamics of free charges and currents (which enter Maxwell's equations directly), but also the dynamics of bound charges and currents (which enter Maxwell's equations through the constitutive relations). As a result, various approximation schemes are typically used.

For example, in real materials, complex transport equations must be solved to determine the time and spatial response of charges, for example, the Boltzmann equation or the Fokker–Planck equation or the Navier-Stokes equations. For example, magnetohydrodynamics, fluid dynamics, electrohydrodynamics, superconductivity, plasma modeling. An entire physical appa-ratus for dealing with these matters has developed. For example, linear response theory, Green–Kubo relations and Green's function (many-body theory).

These complex theories provide detailed formulas for the constitutive relations describing the electrical response of various materials, such as permittivities, permeabilities, conductivities and so forth.

It is necessary to specify the relations between displacement field D and E, and the magnetic H-field H and B, before doing calculations in electromagnetism (i.e. applying Maxwell's macroscopic equations). These equations specify the response of bound charge and current to the applied fields and are called constitutive relations.

Determining the constitutive relationship between the auxiliary fields D and H and the E and B fields starts with the definition of the auxiliary fields themselves:

$$\mathbf{D}(\mathbf{r},t) = \varepsilon_0 \mathbf{E}(\mathbf{r},t) + \mathbf{P}(\mathbf{r},t)$$

$$\mathbf{H}(\mathbf{r},t) = \frac{1}{\mu_0} \mathbf{B}(\mathbf{r},t) - \mathbf{M}(\mathbf{r},t),$$

where P is the polarization field and M is the magnetization field which are defined in terms of microscopic bound charges and bound current respectively. Before getting to how to calculate M and P it is useful to examine the following special cases.

Without Magnetic or Dielectric Materials

In the absence of magnetic or dielectric materials, the constitutive relations are simple:

$$\mathbf{D} = \varepsilon_0 \mathbf{E}, \quad \mathbf{H} = \mathbf{B} / \mu_0$$

where ε_0 and μ_0 are two universal constants, called the permittivity of free space and permeability of free space, respectively.

Isotropic Linear Materials

In an (isotropic) linear material, where P is proportional to E, and M is proportional to B, the constitutive relations are also straightforward. In terms of the polarization P and the magnetization M they are:

$$\mathbf{P} = \varepsilon_0 \chi_e \mathbf{E}, \quad \mathbf{M} = \chi_m \mathbf{H},$$

where χ_e and χ_m are the electric and magnetic susceptibilities of a given material respectively. In terms of D and H the constitutive relations are:

$$\mathbf{D} = \varepsilon \mathbf{E}, \quad \mathbf{H} = \mathbf{B} / \mu,$$

where ε and μ are constants (which depend on the material), called the permittivity and permeability, respectively, of the material. These are related to the susceptibilities by:

$$\varepsilon / \varepsilon_0 = \varepsilon_r = (\chi_e + 1), \quad \mu / \mu_0 = \mu_r = (\chi_m + 1)$$

General Case

For real-world materials, the constitutive relations are not linear, except approximately. Calculating the constitutive relations from first principles involves determining how P and M are created from a given E and B. These relations may be empirical (based directly upon measurements), or theoretical (based upon statistical mechanics, transport theory or other tools of condensed matter physics). The detail employed may be macroscopic or microscopic, depending upon the level necessary to the problem under scrutiny.

In general, the constitutive relations can usually still be written:

$$\mathbf{D} = \varepsilon \mathbf{E}, \quad \mathbf{H} = \mu^{-1} \mathbf{B}$$

but ε and μ are not, in general, simple constants, but rather functions of E, B, position and time, and tensorial in nature. Examples are:

- *Dispersion and absorption* where ε and μ are functions of frequency. (Causality does not permit materials to be nondispersive). Neither do the fields need to be in phase which leads to ε and μ being complex. This also leads to absorption.

- *Nonlinearity* where ε and μ are functions of E and B.

- *Anisotropy* (such as *birefringence* or *dichroism*) which occurs when ε and μ are second-rank tensors,

$$D_i = \sum_j \varepsilon_{ij} E_j \quad B_i = \sum_j \mu_{ij} H_j.$$

- Dependence of P and M on E and B at other locations and times. This could be due to *spatial inhomogeneity*; for example in a domained structure, heterostructure or a liquid crystal, or most commonly in the situation where there are simply multiple materials occupying different regions of space. Or it could be due to a time varying medium or due to hysteresis. In such cases P and M can be calculated as:

$$\mathbf{P}(\mathbf{r},t) = \varepsilon_0 \int d^3\mathbf{r}' dt' \; \hat{\chi}_e(\mathbf{r},\mathbf{r}',t,t';\mathbf{E})\mathbf{E}(\mathbf{r}',t')$$

$$\mathbf{M}(\mathbf{r},t) = \frac{1}{\mu_0} \int d^3\mathbf{r}' dt' \; \hat{\chi}_m(\mathbf{r},\mathbf{r}',t,t';\mathbf{B})\mathbf{B}(\mathbf{r}',t'),$$

in which the permittivity and permeability functions are replaced by integrals over the more general electric and magnetic susceptibilities. In homogenous materials, dependence on other locations is known as spatial dispersion.

As a variation of these examples, in general materials are bianisotropic where D and B depend on both E and H, through the additional *coupling constants* ξ and ζ:

$$\mathbf{D} = \varepsilon\mathbf{E} + \xi\mathbf{H}, \quad \mathbf{B} = \mu\mathbf{H} + \zeta\mathbf{E}.$$

In practice, some materials properties have a negligible impact in particular circumstances, permitting neglect of small effects. For example: optical nonlinearities can be neglected for low field strengths; material dispersion is unimportant when frequency is limited to a narrow bandwidth; material absorption can be neglected for wavelengths for which a material is transparent; and metals with finite conductivity often are approximated at microwave or longer wavelengths as perfect metals with infinite conductivity (forming hard barriers with zero skin depth of field penetration).

Some man-made materials such as metamaterials and photonic crystals are designed to have customized permittivity and permeability.

Calculation of Constitutive Relations

The theoretical calculation of a material's constitutive equations is a common, important, and sometimes difficult task in theoretical condensed-matter physics and materials science. In general, the constitutive equations are theoretically determined by calculating how a molecule responds to the local fields through the Lorentz force. Other forces may need to be modeled as well such as lattice vibrations in crystals or bond forces. Including all of the forces leads to changes in the molecule which are used to calculate P and M as a function of the local fields.

The local fields differ from the applied fields due to the fields produced by the polarization and magnetization of nearby material; an effect which also needs to be modeled. Further, real materials are not continuous media; the local fields of real materials vary wildly on the atomic scale. The fields need to be averaged over a suitable volume to form a continuum approximation.

These continuum approximations often require some type of quantum mechanical analysis such as quantum field theory as applied to condensed matter physics. For example, density functional theory, Green–Kubo relations and Green's function.

A different set of *homogenization methods* (evolving from a tradition in treating materials such as conglomerates and laminates) are based upon approximation of an inhomogeneous material by a homogeneous *effective medium* (valid for excitations with wavelengths much larger than the scale of the inhomogeneity).

The theoretical modeling of the continuum-approximation properties of many real materials often rely upon experimental measurement as well. For example, ε of an insulator at low frequencies can be measured by making it into a parallel-plate capacitor, and ε at optical-light frequencies is often measured by ellipsometry.

Thermoelectric and Electromagnetic Properties of Matter

These constitutive equations are often used in crystallography, a field of solid-state physics.

Electromagnetic properties of solids

Property/ effect	Stimuli/response parameters of system	Constitutive tensor of system	Equation
Hall effect	• E = electric field strength (N C⁻¹) • J = electric current density (A m⁻²) • H = magnetic field intensity (A m⁻¹)	ρ = electrical resistivity (Ω m)	$E_k = \rho_{kij} J_i H_j$
Direct Piezoelectric Effect	• σ = Stress (Pa) • P = (dielectric) polarization (C m⁻²)	d = direct piezoelectric coefficient (K⁻¹)	$P_i = d_{ijk} \sigma_{jk}$
Converse Piezoelectric Effect	• ε = Strain (dimensionless) • E = electric field strength (N C⁻¹)	d = direct piezoelectric coefficient (K⁻¹)	$\varepsilon_{ij} = d_{ijk} E_k$
Piezomagnetic effect	• σ = Stress (Pa) • M = magnetization (A m⁻¹)	q = piezomagnetic coefficient (K⁻¹)	$M_i = q_{ijk} \sigma_{jk}$

Thermoelectric properties of solids

Property/effect	Stimuli/response parameters of system	Constitutive tensor of system	Equation
Pyroelectricity	• P = (dielectric) polarization (C m⁻²) • T = temperature (K)	p = pyroelectric coefficient (C m⁻² K⁻¹)	$\Delta P_j = p_j \Delta T$
Electrocaloric effect	• S = entropy (J K⁻¹) • E = electric field strength (N C⁻¹)	p = pyroelectric coefficient (C m⁻² K⁻¹)	$\Delta S = p_i \Delta E_i$
Seebeck effect	• E = electric field strength (N C⁻¹ = V m⁻¹) • T = temperature (K) • x = displacement (m)	β = thermopower (V K⁻¹)	$E_i = -\beta_{ij} \dfrac{\partial T}{\partial x_j}$

| Peltier effect | • E = electric field strength (N C^{-1})

 • J = electric current density (A m^{-2})

 • q = heat flux (W m^{-2}) | Π = Peltier coefficient (W A^{-1}) | $q_j = \Pi_{ji} J_i$ |

Photonics

Refractive index

The (absolute) refractive index of a medium n (dimensionless) is an inherently important property of geometric and physical optics defined as the ratio of the luminal speed in vacuum c_0 to that in the medium c:

$$n = \frac{c_0}{c} = \sqrt{\frac{\varepsilon \mu}{\varepsilon_0 \mu_0}} = \sqrt{\varepsilon_r \mu_r}$$

where ε is the permittivity and ε_r the relative permittivity of the medium, likewise μ is the permeability and μ_r are the relative permmeability of the medium. The vacuum permittivity is ε_0 and vacuum permeability is μ_0. In general, n (also ε_r) are complex numbers.

The relative refractive index is defined as the ratio of the two refractive indices. Absolute is for on material, relative applies to every possible pair of interfaces;

$$n_{AB} = \frac{n_A}{n_B}$$

Speed of light in matter

As a consequence of the definition, the speed of light in matter is

$$c = 1 / \sqrt{\varepsilon \mu}$$

for special case of vacuum; $\varepsilon = \varepsilon_0$ and $\mu = \mu_0$,

$$c_0 = 1 / \sqrt{\varepsilon_0 \mu_0}$$

Piezooptic effect

The piezooptic effect relates the stresses in solids σ to the dielectric impermeability a, which are coupled by a fourth-rank tensor called the piezooptic coefficient Π (units K^{-1}):

$$a_{ij} = \Pi_{ijpq} \sigma_{pq}$$

Transport Phenomena

Definitions

Definitions (thermal properties of matter)

Quantity (Common Name/s)	(Common) Symbol/s	Defining Equation	SI Units	Dimension
General heat capacity	C = heat capacity of substance	$q = CT$	J K^{-1}	[M][L]^2[T]$^{-2}$[Θ]$^{-1}$
Linear thermal expansion	• L = length of material (m) • α = coefficient linear thermal expansion (dimensionless) • ε = strain tensor (dimensionless)	$\partial L / \partial T = \alpha L$ $\varepsilon_{ij} = \alpha_{ij}\Delta T$	K^{-1}	[Θ]$^{-1}$
Volumetric thermal expansion	β, γ • V = volume of object (m^3) • p = constant pressure of surroundings	$(\partial V / \partial T)_p = \gamma V$	K^{-1}	[Θ]$^{-1}$
Thermal conductivity	$\kappa, K, \lambda,$ • $\mathbf{A}$ = surface cross section of material (m^2) • P = thermal current/power through material (W) • ∇T = temperature gradient in material (K m^{-1})	$\lambda = -P / (\mathbf{A} \cdot \nabla T)$	W m^{-1} K^{-1}	[M][L][T]$^{-3}$[Θ]$^{-1}$
Thermal conductance	U	$U = \lambda / \delta x$	W m^{-2} K^{-1}	[M][T]$^{-3}$[Θ]$^{-1}$
Thermal resistance	R Δx = displacement of heat transfer (m)	$R = 1 / U = \Delta x / \lambda$	m^2 K W^{-1}	[M]$^{-1}$[L][T]^3[Θ]

Definitions (Electrical/magnetic properties of matter)

Quantity (Common Name/s)	(Common) Symbol/s	Defining Equation	SI Units	Dimension
Electrical resistance	R	$R = V / 1$	Ω = V A^{-1} = J s C^{-2}	[M] [L]^2 [T]$^{-3}$ [I]$^{-2}$
Resistivity	ρ	$\rho = RA / l$	Ω m	[M]^2 [L]^2 [T]$^{-3}$ [I]$^{-2}$
Resistivity temperature coefficient, linear temperature dependence	α	$\rho - \rho_0 = \rho_0 \alpha (T - T_0)$	K^{-1}	[Θ]$^{-1}$

Electrical conductance	G	$G = 1/\mathcal{R}$	$S = \Omega^{-1}$	$[T]^3 [I]^2 [M]^{-1}$ $[L]^{-2}$
Electrical conductivity	σ	$\sigma = 1/\rho$	$\Omega^{-1}\,m^{-1}$	$[I]^2 [T]^3 [M]^{-2}$ $[L]^{-2}$
Magnetic reluctance	$R, R_m, \mathcal{R}$	$R_m = \mathcal{M}/\Phi_B$	$A\,Wb^{-1} = H^{-1}$	$[M]^{-1}[L]^{-2}[T]^2$
Magnetic permeance	$P, P_m, \Lambda, \mathcal{P}$	$\Lambda = 1/R_m$	$Wb\,A^{-1} = H$	$[M][L]^2[T]^{-2}$

Definitive Laws

There are several laws which describe the transport of matter, or properties of it, in an almost identical way. In every case, in words they read:

Flux (density) is proportional to a gradient, the constant of proportionality is the characteristic of the material.

In general the constant must be replaced by a 2nd rank tensor, to account for directional dependences of the material.

Property/effect	Nomenclature	Equation
Fick's law of diffusion, defines diffusion coefficient D	<ul><li>D = mass diffusion coefficient ($m^2\,s^{-1}$)</li><li>J = diffusion flux of substance ($mol\,m^{-2}\,s^{-1}$)</li><li>$\partial C/\partial x$ = (1d)concentration gradient of substance ($mol\,dm^{-4}$)</li></ul>	$$J_j = -D_{ij}\frac{\partial C}{\partial x_i}$$
Darcy's law for porous flow in matter, defines permeability κ	<ul><li>κ = permeability of medium (m^2)</li><li>μ = fluid viscosity ($Pa\,s$)</li><li>q = discharge flux of substance ($m\,s^{-1}$)</li><li>$\partial P/\partial x$ = (1d) pressure gradient of system ($Pa\,m^{-1}$)</li></ul>	$$q_j = -\frac{\kappa}{\mu}\frac{\partial P}{\partial x_j}$$
Ohm's law of electric conduction, defines electric conductivity (and hence resistivity and resistance)	<ul><li>V = potential difference in material (V)</li><li>I = electric current through material (A)</li><li>R = resistance of material (Ω)</li><li>$\partial V/\partial x$ = potential gradient (electric field) through material ($V\,m^{-1}$)</li><li>J = electric current density through material ($A\,m^{-2}$)</li><li>σ = electric conductivity of material ($\Omega^{-1}\,m^{-1}$)</li><li>ρ = electrical resistivity of material ($\Omega\,m$)</li></ul>	<ul><li>Simplist form is:</li></ul> $$V = IR$$ <ul><li>More general forms are:</li></ul> $$\frac{\partial V}{\partial x_i} = \rho_{ji}J_i \rightleftharpoons J_j = \sigma_{ji}\frac{\partial V}{\partial x_i}$$

Fourier's law of thermal conduction, defines thermal conductivity λ	• λ = thermal conductivity of material (W m^{-1} K^{-1}) • q = heat flux through material (W m^{-2}) • $\partial T/\partial x$ = temperature gradient in material (K m^{-1})	$q_i = -\lambda_{ij} \dfrac{\partial T}{\partial x_j}$
Stefan–Boltzmann law of black-body radiation, defines emmisivity ε	• I = radiant intensity (W m^{-2}) • σ = Stefan–Boltzmann constant (W m^{-2} K^{-4}) • T_{sys} = temperature of radiating system (K) • T_{ext} = temperature of external surroundings (K) • ε = emissivity (dimensionless)	• For a single radiator: $I = \varepsilon\sigma T^4$ For a temperature difference: $I = \varepsilon\sigma(T_{ext}^4 - T_{sys}^4)$ $0 \le \varepsilon \le 1$ • $\varepsilon = 0$ for perfect reflector • $\varepsilon = 1$ for perfect absorber (true black body)

Kinematics Equations

Kinematics equations refers to the constraint equations of a mechanical system such as a robot manipulator that define how input movement at one or more joints specifies the configuration of the device, in order to achieve a task position or end-effector location. Kinematics equations are used to analyze and design articulated systems ranging from four-bar linkages to serial and parallel robots.

Kinematics equations are constraint equations that characterize the geometric configuration of an articulated mechanical system. Therefore, these equations assume the links are rigid and the joints provide pure rotation or translation. Constraint equations of this type are known as holonomic constraints in the study of the dynamics of multi-body systems.

Loop Equations

The kinematics equations for a mechanical system are formed as a sequence of rigid transformations along links and around joints in a mechanical system. The principle that the sequence of transformations around a loop must return to the identity provides what are known as the *loop equations*. An independent set of kinematics equations is assembled from the various sets of loop equations that are available in a mechanical system.

Transformations

In 1955, Jacques Denavit and Richard Hartenberg introduced a convention for the definition of the joint matrices [Z] and link matrices [X] to standardize the coordinate frames for spatial linkages.

This convention positions the joint frame so that it consists of a screw displacement along the Z-axis

$$[Z_i] = \begin{bmatrix} \cos\theta_i & -\sin\theta_i & 0 & 0 \\ \sin\theta_i & \cos\theta_i & 0 & 0 \\ 0 & 0 & 1 & d_i \\ 0 & 0 & 0 & 1 \end{bmatrix},$$

and it positions the link frame so it consists of a screw displacement along the X-axis,

$$[X_i] = \begin{bmatrix} 1 & 0 & 0 & a_{i,i+1} \\ 0 & \cos\alpha_{i,i+1} & -\sin\alpha_{i,i+1} & 0 \\ 0 & \sin\alpha_{i,i+1} & \cos\alpha_{i,i+1} & 0 \\ 0 & 0 & 0 & 1 \end{bmatrix}.$$

The kinematics equations are obtained using a rigid transformation [Z] to characterize the relative movement allowed at each joint and separate rigid transformation [X] to define the dimensions of each link.

The result is a sequence of rigid transformations alternating joint and link transformations from the base of the chain around a loop back to the base to obtain the loop equation,

$$[Z_1][X_1][Z_2][X_2]\ldots[X_{n-1}][Z_n] = [I].$$

The series of transformations equate to the identify matrix because they return to the beginning of the loop.

Serial Chains

The kinematics equations for a serial chain robot are obtained by formulating the loop equations in terms of a transformation [T] from the base to the end-effector, which is equated to the series of transformations along the robot. The result is,

$$[T] = [Z_1][X_1][Z_2][X_2]\ldots[X_{n-1}][Z_n],$$

These equations are called the kinematics equations of the serial chain.

Parallel Chains

The kinematics equations for a parallel chain, or parallel robot, formed by an end-effector supported by multiple serial chains are obtained from the kinematics equations of each of the supporting serial chains. Suppose that m serial chains support the end-effector, then the transformation from the base to the end-effector is defined by m equations,

$$[T] = [Z_{1,j}][X_{1,j}][Z_{2,j}][X_{2,j}]\ldots[X_{n-1,j}][Z_{n,j}], \quad j = 1,\ldots,m.$$

These equations are the kinematics equations of the parallel chain.

Forward Kinematics

The kinematics equations of serial and parallel robots can be viewed as relating parameters, such as joint angles, that are under the control of actuators to the position and orientation [T] of the end-effector.

From this point of view the kinematics equations can be used in two different ways. The first called *forward kinematics* uses specified values for the joint parameters to compute the end-effector position and orientation. The second called *inverse kinematics* uses the position and orientation of the end-effector to compute the joint parameters values.

Remarkably, while the forward kinematics of a serial chain is a direct calculation of a single matrix equation, the forward kinematics of a parallel chain requires the simultaneous solution of multiple matrix equations which presents a significant challenge.

- Engineering Mathematics Degree Courses UK; list of engineering mathematics courses in the United Kingdom. Whatuni.com website, accessed 9 Dec 2012.

References

- Encyclopaedia of Physics (2nd Edition), R.G. Lerner, G.L. Trigg, VHC publishers, 1991, ISBN (Verlagsgesellschaft) 3-527-26954-1, ISBN (VHC Inc.) 0-89573-752-3

- Kay, J.M. (1985). Fluid Mechanics and Transfer Processes. Cambridge University Press. pp. 10 & 122–124. ISBN 9780521316248.

- Jørgen Rammer (2007). Quantum Field Theory of Nonequilibrium States. Cambridge University Press. ISBN 978-0-521-87499-1.

- Clifford Truesdell & Walter Noll; Stuart S. Antman, editor (2004). The Non-linear Field Theories of Mechanics. Springer. p. 4. ISBN 3-540-02779-3. CS1 maint: Multiple names: authors list (link)

- O. C. Zienkiewicz; Robert Leroy Taylor; J. Z. Zhu; Perumal Nithiarasu (2005). The Finite Element Method (Sixth ed.). Oxford UK: Butterworth-Heinemann. p. 550 ff. ISBN 0-7506-6321-9.

- AC Gilbert (Ronald R Coifman, Editor) (May 2000). Topics in Analysis and Its Applications: Selected Theses. Singapore: World Scientific Publishing Company. p. 155. ISBN 981-02-4094-5.

- Edward D. Palik; Ghosh G (1998). Handbook of Optical Constants of Solids. London UK: Academic Press. p. 1114. ISBN 0-12-544422-2.

- Paul, Richard (1981). Robot manipulators: mathematics, programming, and control : the computer control of robot manipulators. MIT Press, Cambridge, MA. ISBN 978-0-262-16082-7.

Permissions

All chapters in this book are published with permission under the Creative Commons Attribution Share Alike License or equivalent. Every chapter published in this book has been scrutinized by our experts. Their significance has been extensively debated. The topics covered herein carry significant information for a comprehensive understanding. They may even be implemented as practical applications or may be referred to as a beginning point for further studies.

We would like to thank the editorial team for lending their expertise to make the book truly unique. They have played a crucial role in the development of this book. Without their invaluable contributions this book wouldn't have been possible. They have made vital efforts to compile up to date information on the varied aspects of this subject to make this book a valuable addition to the collection of many professionals and students.

This book was conceptualized with the vision of imparting up-to-date and integrated information in this field. To ensure the same, a matchless editorial board was set up. Every individual on the board went through rigorous rounds of assessment to prove their worth. After which they invested a large part of their time researching and compiling the most relevant data for our readers.

The editorial board has been involved in producing this book since its inception. They have spent rigorous hours researching and exploring the diverse topics which have resulted in the successful publishing of this book. They have passed on their knowledge of decades through this book. To expedite this challenging task, the publisher supported the team at every step. A small team of assistant editors was also appointed to further simplify the editing procedure and attain best results for the readers.

Apart from the editorial board, the designing team has also invested a significant amount of their time in understanding the subject and creating the most relevant covers. They scrutinized every image to scout for the most suitable representation of the subject and create an appropriate cover for the book.

The publishing team has been an ardent support to the editorial, designing and production team. Their endless efforts to recruit the best for this project, has resulted in the accomplishment of this book. They are a veteran in the field of academics and their pool of knowledge is as vast as their experience in printing. Their expertise and guidance has proved useful at every step. Their uncompromising quality standards have made this book an exceptional effort. Their encouragement from time to time has been an inspiration for everyone.

The publisher and the editorial board hope that this book will prove to be a valuable piece of knowledge for students, practitioners and scholars across the globe.

Index